Integrated Management and Biocontrol of Vegetable and Grain Crops Nematodes

Integrated Management of Plant Pests and Diseases

Published:

Volume 1
General Concepts in Integrated Pest and Disease Management
edited by A. Ciancio and K.G. Mukerji

Forthcoming:

Volume 3
Integrated Management of Diseases Caused by Fungi,
Phytoplasma and Bacteria
edited by A. Ciancio and K.G. Mukerji

Volume 4
Integrated Management of Fruit Crops Nematodes
edited by A. Ciancio and K.G. Mukerji

Volume 5
Integrated Management of Insect Borne Diseases
edited by A. Ciancio and K.G. Mukerji

Integrated Management and Biocontrol of Vegetable and Grain Crops Nematodes

Edited by

A. Ciancio
C.N.R., Bari, Italy

and

K. G. Mukerji
University of Delhi, India

A C.I.P. Catalogue record for this book is available from the Library of Congress.

ISBN 978-1-4020-6062-5 (HB)
ISBN 978-1-4020-6063-2 (e-book)

Published by Springer,
P.O. Box 17, 3300 AA Dordrecht, The Netherlands.

www.springer.com

Printed on acid-free paper

CONTENTS

SECTION 2 - Crops Ecology and Control

SECTION 4 - Data Analysis and Knowledge-based Applications

CONTRIBUTORS

Manjula Bakhetia
Centre for Plant Sciences
University of Leeds
Leeds, LS2 9JT, UK

Anwar L. Bilgrami
Department of Entomology
Rutgers University
New Brunswick, NJ 08901, USA

P. Cadet
IRD, UMR CBGP Campus de
Baillarguet, CS30016, 34988
Montferrier-sur-Lez Cedex, France

Wayne L. Charlton
Centre for Plant Sciences
University of Leeds
Leeds, LS2 9JT, UK

Aurelio Ciancio
Istituto per la Protezione delle Piante
CNR, Bari, Italy

Norma Coronel
Laboratorio de Nematología. Centro de
Zoología Aplicada, Casilla de correo
122. (5000) Córdoba, Argentina

Giovanna Curto
Servizio Fitosanitario, Regione
Emilia-Romagna,
Laboratorio di Nematologia, 40128
Bologna, Italy

Keith G. Davies
Rothamsted Research, Harpenden,
Hertfordshire, AL5 2JQ, UK

Marcelo E. Doucet
Laboratorio de Nematología
Centro de Zoología Aplicada,
Casilla de correo 122. (5000)
Córdoba, Argentina

Mireille Fargette
IRD, UMR CBGP, Campus de
Baillarguet, CS30016, 34988
Montferrier-sur-Lez Cedex,
France

Javier Franco
Fundación PROINPA, Cochabamba,
Bolivia

Simon Gowen
Department of Agriculture,
School of Agriculture, Policy
and Development, Reading
The University of Reading
RG6 6AR, UK

Leopoldo Hildalgo-Diaz
Centro Nacional de Sanidad
Agropecuaria (CENSA), Apdo 10,
San José de las Lajas, La Habana,
Cuba

Hans-Borje Jansson
Laboratory of Plant Pathology,
Multidisciplinary Institute for
Environmental Studies (MIES)
"Ramon Margalef"
University of Alicante
03080 Alicante, Spain

Brian R. Kerry
Rothamsted Research, Harpenden, Hertfordshire, AL5 2JQ, UK

Paola Lax
Laboratorio de Nematología. Centro de Zoología Aplicada, Casilla de correo 122. (5000) Córdoba, Argentina

Catherine J. Lilley
Centre for Plant Sciences
University of Leeds, Leeds
LS2 9JT, UK

Luis V. Lopez-llorca
Laboratory of Plant Pathology, Multidisciplinary Institute for Environmental Studies (MIES) "Ramon Margalef"
University of Alicante
03080 Alicante, Spain

J. G. Maciá-Vicente
Laboratory of Plant Pathology, Multidisciplinary Institute for Environmental Studies (MIES) "Ramon Margalef", University of Alicante
03080 Alicante, Spain

G. Main
Fundación PROINPA, Cochabamba, Bolivia

Thierry Mateille
IRD, UMR CBGP, Campus de Baillarguet, CS30016, 34988 Montferrier-sur-Lez Cedex, France

Julie M. Nicol
International Wheat and Maize Improvement Center (CIMMYT), Wheat Program, PO Box 39, Emek, 06511, Ankara, Turkey
e-mail: j.nicol@cgiar.org

Gregory R. Noel
U.S. Department of Agriculture, Agricultural Research Service

Cesar Ornat
Departament d'Enginyeria Agroalimentària i Biotecnologia, Universitat Politecnica de Catalunya, ESAB-EUETAB
08860 Castelldefels, Barcelona, Spain

Barbara Pembroke
The University of Reading, Department of Agriculture, School of Agriculture, Policy and Development, Reading, RG6 6AR, UK

Antoon Ploeg
Department of Nematology, University of California, 1463 Boyce Hall, Riverside, CA 92521, USA

Roger Rivoal
UMR INRA/ENSAR, Biologie des Organismes et des Populations Appliquée à la Protection des Plantes (BiO3P),
BP 35327, 35653 Le Rheu, France

A. F. Robinson
USDA – ARS, 2765 F & B Road,
College Station, TX 77845, USA

F. J. Sorribas
Departament d'Enginyeria
Agroalimentària i Biotecnologia,
Universitat Politecnica de Catalunya,
ESAB-EUETAB
08860 Castelldefels, Barcelona,
Spain

Peter E. Urwin
Centre for Plant Sciences
University of Leeds, Leeds
LS2 9JT, UK

PREFACE

This series originated during a visit of prof. Mukerji to the Plant Protection Institute of CNR at Bari, Italy, in November 2005. Both editors agreed to produce a series of five volumes focusing, in a multi-disciplinary approach, on recent advances and achievements in the practice of crop protection.

This Volume deals with nematodes parasitic on plants. Nematodes inhabit the earth since almost half a billion years and will probably remain for an even longer time in the future. They represent a very successful, diversified and specialised animal group, present in all ecological niches in nature. Only a small fraction of species is actually described and, among them, only a reduced number is known to feed on plants. Among them, however, a few species exert an heavy economic impact on crops, representing a severe limiting factor for agricultural productions.

This statement explains the attention devoted in last decades to plant parasitic nematodes, and the efforts deployed for their control. As for other disciplines concerning plant protection, nematology is now in a mature stage in which the optimism initially underlining the widespread use of chemicals and fumigants lent space to a more pragmatic, comprehensive and integrated vision of control.

Although a wide literature already covers chemical or biological control, there is a need for a more holistic vision of management. In this series we attempted to fill this gap, organizing the review in two fields, concerning nematodes of annual (this Volume) and prerennial (Volume 4) crops. We aimed at providing an informative coverage for a broad range of agricultural systems which coexist in the world today, focusing on the solutions suitable for the corresponding economies. Chapters then range from an "anthropological" view of nematode problems and solutions, suitable in self-consumption systems from West Africa and South America, to more technological solutions, suitable in industrialised agricultural systems, based on standardization of management practices, i.e. North and South America extensive productions or consumers-based and policy-oriented sustainable crops, as in the case of mediterranean regions.

In the first chapter, the potentials of predatory nematodes applications is revised. This chapter focuses on biological control attributes and other important characters of predatory species, in reference to their ecology, culture, conservation and feeding abilities. Biological control potentials are discussed in detail for main predator groups, like dorylaims, nygolaims and diplogasterids.

In the following chapter, the integration of biological control with other management tools is reviewed, with reference to crop rotations, antagonistic crops, resistant cultivars, soil solarization, biofumigation and nematicides. The combination of biological control agents with methods to increase microbial abundance and/or activity is also discussed, considering organic amendments, green manures and companion crops. A brief section is also dedicated to the exploitation of genes from natural enemies and to improve formulations and application methods.

The interactions and mode of action of nematophagous fungi are revised in chapter 3. Nematophagous fungi have been extensively studied and a large amount of data is available in the literature. However, there is a need to estabilish the actual horizon for this research field, since only a few species appear suitable for practical, field applications. In this chapter, the authors review the biology, taxonomy and phylogeny of species, focusing on chemotaxis and host adhesion, signalling and differentiation, biochemistry, genomics and proteomics. These are key topics for understanding the potentials of nematophagous fungi, since host recognition and adhesion are fundamental steps in the infection process. The soil and rhizosphere environments are also reviewed, with reference to several aspects concerning the behaviour as root endophytes, the rhizosphere dynamics and biological control efficacy, as well as the role of root exudates, detection and quantification.

In the following chapters, problems and solution applied on a regional scale in management follow, illustrating some case studies ranging from West Africa to North and South America. A wider view of the interactions among nematodes and biological control agents is given in chapter 4. Some nematodes problems of West Africa agricultural systems are revised, as well as the methods locally adopted for management, in a soil conservation approach. The authors show how soil fertility and plant nematodes management fit in the more general problem of protection and conservation of soil. Plant-nematode interactions are discussed in reference to the host plants quality and compatibility, host resistance and antagonistic interactions. The ecology and management of nematode communities are reviewed in the light of the multitrophic relationships occurring in soil and of the nematode-antagonists specificities. A complex, holistic soil health vision is given, aiming at the identification of flexible and adaptive approaches for management.

In chapter 5, further regional and specific agricultural issues dealing with food production are reviewed, in reference to management of tuber and grain crop nematodes in the Andes region. Andean tuber crops are important sources of food for local rural communities and include species like oca (*Oxalis tuberosa*), mashwa (*Tropaeolum tuberosum*) or ullucu (*Ullucus tuberosus*). Andean grain crops like quinoa (*Chenopodium quinoa*) and lupine (*Lupinus mutabilis*) also enrich this diet. Most important nematodes in tuber and grain crops and their role in rotation are reviewed, i.e. potato cyst and rosary nematodes or other species parasitic on oca, with a section on the integrated management practices suitable for the region.

Given the worldwide importance of soybean, the two following chapters deal with the soybean cyst nematode, *Heterodera glycines*, in North and South America. In chapter 6 the management strategies adopted to control this species in the USA are revised, whereas in chapter 7 a detailed description of damages caused by *H. glycines* in Argentina is provided. The authors describe the nematode life-cycle, the occurrence of races and populations, the plant-nematode relationships, the histological alterations and the response of cultivars to parasitism, as well as the relationship of *H. glycines* with the environment. Further sections also illustrate the losses and management strategies, based on early nematode detection or on the identification of races, coupled with chemical control, crop rotation and further preventive measures against cysts dispersal.

Keeping the pace to provide a broad nematological *excursus*, in the following chapter the management of nematodes attacking cotton in North America is reviewed. For several species of significant impact, like *Meloidogyne* spp., *Rotylenchulus reniformis*, *Belonolaimus longicaudatus* and *Hoplolaimus columbus*, the geographical distribution and impact are shown, together with symptomatology, biology, epidemiology, life cycle and interactions. Plant genetic variability is shown as the most important tool for management, whereas sampling and economic thresholds are described as key practices in field damage estimation. For the cited species, control methods are reviewed with focus on natural and physical factors, conventional and novel nematicides, recoverable yield potentials, biological and cultural control and crop rotation. Sanitation practices and weed management are also reviewed. Finally, actual data on genetic resistance to nematodes in cotton are provided, focusing on resistance and tolerance mechanisms, resistant and tolerant cultivars and resistance sources.

Although the potentials of the DNA "revolution" did not yet climb to its *optimum* in the field of biological control and pest management, some interesting tools could soon leave the laboratory to reach a field application status. An elegant approach based on the mechanisms of RNA interference, showing a potential for management, is described in chapter 9. The biotechnological control of plant parasitic nematodes is not yet a field practice indeed, but the RNAi mechanism could soon turn out as a further tool in some niche farming. The mechanism of RNAi in nematodes is described, and RNAi with plant parasites is discussed, either for the uptake of double strand RNA and for the comparison of reported strategies. The authors illustrate dsRNA plant delivery to target nematodes genes and the feeding strategy of sedentary endoparasites, showing how and why in planta RNAi may provide the basis for a biotechnological strategy aimed at nematodes control.

In a more detailed approach to biological control, the potential use of *Pasteuria* spp. is discussed in chapter 10. The authors describe the life cycle and development of one of the nematode antagonists most studied thus far, discussing phenotypic and molecular characters dealing with taxonomy, host range, mass production and *in vitro* and *in vivo* culture. Further topics concern the distribution of *Pasteuria* in natural systems, the association with nematode suppressive soils, as well as further biological and ecological features.

Two chapters then follow, describing management through nematicidal plants. In chapter 11, sustainable methods available for management of the sugarbeet cyst nematode *Heterodera schachtii* in Northern Italy are reviewed. Chemical and agronomic control and biological control methods relying on nematicidal intercrops based on Brassicaceae are described. Application for biological control of *H. schachtii* with suppressive intercrops are included in some production schemes in Northern Italy. Plant species, intercrops management, promotion of *H. schachtii* biological control are revised, together with exploitation of resistance and tolerance. In the following chapter, further data on the biofumigation based management of plant-parasitic nematodes are provided, concerning the Brassica biofumigation mechanism and other nematodes groups, including root-knot species. Non-Brassica based biofumigation practices are also reviewed.

In chapter 13, global knowledge and its application for the integrated management of wheat nematodes are reviewed. The importance of cereals and wheat in the world and the distribution of cereal nematodes, species and pathotypes, are discussed. Cereal cyst, root lesion and other cereal nematodes, including root knot, stem and seed gall species are reviewed, concerning their life cycle, symptoms of damage and yield losses. Integrated control of cereal nematodes, including chemicals, cultural practices i.e. grass-free rotations and fallowing with cultivation, are revised, together with irrigation, sowing, trap and mixed cropping, organic amendments and inorganic fertilizers. Resistance/tolerance and biological control are discussed, in reference to true IPM investigations, for each nematode group.

The following chapter deals with the integrated management of root-knot nematodes in mediterranean horticultural crops. The symptoms, biology, ecology and yield losses caused by *Meloidogyne* spp. are described. Root-knot nematodes management through plant resistance, crop rotation, trap crops, fallowing and tillage are reviewed, together with biological methods, biofumigation and chemicals based management.

Finally, in the last chapter, the basic application of modeling to the study of nematode parasitic bacteria, including *Pasteuria* spp. and other Gram negative species, is reviewed, focusing on potentials in nematodes regulation. Some basic systems like the Lotka-Volterra and Anderson and May models are described, with a further description of requirements for modeling *Pasteuria* regulation, and a final discussion on experimental and practical issues, required in this kind of studies.

In conclusion, our attempts to provide an *excursus* on nematode management solutions available worldwide in a broad range of agricultural systems yielded a comprehensive compilation. Thanks to the efforts and will of many nematologists investigating and applying advanced solutions in their long term research and field practices, we hope we were able to provide a further tool, useful in the environment friendly and sustainable menagement of plant parasitic species. Our hope is that the contributions of this volume, even if confined to some paramount examples, will result useful and helpful for any interested readers also outside these boundaries, inspiring and supporting all research efforts invested in their field and laboratory work.

A. Ciancio
K. G. Mukerji

Section 1

Nematodes in Biological Control

1

ANWAR L. BILGRAMI

BIOLOGICAL CONTROL POTENTIALS OF PREDATORY NEMATODES

Department of Entomology, Rutgers University, New Brunswick, NJ 08901, USA

Abstract. Biological control potential of predatory nematodes is evaluated and discussed in the following chapter. Attributes of a successful biological control agent such as mass production, reproductive potential, longevity, compatibility with agrochemicals, safety to non-target organisms, prey search ability, environmental adaptability, dispersal and persistence capabilities etc., are enumerated. Prey searching and feeding mechanisms, prey preferences, ecology, biology and conservation of predatory nematodes and prey resistance and susceptibility to predation are elaborated and supplemented with the list of plant-parasitic nematodes recorded as prey to various species of predatory nematodes.

1. INTRODUCTION

Air, land and water are alarmingly polluted to the extent that several sensitive species are becoming extinct at the rate never experienced before on earth. Pesticides and chemicals, inextricably associated with us from fabric to food, pose a major threat to our lives. Their adverse effects on human and animal populations, pest resistance and continued ravage on one third of food produced worldwide, call for including nature's own enemies to manage plant pests, including phytoparasitic nematode populations.

Biological control of pests is as old as agriculture. Centuries ago, ducks were used to consume pests, a technique still adopted in China. The first known biological control strategy was implemented in 1762, when a Mynah bird was taken from India to Mauritius to control locusts. However, the landmark in biological control was achieved in 1880 when ladybird beetles were used to control scale insects in citrus plantations. Biological control may be defined as the "*action of parasites, predators and pathogens in maintaining other organism's population density at a lower average than would occur in their absence*". It may be elaborated further as "*any condition under which or practice whereby, survival or activity of a pathogen or pests is reduced through the agency of other living organism*". This is referred to as "*Natural biological control*" since it involves predators and pests without human intervention. However, if "*the use of predators and parasites are induced to multiply and disseminated by human efforts*" it would be referred to as '*induced inundated biological control*". With the advent of biotechnology, the concept of biological

A. Ciancio & K. G. Mukerji (eds.), Integrated Management and Biocontrol of Vegetable and Grain Crops Nematodes, 3–28.

control needs to be redefined as *"the use of nature and/or modified organisms, genes or gene products to regulate or reduce pests in favour of human and animal populations, and agricultural crops besides protecting other beneficial organisms"*. Biological control measures are, therefore, both corrective (e.g., chemical) and preventive (e.g., cultural) in nature. Preventive, as they help evading the disease and corrective because if the disease is already set in, it corrects the malady by reducing pest populations.

The advocacy of nematode biocontrol dates back to several decades, but its usefulness was brought to sharp focus only recently. Initial research by Linford and Oliviera (1937, 1938), though empirical, generated interest in using amendments to control plant-parasitic nematodes. How soil microorganisms/organic amendments reduce plant-parasitic nematodes may provide basic informations to understand nematode biological control. Two hypothesis were proposed to explain why organic amendments mostly help in reducing plant-parasitic nematodes: (1) the products released by amendments are directly toxic to plant-parasitic nematodes and (2) the organic compounds initiate a succession of events which favour the populations of indigenous biological control agents.

Biological control achieved success in Entomology, Plant Pathology and Insect Nematology. Little is achieved with phytoparasitic nematodes except predaceous and parasitic fungi, which contributed 73% of the total research efforts (Table 1). Predatory nematodes attracted 13% research effort whereas the other organisms ranged between 1–6%.

Table 1. Research on different biological control agents with reference to predatory nematodes.

Biocontrol organism	*Research efforts (%)*[1]
Predaceous Fungi	56
Parasitic Fungi	17
Predaceous Nematodes	13
Bacteria	6
Tardigrades	1
Protozoa	<1
Rickettsiae	<1
Collembola	<1
Viruses	<1
Turbellarians	<1
Mites	<1
Enchytraeids	<1

[1]Assessed from examination of 1000 papers in the year 2002.

Research carried out during the last 10–15 years generated interest in evaluating the role of predatory nematodes as nematode biocontrol agents. The use of predatory nematodes is challenging because both predatory and parasitic nematodes are small in size and inhabit soil. Biology, behaviour, food and feeding preferences, prey relationships, together with other ecological parameters are important to fully evaluate their biological control potentials.

2. BIOLOGICAL CONTROL ATTRIBUTES

Effectiveness of predatory nematodes as biocontrol agents depends upon the following characteristics.

Culture: predatory nematodes to be used as biocontrol agent should be easily and cheaply culturable on commercial scale (e.g., diplogasterid predators).

Reproductive potential: predatory species must have a high reproductive rate in order to maintain population at higher densities (i.e., diplogasterid predators). Occasionally, high reproduction adversely affects efficiency, due to high energy requirement. Thus, a judicious balance between reproductive and infective potentials of predatory nematodes needs to be achieved.

Longevity: as a successful biocontrol agent, a predatory nematode should be characterized by significant longevity and stability, so that it can be stored without appreciable loss of its predation capacity.

Application: compatibility of predatory nematodes with agrochemicals and standard farm practices is extremely important in order to achieve successful application and significant control.

Safety to non-target organisms: Although most biocontrol agents are non-pathogenic, the safety of non-target organisms e.g., plants, humans and other beneficial organisms must be kept in mind. The ability of nematodes to avoid predation on organisms other than the target would contribute to its success as an efficient biocontrol agent.

Searching capability: prey searching ability is an important attribute that affects predator's mobility, predation and biocontrol potential. Predatory nematodes possessing efficient searching ability (e.g., diplogasterids) would be more effective as biocontrol agents than those lacking prey search ability (e.g., mononchs).

Environmental adaptability: Any predator that adapts and tolerates existing and changing environmental conditions as well as any species capable of ecological and temporal compatibility would result best fit to act as an efficient biocontrol agent.

Temporal compatibility: perfect temporal compatibility between predatory and pest nematodes is another important attribute that contributes towards the success of biological control. Temporal compatibility synchronizes predator-prey life cycles and eliminates time gaps that allow pests to escape predation.

Dispersal and persistence capability: ability to disperse, persist, survive and reproduce under adverse conditions including absence of prey are ideal candidates for nematode management. Dual feeding habits (e.g., diplogasterid feeding on prey and bacteria) help predators to thrive equally well on alternate food (e.g., diplogasterid predators feed on bacteria in prey absence). Ability of predatory nematodes to reduce parasitic-nematode population within a short time span is also important for biocontrol (e.g., diplogasterid predators). To reduce pest populations below noxious levels, high predation ability and long predators persistence enhance biocontrol efficiency.

Broad spectrum efficiency: monophagous (an undesirable biological control trait) predators may be efficient regulators, but they may allow development and establishment of other noxious nematode species. It is desirable that predatory nematodes should possess a broad host range, in order to harm a diverse spectrum of noxious nematodes.

Capacity to produce toxic metabolites: Predatory nematodes that produce toxic secretions to kill or inactivate pest organisms (e.g., *Seinura* injects toxic substances to inactivate its prey) possess yet another attribute of a successful biocontrol agent.

Hyperparasitism: this trait significantly affects biological control potential. Predatory nematodes (e.g., mononchs) resorting to cannibalism in absence of preys (an example of hyperparasitism) can never represent an optimal good choice as other nematodes biocontrol agent. Cannibalism is a condition in which predators feed on conspecific individuals, thus reducing biological control potential.

2.1. Prey Capturing and Feeding Abilities

Predatory feeding is divided into different phases (Fig. 1), namely encounter with prey, attack response, attack, extra corporeal digestion and ingestion (Bilgrami & Jairajpuri, 1989b).

Encounter with prey: this phase is established either by willful movements of predators in response to kairomones emitted by the prey (diplogasterid, dorylaim or nygolaim predators) (Bilgrami & Jairajpuri, 1988a; Bilgrami, & Pervez, 2000; Bilgrami, Pervez, Yoshiga, & Kondo, 2000; Bilgrami, Pervez, Kondo, & Yoshiga, 2001) or by a chance contact (e.g., mononchs) (Grootaert & Maertens, 1976) (Fig. 1). Cutting and sucking type (e.g., *Mononchoides, Butlerius*) or stylet-bearing predators (e.g., *Mesodorylaimus, Aquatides*) establish contacts with the prey in response to attractants (Bilgrami, 1997). Predator attraction towards prey and aggregation around the feeding sites suggest an important role of prey secretions in establishing predator-prey contacts. Unlike other predatory groups, diplogasterids are attracted towards bacteria (Bilgrami & Jairajpuri, 1988a).

Attack response: Attack response is generated as a result of head probing, feeding apparatus movements and oesophageal pulsations. Prey contacts at right angles are necessary to initiate an attack (Fig. 1) as glancing contacts or contact other than right angles do not result in successful attacks. Attack response varies from predator to predator, it may be aggressive as in *Prionchulus* or *Mylonchulus*, vigorous but confined in *Labronema*, gradual and restricted in *Aquatides* or *Dorylaimus*. Attacks always followed probing of prey, which may be limited to a rapid side-to-side lip rubbing (*Mononchus*), vigorous (*Mononchoides*) or just an head shaking and lip rubbing against prey's body (*Butlerius*).

Attack: predators cut or penetrate the prey cuticle by side-to-side lip rubbing with simultaneous movements of the feeding apparatus (Bilgrami & Jairajpuri, 1989b). If a predator fails to puncture the cuticle it searches another spot on the prey body or reverts back to search for another prey. The prey is attacked by the stylet (e.g., *Mesodorylaimus, Discolaimus, Seinura*), mural tooth (e.g., *Aquatides*), dorsal tooth (e.g., *Mylonchulus*), onchia (*Actinolaimus*), teeth (e.g., *Ironus*) or by the combined actions of a movable dorsal tooth and high esophageal suction (e.g., *Mononchoides, Butlerius). Aquatides* and *Dorylaimus* puncture prey cuticle by gradual and intermittent thrusting of the stylet (Shafqat, Bilgrami, & Jairajpuri, 1987) whereas *Labronema* achieves puncturing through quick stylet movements (Wyss & Grootaert, 1977) (Fig. 2). *Diplenteron* (Yeates, 1969), *Butlerius* (Grootaert, Jaques, & Small, 1977) or *Mononchoides* (Bilgrami & Jairajpuri, 1989b) use their movable dorsal tooth and esophageal suction to slit open the prey cuticle (Fig. 2). *Mononchus*, *Iotonchus* and other mononchs engulf and swallow their prey whole or shred their body prior to feeding (Fig. 2) (Bilgrami, Ahmad, & Jairajpuri, 1986). *Dorylaimus* needs 6-8 thrusts to penetrate the prey cuticle (Shafqat et al., 1987) whereas *Labronema* and *Aquatides* requires 5–6 stylet thrusts (Wyss & Grootaert, 1977). *Mesodorylaimus* needs fewer stylet thrusts (6–9) than *Aporcelaimellus* (7–12) (Khan, Bilgrami, & Jairajpuri, 1991) to perforate the prey cuticle. *Seinura* injects toxic esophageal secretions to paralyze its prey (Hechler, 1963). Other stylet bearing predators disorganize internal body organs of prey to make them immobile before initiating feeding. *Mononchs* inactivate their prey by holding them firmly with the buccal armature and high esophageal suction.

Extracorporeal digestion: stylet bearing predators partially digested their food outside the oesophagus prior to ingestion since their lumen is too narrow to ingest large food particles, a phenomenon known as extracorporeal digestion in plant parasitic (Wyss, 1971) and predatory nematodes (Bilgrami & Jairajpuri, 1989b). Mononchs do not pre-digest food since they can swallow a prey whole or ingest its pieces through the wide oral aperture. In contrast, diplogasterid predators partially digest food molecules prior to ingestion, by releasing esophageal gland secretions (Bilgrami & Jairajpuri, 1989b). Complex food globules are broken down into small particles before they are ingested through stylet lumen en route to the intestine. *Diplenteron*, *Dorylaimus*, *Aquatides*, *Seinura*, *Mononchoides* and other predators show extra corporeal digestion of food molecules.

Ingestion: most species of mononchs engulf prey or ingest it after shredding into pieces (e.g., *Iotonchus*) (Fig. 2) but few (e.g., *Mylonchulus*) feeds by cutting and sucking their prey (Bilgrami et al., 1986). Swallowing of prey is supported by the esophageal muscle contractions that pull prey into the buccal cavity through vertically positioned plates. Some individuals show periods of inactivity after devouring an entire prey, while others initiate further attacks. Stylet bearing predators cannot engulf their prey or shred it into pieces, but penetrate and rupture the internal body structures by making sideways movements of the feeding apparatus. Prey contents are ingested through the esophago-intestinal junction by simultaneous relaxation and contraction of the esophageal bulb. Once the contents are ingested, predators detach their lips from the prey, retract feeding apparatus and move in search of another prey. Stylet bearing predators also feed upon the eggs of other nematode species but not conspecific eggs. When in contact with conspecific eggs, these predators probe in an exploratory fashion by making side-to-side lip rubbing but cause no harm to the eggs (Esser, 1987). Diplogasterids could devour intact first stage juveniles of small prey nematodes (e.g., *Acrobeloides* or *Cephalobus*) but must cut larger preys into pieces to feed. The process of food ingestion in *Neoactinolaimus, Ironus* or *Thalassogenus* is identical.

Predators struggle among themselves to feed if their number exceeds 3 at a feeding site. Aggregation at feeding sites is common in dorylaim (Bilgrami et al., 2000), nygolaim (Bilgrami et al., 2001) and diplogasterid predators (Bilgrami & Jairajpuri, 1988a) (Fig. 2). Up to eight diplogasterid predators were found aggregated at the feeding site. Aggregation is most pronounced at low prey densities, allowing predators to quickly finish feeding before hunting other preys.

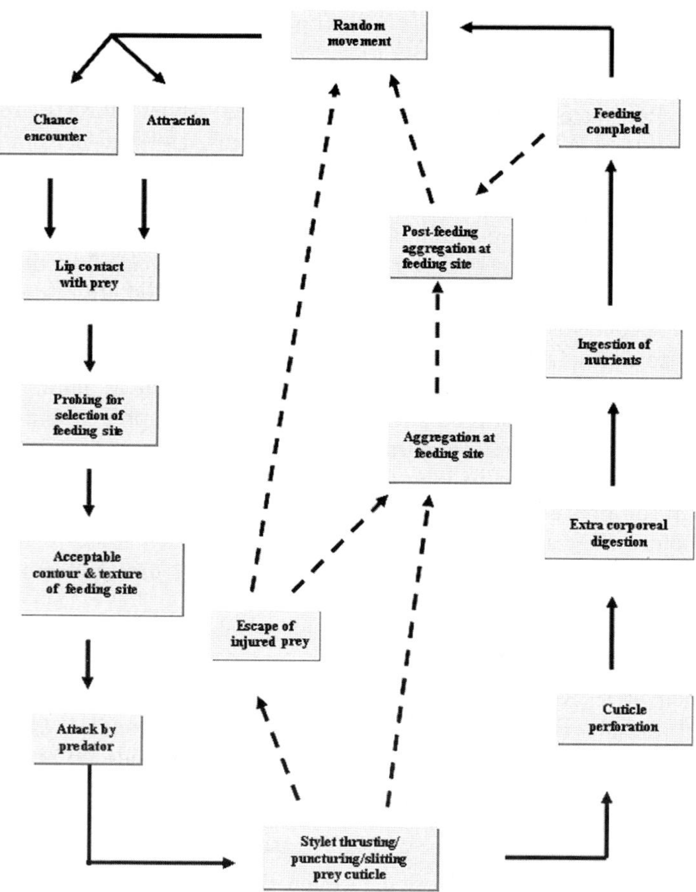

Figure 1. Capturing and feeding mechanisms in predatory nematodes.

2.2. Prey Resistance and Susceptibility to Predation

Successful biological control could be achieved if predators possess high strike rate and prey nematodes are highly susceptible. Cohn and Mordechai (1974), Grootaert et al. (1977), Small and Grootaert (1983) and Bilgrami and Jairajpuri (1989a) differentiated prey nematodes depending upon their abilities to resist predation. The ability of prey nematodes to defend themselves against predator's onslaught varies from species to species.

Resistance to predation is due to the coarse body annulations (e.g., *Hemicriconemoides*), thick or double body cuticles (e.g., *Hoplolaimus*), gelatinous matrix, toxic body repellents (e.g., *Helicotylenchus*) or rapid undulatory body movements (e.g., *Rhabditis*). Bilgrami and Jairajpuri (1989a) proposed the following equations to determine predator strike rate and prey resistance and susceptibility to predation.

Strike rate of predators (%) $SR = (EA/A) \cdot 100$

Prey resistance (%) $PR = (EA\text{-}AW)/EA \cdot 100$

Prey susceptibility (%) $PS = 100 - PR$

Where: SR = strike rate of predators; PR = prey resistance; PS = prey susceptibility; EA = number of encounters of predators with prey resulting into attack; AW = number of attacks by predators resulting into prey wounding; A = total number of encounters with the prey.

2.3. Prey Preference

Prey preference is a key feature for the selection of a biological control agent. A broad or indiscriminate host/prey range, as is the case for many predators, can be an undesirable feature in a predator intended for field release. A highly specific predator, on the other hand, limits its effectiveness against target species and mass culturing. Prey preference is determined either from the chance observations in petri dishes or from the analysis of preserved materials. Mononchs are broad in prey specificity as they engulf all types of organisms including nematodes, rotifers, protozoa, oligochetes and other invertebrates (Table 2) (Bilgrami et al., 1986). They are rapacious, with reports of a single individual mononch killing up to 83 cyst nematode (*Heterodera*) per day; another individual ingested 1332 preys over its life span (Steiner & Heinly, 1922).

In a recent study Bilgrami, Gaugler, and Brey (2005) showed prey preference of a predatory nematode in choice and no choice experiments (Fig. 3). They proposed method to determine coefficient of preference based on predators rejection or acceptance of a prey and prey choices they were given. Coefficient of preference is based on the probabilities of success (prey accepted = proportion of one prey killed higher than the other in a prey combination) and failures after prey rejected (proportion of one prey killed less than the other in a prey combination in relation to the number of events, number of combinations for one species i.e., ten) (Table 3) (Bilgrami et al., 2005). Prey preferences were designated as positive (more prey killed) or negative (fewer prey killed) for the sake of convenience and comparison. Prey rejected does not mean that no prey was killed or eaten. Coefficient of preference, referred to as positive (prey accepted) and negative (prey rejected) ranged between 0 to + 1 and –1 to 0 respectively. Prey species having coefficient of preference approaching + 1 were highly accepted and those approaching – 1 as rejected. Based on Table 3, the coefficient of preference (Table 4) for each species was calculated as follows:

$$\frac{\text{Mean prey accepted (\%)} - \text{Mean prey rejected (\%)}}{\text{Mean prey accepted (\%)} + \text{Mean prey rejected (\%)}}$$

Table 2. Analysis of intestinal contents of mononchs (from Bilgrami et al.*, 1986).*

Predators	*Observed*	*Total*	*Specimens containing prey*[*]						
			D	T	F[**]	NI	C	MO	MG
Parahadronchus	164	112 (68%)	42	48	68	48	21	38	14
Mononchus	198	87 (44%)	22	24	55	33	19	10	23
Miconchus	34	15 (44%)	10	8	15	6	3	4	3
Clarkus	62	26 (42%)	4	6	17	8	4	8	8
Prionchulus	105	32 (30%)	18	20	22	24	16	12	0
Sporonchulus	59	16 (27%)	4	6	16	8	7	3	6
Coomansus	24	5 (21%)	4	5	4	4	2	1	0
Iotonchus	173	75 (43%)	50	49	70	52	24	24	20
Mylonchulus	190[***]	0	0	0	0	0	0	0	0
Total	816	368 (43%)	154	166	277	184	96	100	74

* D = Dorylaims; T = Tylenchs; F = free living; NI = not identified; C = cuticular parts; MO = mononchs of other genera; MG = mononchs of same genera.
** Includes monohysterid, diplogasterid and rhabditid nematodes.
*** Not included in the total as no specimen of this genus had prey in the intestine.

2.4. Ecology

Ecological studies revealed significant generic diversity in predatory nematodes (Bilgrami et al., 2000; Bilgrami, Khan, Kondo & Yoshiga, 2002; Bilgrami, Wenju, Wang, & Qi, 2003). Diversity up to 32% was recorded in the presence of other nematode species (Bilgrami et al., 2003). At the nematode community level, plant-parasitic nematodes dominated but predatory species constituted maximum biomass (Bilgrami et al., 2000, 2003). The positive correlation of predators with plant-parasitic species suggested that the latter represent a suitable food source for predatory nematodes. Such a correlation also indicates a role of predatory nematodes as effective biocontrol agents. In another study Bilgrami et al. (2000) showed predominance of predatory over plant-parasitic and bacteriophagous nematodes in a deciduous forest. Predatory nematodes constituted a major component of the nematode community due to their abundance, moderate to high density and maximum biomass. Positive correlation between predatory and other nematode species suggested density dependent regulation.

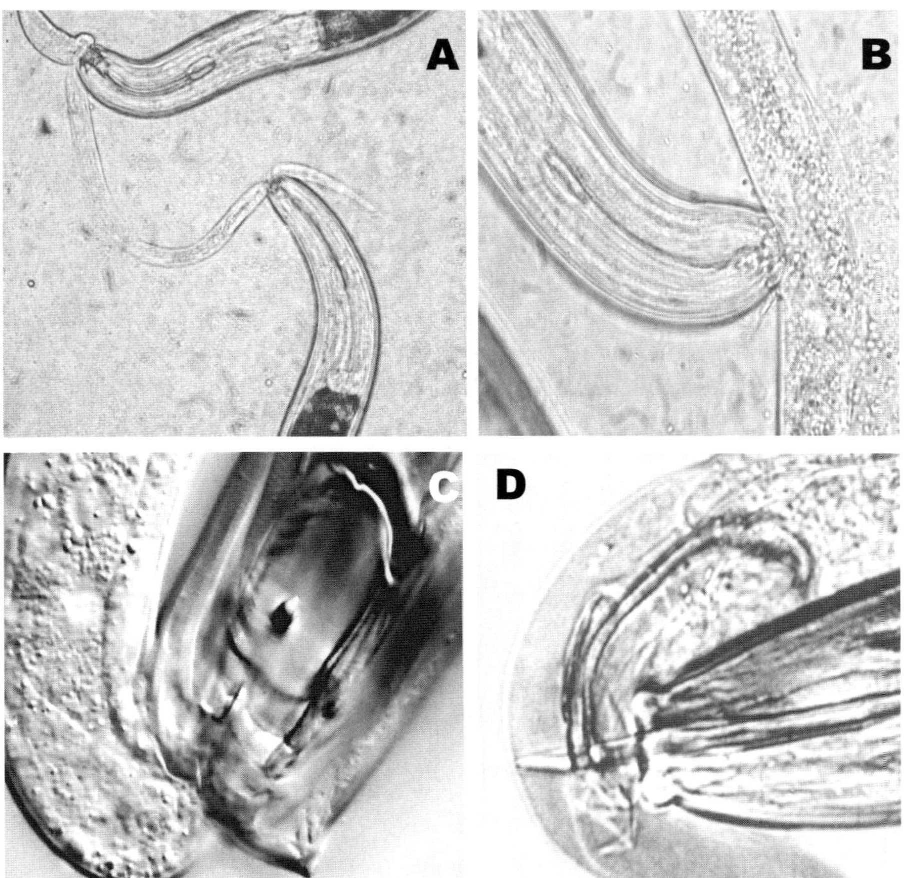

Figure 2. Predatory nematodes feeding activities. (A) two individuals of Mononchoides gaugleri *(diplogasterid) feeding on the same prey. (B)* M. gaugleri *feeding on a prey. (C)* Anatonchus tridentatus *(mononch) ingesting* Panagrellus redivivus. *(D)* Labronema vulvapapillatum *(stylet bearing predator) sucking prey contents.*

Entomopathogenic nematodes feed on specific symbiotic bacteria within the host cadaver. Diplogasterid predators differ from entomopathogenic species in one fundamental way: under natural conditions, they feed on bacteria besides preying nematodes (Pillai &Taylor, 1968; Yeates, 1969; Jairajpuri & Bilgrami, 1990; Yeates, Bongers, De Goede, Freckman, & Georgieva, 1993). The ability of diplogasterids to "switch" between predator and microbivore feeding modes rests in the anticipated ability to survive periods of low prey densities. Switching behaviour buffers predator populations, and thereby serves as a "powerful stabilizing mechanism" (Hassell, 1978).

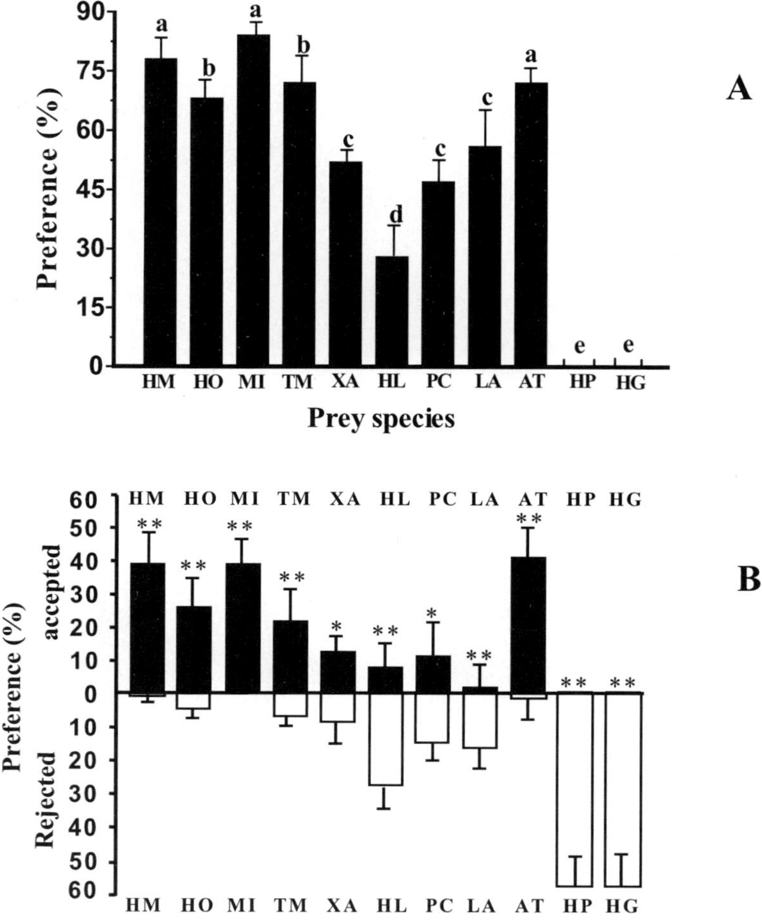

Figure 3. Prey preference by Mononchoides gaugleri *in no-choice (A) and choice experiments (B).* HM = Heterodera mothi *J_2;* HO = Hirschmanniella oryzae*;* MI = Meloidogyne incognita *J2;* TM = Tylenchorhynchus mashhoodi*;* XA = Xiphinema americanum*;* HL = Helicotylenchus indicus*;* PC = Paratrichodorus christei*;* LA = Longidorus attenuatus*;* AT = Anguina tritici *J_2;* HI = Hoplolaimus indicus*;* HG = Hemicriconemoides mangiferae. *Vertical lines on the bars show ± SD. Different letters show significant differences between preference (A) and prey accepted and rejected (B).* ***Prey accepted significantly different from prey rejected.* **Prey accepted not significantly different from prey rejected (B). Adapted from Bilgrami* et al. *(2005).*

Dauer juveniles are metabolically active and motile, non-aging but developmentally arrested. Environmental stresses induce formation of the “dauer juvenile” that enhances the tolerance to moisture, temperature, and chemicals extreme conditions. Only predatory diplogasterids – the cutting and sucking type of predators – have such a resting stage. It shares strong similarities with that of entomopathogenic nematodes in being induced when conditions are unfavorable and in possessing enhanced survival abilities. Most other differences remain uncertain, as in sharp contrast to the dauer juveniles of entomopathogenic species, diplogasterids dauers received thus far little attention. It is hypothesized that diplogasterid dauer juveniles possess some degree of tolerance to anhydrobiotic conditions too.

Table 3. Predatory nematodes preference for prey species in choice experiments.

Prey	*prey accepted or rejected (%)*										
	HM	HO	MI	TM	XA	HL	PC	LA	AT	HI	HG
HM		–16	+04	–16	–48	–56	–32	–52	+04	–84	–88
HO	+16		+24	–12	–44	–32	–28	–04	+12	–72	–76
MI	–04	–24		–16	–44	–52	–28	–40	–16	–84	–88
TM	+16	+12	+16		–12	–28	–08	–08	+28	–76	–80
XA	+48	+44	+44	+12		–24	–12	–08	+40	–56	–60
HL	+56	+28	+52	+28	+24		+12	+28	+56	–44	–36
PC	+32	+28	+28	+16	+12	–12		–16	+44	–44	–40
LA	+52	+04	+40	+08	+08	–28	+16		+48	–36	–32
AT	+04	–12	+16	–28	–40	–56	–44	–48		–92	–88
HI	+84	+72	+84	+76	+44	+44	+44	+36	+92		00
HG	+88	+76	+88	+80	+40	+36	+40	+32	+88	00	

Mean prey accepted or rejected for each species calculated from ten combinations of two prey species. Predator preference = difference in the proportion of two prey species killed in a combination. Proportion of one prey (e.g., HM in a column) killed higher than the other (e.g., HO in a row) in a combination designated as positive (+16%) and referred to as prey accepted, whereas proportion of one prey (e.g., HO in a column) killed less than the other (e.g., HM in a row) in a combination is designated as negative (–16%) and referred to as prey rejected. HM = *Heterodera mothi* (juveniles); HO = *Hirschmanniella oryzae*; MI = *Meloidogyne incognita* (juveniles); TM = *Tylenchorhynchus mashhoodi*; XA = *Xiphinema americanum*; HL = *Helicotylenchus indicus*; PC = *Paratrichodorus christei*; LA = *Longidorus attenuatus*; AT = *Anguina tritici* (juveniles); HI= *Hoplolaimus indicus*; HG = *Hemicriconemoides mangiferce*. Adapted from Bilgrami et al. (2005).

2.5. Culture

Efficacy studies largely reflect lack of *in vitro* production methodology (Bilgrami & Brey, 2005). With few exceptions, predators are reared using *in vivo* methods, which require maintaining concurrent prey cultures, thereby greatly reducing efficiency. The ability to mass rear entomopathogenic nematodes was the catalyst driving their development (Gaugler & Han, 2002). Ease of culture here is due to the ability of entomopathogenic species to feed on symbiotic bacteria, leading ultimately to rearing in 80 000-liter bioreactors (Georgis, 2002).

Table 4. Coefficient of preference for prey nematodes of Mononchoides gaugleri.

Prey species	*Coefficient of preference* [1]	*Combinations preferred* [2]
Meloidogyne incognita	1.00	10
Heterodera mothi	0.92	09
Anguina tritici	0.92	09
Hirschmanniella oryzae	0.67	07
Tylenchorhynchus mashhoodi	0.51	06
Xiphinema americanum	0.19	05
Paratrichodorus christei	–0.15	03
Longidorus attenuatus	–0.42	03
Helicotylenchus indicus	–0.57	02
Hoplolaimus indicus	0.00	00
Hemicriconemoides mangiferae	0.00	00

[1]Preference is measured on the scale of 0 to + 1 for prey accepted and 0 to – 1 for prey rejected.
[2]Number of combinations a prey was killed more than other species. Adapted from Bilgrami et al. (2005).

Diplogasterids can be reared on either prey nematodes or bacteria, both by *in vivo* or *in vitro* methods, since they are facultative and biphasic. *Diplenteron colobocercus*, *B. degressei*, *M. fortidens*, *M. longicaudatus* and *M. gaugleri* have been successfully maintained on *Caenorhabditis*, *Rhabditis*, *Panagrellus*, *Cephalobus*, bacteria or on a combination of nematode and bacteria for multiple generations over a period of several months. In a study on reproductive capacity of *Mononchoides,* cultures with 25 adult female nematodes per 5.5-cm agar Petri dish were started with *E. coli*. After 20 days at 30°C, culture plates averaged an impressive 10 376 individuals. The oviposition rate was 8–10 eggs day^{-1} female^{-1} (Siddiqi et al., 2003).

Mononchs possess significant potential to reduce populations of phytoparasitic nematode under field conditions, but they were never considered as a good biocontrol agent. These predators are fastidious to culture due to their localized distribution in field, long life cycles and low rate of fecundity. In contrast, stylet bearing predators appear as better biocontrol agents since they are widely distributed and occur naturally at high densities. However, their long life cycle and culture conditions hinder any practical application. Pillai and Taylor (1968) cultured diplogasterids on a dixenic culture of bacteria and *Aphelenchus avenae*. Prey nematodes and bacteria have supported growth and development of diplogasterid predators, although some appeared to provide better nematode reproduction than others.

2.6. *Conservation*

Predatory nematode conservation under natural conditions could make their practical utilization possible (Bilgrami & Brey, 2005). As compared to insects and other beneficial predatory nematodes, conservation is simple and cost-effective. Their population and predatory activities may be stimulated to counter parasitic nematode populations in the field. More studies are needed to develop methods for predatory nematodes conservation under natural habitats.

Neem (*Azadirachta indica*) products e.g., leaf powder, sawdust and oilseed cake, used as organic amendments, showed encouraging results in maintaining and conserving predatory nematode densities in the field (Akhtar, 1995; Akhtar & Mahmood, 1993). Mulching may be another option to improve conservation of predatory nematodes in the field. Mulching was found effective in stabilizing a *Iotonchus tenuicaudatus* population feeding on *Tylenchulus semipenetrans* and *Helicotylenchus dihystera* in orange orchards (Rama & Dasgupta, 1998). More studies are needed on the role of organic soil amendments and nitrogenous compounds in predatory nematode conservation.

3. BIOLOGICAL CONTROL POTENTIALS

Predatory nematodes belong to the Orders Mononchida, Diplogasterida, Rhabditida, Aphelenchida and super families, Dorylaimoidea, Nygolaimoidea, Actinolaimoidea and families Ironidae, Oncholaimidae, Monohysteridae and Thalassogeneridae etc. They show different types of feeding apparatus, and modes of prey searching, catching and feeding mechanisms. Predators of the order Mononchida possess a well developed buccal cavity with strong buccal musculature, tooth, teeth and denticles. They are commonly known as mononchs which feed by cutting, sucking and engulfing an intact prey (e.g., *Mononchus*, *Mylonchulus, Iotonchus* etc.) (Bilgrami et al., 1986; Jairajpuri & Bilgrami, 1990). As a result of their inability to perceive prey secretions, their contacts with prey depend on chance encounters. Species belonging to Aphelenchida, Dorylaimoidea and Nygolaimoidea are commonly known as aphelench, dorylaim and nygolaim predators.

Feeding apparatus in dorylaim predators (e.g., *Mesodorylaimus*) is a stylet provided with a lumen. Nygolaim predators (e.g., *Aquatides*) have a feeding apparatus called mural tooth, which is solid. Aphelench predators (e.g. *Seinura*) are provided with a pointed stylet with a lumen for ingestion. Feeding in aphelench, dorylaim and nygolaim predators is piercing and sucking type. Members belonging to Diplogasterida are commonaly referred to as diplogasterid predators (e.g., *Mononchoides*) and possess a strong buccal cavity with dorsal movable tooth. Their feeding apparatus is cutting and sucking type. Other nematode groups e.g., pelagonematids, actinolaimids, ironids, monohysterids and enoplids also include predatory species which possess cutting, sucking or piercing types of feeding. However, little is known about predation abilities and role in nematode management.

Predatory nematodes like *Seinura paynei* have been recovered from mushroom substrates feeding on free living nematodes e.g., *Acrobeloides* and *Bursilla* (Grewal, Siddiqi, & Atkey, 1991). The widespread distribution of *Seinura* and their feeding on nematodes in mushroom substrates suggest that these predators may also control populations of *Aphelenchoides,* parasitic on mushrooms (Grewal et al., 1991). However, more studies are needed to understand true predatory potential of aphelenchid species.

3.1. Biocontrol Potential of Mononchs

Prospects for use of mononchs to control plant-parasitic nematodes were speculated by Cobb (1917; 1920) and Steiner and Henley (1922). Thorne (1927) thought otherwise, considering mononchs unable to control nematode populations. Cassidy (1931), however, reported partial control under suitable conditions using *Iotonchus brachylaimus* as predator.

Further studies were made by Mulvey (1961), Esser (1963) and Ritter and Laumond (1975). Mononchs feed on a variety of soil microorganisms including nematodes (Table 5). According to Webster (1972) and Jones (1974) non-specific predators like mononchs exert only partial control and the possibility of these being successful agents of biological control are remote.

Table 5. List of plant-parasitic nematodes recorded as preys of mononchs.

Predators	*Prey nematodes*	*References*
Anatonchus amiciae	*Tylenchus, Xiphinema*	Coomans and Lima (1965)
A. ginglymodontus	*Meloidogyne hapla* (juv.)	Szczygiel (1966; 1971)
A. tridentatus	*Paratylenchus macrophallus, Aphelenchus,*	Mulvey (1961),
	Longidorus, Pratylenchus	Banage (1963)
Clarkus mulveyi	*Tylenchorhynchus nudus, Helicotylenchus multicinctus, Rotylenchulus reniformis, M. incognita* (juv.).	Mohandas and Prabhoo (1980)
C. papillatus	*Tylenchus, Tylenchulus semipenetrans,*	Cobb (1917), Menzel (1920)
	Tylochephalus auriculatus, Heterodera schachtii (juv.), *Hemicriconemoides, Aphelenchoides, M. hapla* (juv.)	Steiner and Heinley (1922)
C. sheri	*Tylenchorhynchus, Aphelenchus*	Bilgrami et al. (1986)
Coomansus indicus	*Pratylenchus, Tylenchorhynchus, Hemicriconemoides, Xiphinema*	Bilgrami et al. (1986)
Iotonchus acutus	*Trichodorus obtusus, R. robustus,*	Cobb (1917),
	Xiphinema americanum	Thorne (1932)

I. amphigonicus	*H. schachtii (*juv.)	Thorne (1924)
I. antidontus	*Tylenchorhynchus*	Bilgrami et al. (1986)
I. basidontus	*Tylenchorhynchus*	Bilgrami et al. (1986)
I. brachylaimus	*Rhadopholus similis*, *T. semipenetrans*	Cassidy (1981), Mankau (1980)
I. indicus	*Tylenchorhynchus*	Bilgrami et al. (1986)
I. kherai	*T. nudus*, *Hirschmanniella oryzae*, *H. multicinctus*, *R. reniformis*, *Meloidogyne incognita* (juv.), *Xiphinema elongatum*	Mohandas and Prabhoo (1980)
I. longicaudatus	*Hoplolaimus*, *Hirschmanniella*	Bilgrami et al. (1986)
I. monhystera	*T. nudus*, *H. oryzae*, *H. multicinctus*, *R. reniformis*, *M. incognita* (juv.)	Azmi (1983), Bilgrami et al. (1986)
I. nayari	*X. elongatum, H. oryzae, H. multicinctus, R. reniformis, M. incognita* (juv.), *T. nudus*	Mohandas and Prabhoo (1980)
I. parabasidontus	*Hirschmanniella*	Bilgrami et al. (1986)
I. prabhooi	*R. reniformis, M. incognita* (juv.)	Mohandas and Prabhoo (1980),
I. risoceiae	*Pratylenchus*	Bilgrami et al. (1986)
I. shafi	*Hoplolaimus*	Bilgrami et al. (1986)
I. trichuris	*Pratylenchus, Hoplolaimus, Tylenchorhynchus, Xiphinema*	Bilgrami et al. (1986)
I. vulvapapillatus	*Tylenchorhynchus*	Andrassy (1964), Andrassy (1973)
Miconchus aquaticus	*Helicotylenchus*, *Xiphinema*, *Hemicycliophora*	Bilgrami et al. (1986)
M. citri	*Pratylenchus*, *Tylenchorhynchus*	Bilgrami et al. (1986)
M. dalhousiensis	*Aphelenchoides*	Bilgrami et al. (1986)
Mononchus aquaticus	*Tylenchorhynchus mashoodi, H. oryzae, Hoplolaimus indicus, Helicotylenchus indicus, X. americanum, Longidorus, Paralongidorus citri, Paratrichodorus*, *Anguina tritici* (juv.), *M. incognita* (juv.), *Meloidogyne naasi* (juv.), *Heterodera mothi* (juv.), *Rotylenchus fallorobustus, Globodera rostochiensis* (juv.)	Grootaert and Maertens (1976), Grootaert et al. (1977), Small and Grootaert (1983), Bilgrami (1992), Bilgrami et al. (1986)
M. truncatus	*H. schachtii*	Thorne (1927)

(continued)

Table 5 (continued)

Predators	*Prey nematodes*	*References*
M. tunbridgensis	*Aphelenchus avenae, T. semipeterans, Hoplolaimus, Tylenchorhynchus, Hemicriconemoides*	Mankau (1980), Bilgrami et al. (1986)
Mylonchulus agilis	*Helicotylenchus vulgaris, R. fallorobustus, Longidorus caespiticola*	Doucet (1980)
M. brachyuris	*Subanguina radicicola, R. similis*	Cassidy (1931)
M. dentatus	*A. avenae, Helicotylenchus indicus, H. indicus, T. mashhoodi, M. incognita* (juv.), *H. mothi* (juv.), *H. oryzae, T. semipenetrans, Basiria, Xiphinema, Paralongidorus citri, Longidorus*	Jairajpuri and Azmi (1978), Bilgrami and Kulshreshtha (1994)
M. hawaiiensis	*T. nudus, H. oryzae, R. reniformis M. incognita* (juv.)	Mohandas and Prabhoo (1980)
M. minor	*A. tritici* (juv.), *M. incognita* (juv), *T. semipenetrans, X. americanum. R. reniformis*	Kulshreshtha, Bilgrami, and Khan (1993), Choudhary and Sivakumar (2000)
M. parabrachuris	*H. schachtii* (juv)	Thorne (1927)
M. sigmaturus	*H. schachtii* (juv.), *R. similis, T. semipenetrans, Meloidogyne javanica* (juv.), *Subanguina radicicola*	Thorne (1927), Cassidy (1931), Cohn and Mordechai (1973, 1974), Mankau (1982)
Prionchulus muscorum	*Aphelenchus, Hoplolaimus,Tylenchorhynchus, Hemicriconemoides,*	Szczygiel (1971), Arpin (1976)
	Aphelenchus	Bilgrami et al. (1986)
P. punctatus	*A. avenae, M. naasi* (juv.) *G. rostochiensis* (juv.), *R. fallorobustus, Helicotylenchus, A. tritici* (juv.)	Nelmes (1974), Maertens (1975), Grootaert et al. (1977), Small and Grootaert (1983), Small (1979)
Sporonchulus ibitiensis	*Aphelenchus, Aphelenchoides*	Carvalho (1951)
S. vagabundus	*Aphelenchoides, Hemicycliophora, Trichodorus*	Bilgrami et al. (1986)

Predatory nematodes remained neglected until 1974 when Cohn and Mordechai (1974) found correlation between *Mylonchulus* and *Tylenchulus* in pot experiments. Similarly, Small (1979) reported significant reduction in *Globodera* and *Meloidogyne* populations in the presence of *Prionchulus*. Ahmad and Jairajpuri (1982) reported significant correlation between *Parahadronchus* and *Trichodorus* and *Hemicriconemoides* under field conditions. Azmi (1983) indicated increase in *Iotonchus* and reduction in *Helicotylenchus* populations.

Observations on the predation by mononchs viz., factors influencing predation (Bilgrami, Ahmad, & Jairajpuri, 1983); predation (Nelmes, 1974; Small & Grootaert, 1983; Bilgrami, Ahmad, & Jairajpuri, 1984; Kulshreshtha et al., 1993; Bilgrami et al., 1986); predator strike rate, prey resistance and susceptibility to predation (Bilgrami & Jairajpuri, 1989a; Bilgrami, 1992, 1995); relationships with prey trophic groups (Bilgrami, 1992); cannibalism (Bilgrami & Jairajpuri, 1984); and range of prey (Small, 1979, 1987) etc., were made to evaluate predatory potential of mononchs. In a study by Bilgrami et al. (1986) analysis of gut contents of mononchs revealed their voracious feeding on different species of plant-parasitic nematodes. Dorylaim, tylench and bacteriophagous nematodes were found intact within the intestine, while others were present in semi digested conditions (Table 2).

Under natural conditions mononchs feed upon all types of nematodes, besides rotifers and other soil microorganisms. Arpin (1979) and Mahapatra and Rao (1980) found significant correlation between mononchs and free-living but Nelmes and McCulloch (1975) did not find such a correlation. Study made by Bilgrami et al., (1986) showed that more predators (75%) had free-living nematodes in their intestine than tylenchs (45%) or dorylaims (41%) (Table 2).

Any relationship of mononchs with prey nematodes present in the soil could not be determined since observations were made on mounted specimens and not the live populations. It cannot be suggested with certainty that widespread presence of free-living nematodes is either due to any preference or due to the widespread occurrence of free-living nematodes. Of all the mononchs, *Parahadronchus* was the most active predator as 68% of its specimens had prey in its intestine while *Coomansus* was least active with only 21% prey. Eight genera of Tylenchida, six of Dorylaimida, five of Mononchida, three of Rhabditida and one each of Diplogasterida and Monhystera were identified as prey of *Parahadronchus shakily*. Mohandas and Prabhoo (1980) did not find any prey in the intestine of *Mylonchulus* spp.

3.2. Biocontrol Potential of Dorylaim and Nygolaim Predators

Dorylaim, nygolaim and aphelench predators, which have piercing-sucking type of feeding (Bilgrami & Gaugler, 2004), can switch to feeding on bacteria and fungi (Hollis, 1957; Ferris, 1968; Wood, 1973), which presumably enhances their survival when prey nematodes are scarce.

In addition to nematodes (Wyss & Grootaert, 1977; Shafqat et al., 1987; Khan et al., 1991), dorylaim and nygolaim predators also feed on algae and fungi (Hollis, 1957; Ferris, 1968; Wood, 1973; Bilgrami, 1990b) (Table 6). Consequently, they can also be grown on algae and fungi. Their widespread and abundant presence reflects the possibility of controlling nematode populations. It is, however, not known up to what extent and under what conditions nematode populations are reduced, since such an evaluation has never been made. Dorylaim and nygolaim predators occur in all soil types, climates and habitats. The presence of 2, 3 or more genera at one field/place is quite common.

Feeding of *Eudorylaimus obtusicaudatus* on *Heterodera schachtii* eggs and increased population of *Thornia* sp., in the presence of citrus nematode suggests their control potential. *Aporcelaimellus*, *Discolaimus*, *Mesodorylaimus* and *Dorylaimus* (Khan et al., 1991; Khan, Bilgrami, & Jairajpuri, 1995a; Khan, Bilgrami, & Jairajpuri, 1995b; Bilgrami, 1992, 1993, 1995) showed significant predatory potential. They are attracted towards prey and aggregate at the feeding sites in response to prey secretions. Predation rate, feeding, aggregation, and prey search activities are governed by biotic and abiotic factors such as temperature, density, starvation, incubation, etc. These factors affect their chemotactic respons (Bilgrami & Jairajpuri, 1988a), dispersion of prey kairomones (Green, 1980) and rate of predation (Bilgrami, 1997). Reduced predator activity (Bilgrami et al. 1983) and depleted prey attractants (Huettel, 1986) as influenced by temperature extremes are also possible causes of reduced predation.

Similarly to temperature, predatory activities are also affected by starvation (Jairajpuri & Bilgrami, 1990). Starvation of 14 days did not alter predation by *Dorylaimus stagnalis* (Bilgrami et al., 1984; Shafqat et al., 1987) but short-term food deprivation enhanced predation. Bilgrami and Gaugler (2005) observed maximal predation in 6 days starved predators, presumably because food deprivations increased predator ability to detect more prey individuals to kill. Doncaster and Seymour (1974) concluded that starved nematodes could perceive weaker stimuli much faster than when they are well fed, because of decreased minimum response threshold. Stylet bearing predators show density dependent predation (Khan et al., 1991) similar to other group of predators. More predator-prey encounters at higher prey densities always result in the increased rate of predation.

Table 6. Plant-parasitic nematodes as prey of dorylaim, nygolaim and tylenchid predators.

Predators	Prey nematodes	References
Allodorylaimus americanus	*M. incognita* (juv.), *A. tritici* (juv.), *Xiphinema basiri, Longidorus, T. mashoodi, H. oryzae, Aphelenchoides, Basiria, A. avenae, T. semipenetrans, Trichodorus*	Khan et al. (1995a, 1995b)
A. amylovorus	*T. semipenetrans*	Mankau (1982)
A. obscurus	*H. schachtii* (juv.)	Thorne and Swanger (1936)
A. obtusicaudatus	*H. schachtii* (juv.)	Marinari, Vinciguerra, Vovlas and Zullini (1982)
A. nivalis	*M. incognita* (juv.), *H. mothi* (juv.), *X. basiri, Longidorus, T. mashoodi, H. oryzae, H. indicus, Aphelenchoides, Basiria, A. avenae, T. semipenetrans, Trichodorus*	Bilgrami (1993), Khan et al. (1991)
Discolaimus arenicolus	*M. incognita* (juv.)	Yeates et al. (1993)

D. silvicolus	*M. incognita* (juv.), *H. mothi* (juv.), *A. tritici* (juv.), *X. basiri, Longidorus,T. mashoodi, H. oryzae, Aphelenchoides, Basiria, A. avenae, T. semipenetrans, Trichodorus*	Khan et al. (1995a)
Dorylaimus obtusicaudatus	*H. schachtii* (eggs)	Cobb (1929)
D. obscurus	*H. schachtii* (eggs)	Thorne and Swanger (1936)
D. stagnalis	*T. mashoodi, H. oryzae, H. indicus X. americanum, Longidorus, P. citri, A. tritici* (Juv.), *M. incognita* (juv.) *H. mothi* (juv.)	Bilgrami (1992) Shafqat et al. (1987)
Eudorylaimus obtusicaudatus	*H. schachtii*	Esser (1987)
Labronema vulvapapillatum	*A. avenae, A. tritici* (juv.)	Wyss and Grootaert (1977)
	M. naasi (juv.), *G. rostochiensis* (juv.)	Grootaert and Small (1982), Small and Grootaert (1983), Esser (1987)
Mesodorylaimus bastiani	*M. incognita* (juv.), *H. mothi* (juv.)	Bilgrami (1992)
	X. basiri, X. americanum, X. insigne, Longidorus, T. mashoodi, H. oryzae, H. indicus, Aphelenchoides, Basiria, A. avenae, T. semipenetrans, Trichodorus, Paratrichodorus, A. tritici (juv.), *Longidorus, T. mashoodi*	
Pungentus monohystera	*T. semipenetrans*	Mankau (1982)
Seinura celeris	*A. avenae*	Hechler and Taylor (1966)
S. demani	*A. bicaudatus, A. avenae*	Wood (1974)
S. oliveirae	*A. avenae*	Hechler and Taylor (1966)
S. oxura	*A. avenae, Ditylenchus myceliophagus*	Hechler and Taylor (1966), Cayrol (1970)
S. steineri	*A. avenae*	Hechler and Taylor (1966)
S. tenuicaudata	*M. marioni* (juv.), *Pratylenchus pratensis A. avenae, A. parietinus*	Linford and Oliviera (1937),
	D. dipsaci, Heterodera trifolii (juv.), *M. hapla* (juv.), *Neotylenchus linfordi*	Hechler (1963)

(continued)

3.3. Biocontrol Potential of Diplogasterid Predators

Diplogasterid predators remained largely neglected until Yeates (1969) evaluated predatory abilities of *Diplenteron*. Subsequent studies (Goodrich, Hechler, & Taylor, 1969; Grootaert et al., 1977) brought to light more informations on their biology, behaviour, predator-prey relationships, ecology, predation abilities etc. Despite these efforts, diplogasterid predators have received thus far less attention (Bilgrami & Jairajpuri, 1989a; Fauzia, Jairajpuri, & Khan, 1998) (Table 7) than the large and easily studied mononchs, yet they possess more favorable biological control traits.

Bilgrami and Jairajpuri (1988a, 1989a, 1989b) and Bilgrami (1990a, 1997) made the first case for diplogasterid predators by offering detailed accounts on their prey searching, preference, strike rate, and prey resistance and susceptibility to predation. Among the advantages of diplogasterids over mononch predators, these authors listed ease of *in vitro* culture, high rates of reproduction and predation, short life cycle, ability to detect and respond to prey attractants, and rare cannibalism. Diplogasterids further differ from mononch juveniles in possessing greater tolerance to unfavourable environmental conditions (Bilgrami, 1997).

Particularly significant were the observations of Yeates (1969), Grootaert et al., (1977) and Bilgrami (1997) that the diplogasterids *Diplenteron* and *Butlerius* switch to feeding on bacteria in the absence of prey, strongly suggesting an enhanced capability to persist when prey populations are reduced. Switching food resources is therefore a common trait among predaceous diplogasterids. Fauzia et al. (1998) subsequently demonstrated the ability of *Mononchoides* to reduce galling by root knot nematodes in post tests, resulting in improved vegetative growth and increased root mass.

Recently, Bilgrami, Brey, and Gaugler (2007) made first field release of a diplogasterid predator *Mononchoides gaugleri* to determine its effect on existing parasitic nematode populations in a turf grass fields. They reported significant control of plant-parasitic nematodes although the rate of predator persistence was low.

Prey preference is another desirable feature in biological control agents but predators, whether mammalian, reptilian, insect, or nematode, tend to be polyphagous. Mononchs, too, are polyphagous (Bilgrami et al., 1984; Bilgrami 1997). However, diplogasterid predators appear to be more prey-specific as indicated by *Odontopharynx* which attacked and killed six of 17 species presented in a laboratory study (Chitambar & Noffsinger, 1989). Moreover, some prey species were preferred more strongly than others. A strong degree of preference was similarly reported for other diplogasterid predators *Butlerius* and *Mononchoides*. Bilgrami and Jairajpuri (1989a) and Bilgrami et al. (2005) showed that *M. longicaudatus*, *M. fortidens* and *M. gaugleri* preferred endoparasitic over ectoparasitic prey species.

Table 7. List of plant-parasitic nematodes recorded as prey of diplogasterid predators.

	Prey nematodes	*References*
Butlerius degrissei	*A. avenae, A. fragariae*	Grootaert et al. (1977)
	Pratylenchus, G. rostochiensis (juveniles)	Grootaert and Jaques (1979)
	R. robustus	Small and Grootaert (1983)
B. micans	*A. avenae*	Pillai and Taylor (1968)
Fictor anchicoprophaga	*A. avenae*	Pillai and Taylor (1968)
Mononchoides bollingeri	*A. avenae*	Goodrich et al. (1968)
M. changi	*A. avenae*	Goodrich et al. (1968)
M. fortidens	*M. incognita* (juv.), *A. tritici* (juv.)	Bilgrami and Jairajpuri (1988, 1989)
	T. mashoodi, X. americanum, H. indicus, Longidorus, Trichodorus	
M. gaugleri	*M. incognita* (juv.), *A. tritici* (juv.),	Bilgrami et al. (2005)
	H. mothi (juv), *T. mashhoodi, Longidorus, X. americanum, Trichodorus*	
	H. indicus, H. mangiferae, P. christei	
M. longicaudatus	*M. incognita* (juv.), *A. tritici* (juv.) *T. mashoodi, X. americanum, H. indicus, Longidorus, Trichodorus*	Bilgrami and Jairajpuri (1988b, 1989)

4. FUTURE PROSPECTS

Predatory nematodes represent a small amount of the available biomass in the soil, but their presence across so many trophic levels e.g., plant, fungal, bacterial and carrion feeders is vitally important in soil ecosystem processes (Barker & Koenning, 1998). Their future role in nematode management depends greatly on advances made on other control methods, their effectiveness, and the resources provided to establish research programs.

The real possibility of using predatory nematodes in nematode management programs lies in the diplogasterid predators due to their biphasic feeding, high rates of predation and fecundity, short life cycle, ability to search for prey and the presence of resistant juveniles. Diplogasterid predators rarely resort to cannibalism due to their bacteriophagous feeding habits.

Despite remarkable similarities with the attributes of entomopathogenic nematode species, diplogasterids should not be considered as unilateral inundative agents (i.e., repeated applications for short-term control). The flexible bi-phasic feeding behaviour of diplogasterids should endow them with superior persistence; that is, when prey become scarce they should switch to feeding on soil bacteria to maintain themselves. Nematode predators are likely to offer the most promise as augmentative agents in colonization efforts in combination with cultural control tactics, such as rotation, cover cropping, green manuring, organic amendments.

Dorylaim and nygolaim predators are ubiquitous species, occurring in all types of climates and habitats. The presence of two, three or more genera of dorylaims and nygolaims at one field/place is quite usual and their abundance has been estimated to be 200–500 millions/acre (Thorne, 1930). Their widespread and abundant presence, the omnivorous feeding habits, the ability to perceive prey kairomones, and the inverse relationships with prey populations observed in pot trails (Boosalis & Mankau, 1965) indicate their potential as nematode biological control agents.

REFERENCES

Ahmad, N., & Jairjapuri, M. S. (1982). Population fluctuations of the predatory nematode *Parahadronchus shakili* (Jairajpuri, 1969) (Mononchida). *Proceedings of the Symposium on Animal Population, India, 3*, 1–12.

Akhtar, M. (1995). Biological control of the root-knot nematode *Meloidogyne incognita* in tomato by the predatory nematode *Mononchus aquaticus*. *International Pest Control, 37*, 18–19.

Akhtar, M., & Mahmood, I. (1993). Effect of *Mononchus aquaticus* and organic amendments on *Meloidogyne incognita* development on chili. *Nematologia Mediterranea, 21*, 251–262.

Andrassy, I. (1964). Subwasserbematoden aus groben *Gebrigsgegenden ostafrikas*. *Acta Zoologica Academiae Scientiarum Hungaricae, 10*, 1–59.

Andrassy, I. (1973). 100 Neue nematodenarten in der ungarischen fauna. *Opuscula Zoologica,* 9, 187–233.

Arpin, P. (1976). Etude et discussion sur un milieu de culture pour Mononchidae (Nematoda). *Revue de Ecologie et de Biologie du Sol, 13*, 629–634.

Arpin, P. (1979). Ecologie et systematique des nematodes Mononchidae des zones forestières et harpaeces zones climate, temperate humid I. Types de sol et groupements specifiques. *Revue de Nematologie, 4*, 131–143.

Azmi, M. I. (1983). Predatory behaviour of nematodes: Biological control of *Helicotylenchus dihystera* through the predaceous nematode, *Iotonchus monhystera*. *Indian Journal of Nematology, 13*, 1–8.

Banage, W. B. (1963). The ecological importance of free living soil nematodes with special reference to those of moorland soil. *Journal of Animal Ecology, 32*, 133–140.

Barker, K. R., & Koenning, S. R. (1998). Developing sustainable systems for nematode management. *Annual Review of Phytopathology*, 36, 165–205.

Bilgrami, A. L. (1990a). Diplogasterid predators. In M. S. Jairajpuri, M. M. Alam & I. Ahmad (Eds.), *Nematode biocontrol: Aspects and prospects* (pp. 136–142). New Delhi: CBS Publisher & Distributor Pvt. Ltd.

Bilgrami, A. L. (1990b). Dorylaim predators. In M. S. Jairajpuri, M. M. Alam & I. Ahmad (Eds.), *Nematode biocontrol: Aspects and prospects* (pp. 143–148). New Delhi: CBS Publisher & Distributor Pvt. Ltd.

Bilgrami, A. L. (1992). Resistance and susceptibility of prey nematodes to predation and strike rate of the predators, *Mononchus aquaticus, Dorylaimus stagnalis* and *Aquatides thornei. Fundamental and Applied Nematology, 15*, 265–270.

Bilgrami, A. L. (1993). Analyses of relationships between predation by *Aporcelaimellus nivalis* and different prey trophic categories. *Nematologica, 39*, 356–365.

Bilgrami, A. L. (1995). Numerical analysis of the relationship between the predation by *Mesodorylaimus bastiani* (Nematoda: Dorylaimida) and different prey trophic categories. *Nematologia Mediterranea, 23*, 81–88.

Bilgrami, A. L. (1997). *Nematode Biopesticides* (p. 262). Aligarh: Aligarh University Press.

Bilgrami, A. L., & Brey, C. (2005). *Potential of predatory nematodes to control plant-parasitic nematodes* (pp. 447–464). UK: CABI.

Bilgrami, A. L., & Gaugler, R. (2004). Feeding behaviour. In R. Gaugler & A. L. Bilgrami (Eds.), *Nematode behaviour* (pp. 91–125). UK: CABI.

Bilgrami, A. L., & Gaugler, R. (2005). Feeding behaviour of the predatory nematodes *Laimydorus baldus* and *Discolaimus major* (Nematoda: Dorylaimida). *Nematology, 7*, 11–20.

Bilgrami, A. L., & Jairajpuri, M. S. (1984). Cannibalism in *Mononchus aquaticus. Indian Journal of Nematology, 14*, 202–202.

Bilgrami, A. L., & Jairajpuri, M. S. (1988a). Attraction of *Mononchoides longicaudatus* and *M. fortidens* (Nematoda: Diplogasterida) towards prey and factors influencing attraction. *Revue de Nematologie, 11*, 195–202.

Bilgrami, A. L., & Jairajpuri, M. S. (1988b). Aggregation of *Mononchoides longicaudatus* and *M. fortidens* (Nematoda: Diplogasterida) at feeding sites. *Nematologica, 34*, 119–121.

Bilgrami, A. L., & Jairajpuri, M. S. (1989a). Resistance of prey to predation and strike rate of the predators *Mononchoides longicaudatus* and *M. fortidens* (Nematoda: Diplogastgerida). *Revue de Nematologie, 12*, 45–49.

Bilgrami, A. L., & Jairajpuri, M. S. (1989b). Predatory abilities *of Mononchoides longicaudatus* and *M. fortidens* (Nematoda: Diplogasterida) and factors influencing predation. *Nematologica, 35*, 475–488.

Bilgrami, A. L., & Kulshreshtha, R. (1994). Evaluation of predation abilities of *Mylonchulus dentatus. Indian Journal of Nematology, 23*, 191–198.

Bilgrami, A. L., & Pervez, R. (2000). Prey searching and attraction behaviours of *Mesodorylaimus bastiani* and *Aquatides thornei* (Nematoda: Dorylaimida). *International Journal of Nematology, 10*, 199–206.

Bilgrami, A. L., Ahmad, I., & Jairajpuri, M. S. (1983). Some factors influencing predation by Mononchus aquaticus. *Revue de Nematologie, 35*, 475–488.

Bilgrami, A. L., Ahmad, I., & Jairajpuri, M. S. (1984). Observations on the predatory behaviour of *Mononchus aquaticus. Nematologia Mediterranea, 12*, 41–45.

Bilgrami, A. L., Ahmad, I., & Jairajpuri, M. S. (1986). A study on the intestinal contents of some mononchs. *Revue de Nematologie, 9*, 191–194.

Bilgrami, A. L., Brey, C., & Gaugler, R. (2007). First field release of a predatory nematode *Mononchoides gaugleri* (Nematoda: Diplogasterida) to control plant parasitic nematodes. *Nematology, 9*, (in press).

Bilgrami, A. L., Gaugler, R., & Brey, C. (2005). Prey preference and feeding behaviour of the diplogastrid predator *Mononchoides gaugleri* (Nematoda: Diplogastrida). *Nematology, 7*, 333–342.

Bilgrami, A. L., Khan, Z., Kondo, E., & Yoshiga, T. (2002). Generic diversity and community dynamics of nematodes with particular reference to predaceous nematodes at deciduous forest of Henukuma, Saga Prefecture, Japan. *International Journal of Nematology, 12*, 46–54.

Bilgrami, A. L., Wenju, L., Wang, P., & Qi, L. (2003). Generic diversity, population structure and community ecology of plant and soil nematodes. *International Journal of Nematology, 13*, 104–117.

Bilgrami, A. L., Pervez, R., Kondo, E., & Yoshiga, T. (2001). Attraction and aggregation behaviour of *Mesodorylaimus bastiani* and *Aquatides thornei* (Nematoda: Dorylaimida). *Applied Entomology and Zoology, 36*, 243–249.

Bilgrami, A. L., Pervez, R., Yoshiga, T., & Kondo, E. (2000). Feeding, attraction and aggregation behaviour of *Mesodorylaimus bastiani* and *Aquatides thornei* at the feeding site using *Hirschmanniella oryzae* as prey. *International Journal of Nematology, 10*, 207–214.

Boosalis, M., & Mankau, R. (1965). Parasitism and predation of soil microorganisms. In F. Baker & W. Synder (Eds.), *Ecology of soil-borne plant pathogens* (pp. 374–391). Berkely: University of California Press.

Cassidy, G. H. (1931). Some mononchs of Hawaii. *Hawaiian Planters Records, 35*, 305–339.
Chitambar, J. J., & Noffsinger, M. (1989). Predaceous behaviour and life history of *Odontopharynx longicaudata* (Diplogasterida). *Journal of Nematology, 21*, 284–291.
Carvalho, J. C. (1951). Una nova especie de *Mononchus* (*M. ibitiensis* n. sp.). *Bragantia*, 11, 51–54.
Cayrol, J. C. (1970). Action des autres composants de la biocenose du champignon de couche sur le nematode mycophage, *Ditylenchus myceliophagus* J. B. Goodey, 1958 et étude de son anabiose: Forme de survie en conditions defavorables. *Revue d'Ecologie et de Biologie du Sol, 7*, 409–440.
Choudhary, B. N., & Sivakumar, C. V. (2000). Biocontrol potential of *Mylonchulus minor* against some plant-parasitic nematodes. *Annals of Plant Protection Sciences, 8*, 53–57.
Cobb, N. A. (1917). The *Mononchus*: a genus of free living predatory nematodes. *Soil Science, 3*, 431–486.
Cobb, N. A. (1920). Transfer of nematodes (mononchs) from place to place for economic purposes. *Science*, 51, 640–641.
Cobb, N. A. (1929). Nemas of the genus *Dorylaimus* attacking the edges of mites. *Journal of Parasitology, 15*, 284 [Abstract].
Cohn, E., & Mordechai, M. (1973). Biological control of the citrus nematode. *Phytoparasitica, 1*, 32 [Abstract].
Cohn, E., & Mordechai, M. (1974). Experiments in suppressing citrus nematode populations by use of marigold and predaceous nematode. *Nematologia Mediterranea, 2*, 43–53.
Coomans, A., & Lima, B. (1965). Description of *Anatonchus amiciae* n. sp. (Nematoda: Mononchidae) with observations on its juvenile stages and anatomy. *Nematologica, 11*, 413–431.
Doncaster, C. C., & Seymour, M. K. (1973). Exploration and selection of penetration site by Tylenchida. *Nematologica, 19*, 137–145.
Doucet, M. E. (1980). Description d'une nouvelle espèce du genere *Mylonchulus* (Nematoda: Dorylaimida). *Nematologia Mediterranea, 8*, 37–42.
Esser, E. (1963). Nematode interactions in plate of non sterile water-agar. *Proceedings of the Soil and Crop Science Society of Florida, 23,* 121–138.
Esser, E. (1987). Biological control of nematodes by nematodes I. Dorylaims (Nematoda: Dorylaimida). *Nematology Circular, 144*, 4. pp.
Fauzia, M., Jairajpuri, M. S., & Khan, Z. (1998). Biocontrol potential of *Mononchoides longicaudatus* on *Meloidogyne incognita* on tomato plants. *International Journal of Nematology, 8*, 89–91.
Ferris, V. R. (1968). Biometric analysis in the genus *Labronema* (Nematoda: Dorylaimida) with a redescription of *L. thornei* n. sp. *Nematologica, 14*, 276–284.
Gaugler, R., & Haan, R. (2002). Mass production technology. In R. Gaugler (Ed.), *Entomopathogenic Nematology* (pp. 289–310). UK: CABI.
Grewal, P. S., Siddiqi, M. R., & Atkey, P. T. (1991). *Aphelenchoides richardsoni* sp. nov. and *Seinura paynei* sp. nov. from mushrooms in the British Isles and *S. obscura* sp. nov. from India (Nematoda: Aphelenchina). *Afro-Asian Journal of Nematology, 1*, 204–211.
Georgis, R. (2002). The Biosys experiment: an insider's perspective. In R. Gaugler (Ed.), *Entomopathogenic Nematology* (pp. 357–372). UK: CABI.
Goodrich, M., Hechler, H. C., & Taylor, D. B. (1969). *Mononchoides changi* n. sp., and *M. bollingeri* n. sp. (Nematoda: Diplogasterinae) from a waste treatment plant. *Nematologica, 14*, 25–36.
Grootaert, P., Jaques, A., & Small, R. W. (1977). Prey selection in *Butlerius* sp. (Rhabditida: Diplogasteridae). *Mededelingen van de Faculteit Landbouwwetenschappen Rijksuniversiteit Gent, 42*, 1559–1563.
Grootaert, P., & Maertens, D. (1976). Cultivation and life cycle of *Mononchus aquaticus*. *Nematologica, 22*, 173–181.
Grootaert, P. & Wyss, U. (1979). Ultrastructure and function of the anterior feeding apparatus in *Mononchus aquaticus*. *Nematologica,* 25, 163–173.
Grootaert, P. & Small, R. W. (1982). Aspects of the biology of *Labronema vulvapapillatum* (Meyl) (Nematoda: Dorylaimida) in laboratory culture. *Biologisch Jaarboek Dodonaea, 50,* 135–148.
Green, C. D., (1980). Nematode sex attractants. *Helminthological Abstracts Series B, 49*, 81–93.
Hassell, M. P. (1978). *The dynamics of arthropod Predator-Prey systems* (p. 25). Princeton New Jersey: Monographs in Population Biology 13, Princeton University Press.

Hechler, H. C. (1963). Description, developmental biology and feeding habits of *Seinura tenuicaudata* (de Man), a nematode predator. *Proceedings of the Helminthological Society of Washington, 30*, 183–195.

Hechler, H. C., & Taylor, D. P. (1966). The life cycle histories of *Seinura celeries*, *S. olivierae*, *S. oxura* and *S. steineri* (Nematoda: Aphelenchoididae). *Proceedings of the Helminthological Society of Washington, 33*, 71–83.

Hollis, J. P. (1957). Cultural studies with *Dorylaimus ettersbergensis*. *Phytopathology, 47*, 468–473.

Huettel, R. N., (1986). Chemical communicators in nematodes. *Journal of Nematology*, 18, 3–8.

Jairajpuri, M. S., & Azmi, M. I. (1978). Some studies on the predatory behaviour of *Mylonchulus dentatus*. *Nematologia Mediterranea, 6*, 205–212.

Jairajpuri, M. S., & Bilgrami, A. L. (1990). Predatory nematodes. In M. S. Jairajpuri, M. M. Alam, & I. Ahmad (Eds.), *Nematode Bio-control: Aspects and Prospects* (pp. 95–125). New Delhi: CBS Publishers and Distributors Pvt., Ltd.

Jones, F. G. W. (1974). Control of nematode pests, background and outlook for biological control. In D. Price-Jones & M. E. Solomon (Eds.), *Biology in pest and disease control* (pp. 249–268). The British Ecological Society: Oxford Press.,

Khan, Z., Bilgrami, A. L., & Jairajpuri, M. S. (1991). Some studies on the predation abilities of *Aporcelaimellus nivalis* (Nematoda: Dorylaimida). *Nematologica, 37*, 333–342.

Khan, Z., Bilgrami, A. L., & Jairajpuri, M. S. (1995a). Observations on the predation abilities of *Neoactinolaimus agilis* (Dorylaimida: Actinolaimoidea). *Indian Journal of Nematology, 25*, 129–135.

Khan, Z., Bilgrami, A. L., & Jairajpuri, M. S. (1995b). A comparative study on the predation by *Allodorylaimus americanus* n. sp. and *Discolaimus silvicolus* (Nematoda: Dorylaimida) on different species of plant-parasitic nematodes. *Fundamental and Applied Nematology, 18*, 99–108.

Kulshreshtha, R., Bilgrami, A. L., & Khan, Z. (1993). Predation abilities of *Mylonchulus minor* and factors influencing predation. *Annals of Plant Protection Sciences, 1*, 79–84.

Linford, M. B., & Oliviera, J. M. (1937). The feeding of some hollow stylet bearing nematodes. *Proceedings of the Helminthological Society of Washington, 4*, 41–46.

Linford, M. B., & Oliviera, J. M. (1938). Potential agents of biological control of plant-parasitic nematodes. *Phytopathology, 28*, 14.

Maertens, D. (1975). Observations on the life cycle of *Prionchulus punctatus* (Cobb, 1917) and culture conditions. *Biologische Jaarboek Dodonaea, 43*, 197–218.

Mahapatra, N. K., & Rao, Y. S. (1980). Bionomics of *Iotonchus punctatus* (Cobb, 1917) and cultural conditions. *Proceeding of the Indian Academy of Parasitol*ogy, *2*, 85–87.

Mankau, R. (1980). Biological control of nematode pests by natural enemies. *Annual Revue of Phytopathology, 18*, 415–440.

Marinari, A., Vinciguerra, M. T., Vovlas, N., & Zullini, A. (1982). Nematodi delle dune costiere d'Italia. In Quaderni sulla Struttura delle Zoocenosi Terrestri. 3. Ambienti Mediterranei. I. Le Coste Sabbiose, 27–50.

Menzel, R., (1920). Über die Nahrung der freilebenden Nematoden und die Art ihrer Aufnahme. Ein Beitrag zur Kenntnis der Ernährung der Würmer. *Verhandlungen der Naturforschenden Gesellschaft in Basel*, 31, 153–188.

Mohandas, C., & Prabhoo, N. R. (1980). The feeding behaviour and food preference of predatory nematodes (Mononchida) from the soil of Kerala, India. *Revue d'Ecologie et de Biologie du Sol, 17*, 153–160.

Mulvey, R. H. (1961). The Mononchidae: a family of predaceous nematodes I. genus *Mylonchulus* (Enoplida: Nematoda) *Canadian Journal of Zoology, 39*, 665–696.

Nelmes, A. J. (1974). Evaluation of the feeding behaviour of *Prionchulus punctatus* (Cobb) a nematode predator. *Journal of Animal Ecology, 43*, 553–565.

Nelmes, A. J., & McCulloch, J. S. (1975). Number of mononchid nematodes in soil sown to cereals and grasses. *Annals of Applied Biology, 79*, 231–242.

Pillai, J., & Taylor, D. (1968). Biology of *Paroigolaimella bernensis* and *Fictor anchicoprophaga* (Diplogasteriniae) in laboratory culture. *Nematologica, 14*, 159–170.

Poinar, G. O., Triggiani, O., & Merritt, R. (1976). Life history of *Eudiplogaster aphodii,* a facultative parasite of *Aphodius fimetarius*. *Nematologica, 22*, 79–86.

Rama, K., & Dasgupta, M. K. (1998). Biocontrol of nematodes associated with mandarin orange decline by the promotion of predatory nematode *Iotonchus tenuicaudatus* (Kreis, 1924). *Indian Journal of Nematology, 28*, 118–124.

Ritter, M., & Laumond, C., (1975). Review of the use of nematodes in biological control programs against parasites and pests of cultivated plants. *Bulletin des Recherches Agronomiques de Gembloux,* 43, 331–334.

Shafqat, S., Bilgrami, A. L., & Jairajpuri, M. S. (1987). Evaluation of the predatory behaviour of *Dorylaimus stagnalis* (Nematoda: Dorylaimida). *Revue de Nematologie, 10,* 455–461.

Siddiqi, M. R., Bilgrami, A. L. & Tabassum, K. (2003). Description and biology of *Mononchoides gaugleri* sp. n. (Nematoda: Diplogasterida). *International Journal of Nematology,* 14, 124–129.

Small, R. W. (1979). The effects of predatory nematodes on populations of plant-parasitic nematodes in pots. *Nematologica, 25*, 94–103.

Small, R. W. (1987). A review of the prey of predatory soil nematodes. *Pedobiologia, 30*, 179–206.

Small, R. W., & Grootaert, P. (1983). Observations on the predatory abilities of some soil dwelling predatory nematodes. *Nematologica, 29*, 109–118.

Steiner, G., & Heinley, H. (1922). The possibility of control of *Heterodera radicicola* and other plant injurious nematodes by means of predatory nemas, especially *Mononchus papillatus*. *Journal of Washington Academy of Science, 12*, 367–385.

Szczygiel, A., (1966). Studies on the fauna and population dynamics of nematodes occurring on strawberry plantation. *Ekologia Polska,* 14, 651–709.

Szczygiel, A. (1971). Wystepowanie drapieznych nicieni zrodzniy Mononchidae w glebach uprawnych w polsce. *Zeszyty Problemowe Postepow Nauk Rolniczych, 121*, 145–158.

Thorne, G. (1924). Utah nematodes of the genus *Mononchus*. *Transactions of the American Microscopy Society, 43*, 157–171.

Thorne, G. (1927). The life history, habits and economic importance of some mononchs. *Journal of Agricultural Research, 34*, 265–286.

Thorne, G. (1930). Predaceous nemas of the genus *Nygolaimus* and new genus *Sectonema*. *Journal of Agricultural Research, 44*, 445–446.

Thorne, G. (1932). Specimens of *Mononchus acutus* found to contain *Trichodorus obtusus*, *Tylenchulus robustus* and *Xiphinema americanum*. *Journal of Parasitology, 19*, 90.

Thorne, G., & Swanger, H. H. (1936). A monograph of the nematode genera *Dorylaimus*, *Aporcelaimus*, *Dorylaimoides* and *Pungentus*. *Capita Zoologica, 6*, 1–233.

Webster, J. M. (1972). Nematodes and biological control. In J. M. Webster (Ed.). *Economic nematology*. Academic Press: New York. NY, pp. 469–496.

Wood, F. H. (1973). Nematode feeding relationships. Relationships of soil dwelling nematodes. *Soil Biology Biochemistry*, 5, 593–601.

Wood, F. H. (1974). Biology of *Seinura demani* (Nematoda: Aphelenchoididae). *Nematologica, 20*, 347–353.

Wyss, U. (1971). Der Mechanismus der Nahrungsaufnahme bei *Trichodorus similes*. *Nematologica, 7*, 508–518.

Wyss, U., & Grootaert, P. (1977). Feeding mechanisms of *Labronema* sp. *Mededelingen van de Faculteit Landbouwwetenschappen Rijksuniversiteit Gent, 42*, 1521–1527.

Yeates, G. W. (1969). Predation by *Mononchoides potohikus* in laboratory culture. *Nematologica, 15*, 1–9.

Yeates, G. W., Bongers, T., De Goede, R. G. M., Freckman, D. W., & Georgieva, S. S. (1993). Feeding habits in soil nematode families and genera-An outline for soil ecologist. *Journal of Nematology, 25*, 315–331.

2

L. HILDALGO-DIAZ[1] AND B. R. KERRY[2]

INTEGRATION OF BIOLOGICAL CONTROL WITH OTHER METHODS OF NEMATODE MANAGEMENT

[1]*Centro Nacional de Sanidad Agropecuaria (CENSA), Apdo 10, San José de las Lajas, La Habana, Cuba*
[2]*Rothamsted Research, Harpenden, Hertfordshire, AL5 2JQ, UK*

Abstract. This chapter describes measures used to improve the performance of biological control agents for nematode management. Suppressive soils have been associated with the continuous cultivation of nematode-susceptible crops, which support increases in the natural enemy community. Soils that become suppressive to nematode pests and the agronomic practices that may destroy such natural control and lead to increased nematode infestations are discussed. Biological control alone is often inadequate to maintain nematode populations below their economic threshold and must be integrated with other management methods. Methods that decrease nematode infestations in soil or increase the activity of microbial agents are reviewed and some examples given where their combination with agents applied to soil have enhanced the efficacy of biological control. There may be problems for growers with the delivery of such integrated control strategies unless they receive adequate support from extension services, which may be absent in many countries. Hence, the exploitation of natural enemies as a source of genes and compounds with anti-nematode properties, which could be used in chemical and genetic interventions may provide alternative approaches for nematode management.

1. INTRODUCTION

The practical development of biological control methods for plant parasitic nematodes has depended on the use of microbial agents (Stirling, 1991). Only predatory nematodes have also been seriously considered as potential agents and their importance in agriculture is still unclear (Khan & Kim, 2007) but difficulties in producing sufficient inoculum in mass culture will probably restrict their use. As no organism has provided adequate control when applied alone, this chapter describes a range of measures that may be used to improve the performance of biological control agents for nematode management. Soils that become suppressive to nematode pests because they have supported an increase in natural enemy populations have provided sustainable control of some pest species (Kerry & Crump, 1998) and discussion here is limited to the agronomic practices that may destroy such natural control and lead to increased nematode infestations. Suppressive soils

A. Ciancio & K. G. Mukerji (eds.), Integrated Management and Biocontrol of Vegetable and Grain Crops Nematodes, 29–49.

have been associated with the continuous cultivation of nematode-susceptible crops, which support increases in the natural enemy community; the development of suppressive soils is to some extent dependent on the appropriate crop sequence (Gair, Mathias, & Harvey, 1969; Kerry, 1995). The research discussed in this chapter concerns the application of bacteria and fungi as soil inoculants as part of integrated pest management strategies. The use of plant-derived biocidal compounds is treated as chemical control (see Chapters 11–12, this volume) and is not discussed here.

Microbial agents may be antagonistic and produce bioactive compounds that kill or affect the development of nematodes, or be parasitic/pathogenic and destroy nematodes following their colonisation, or they may compete for resources; some organisms have more than one mode of action (Kerry, 2000). Bacteria and fungi that parasitise nematodes may depend solely on their hosts for nutrition (obligate parasites, such as *Pasteuria penetrans*) or have a saprotrophic phase in their life cycle (facultative parasites, such as *Pochonia chlamydosporia*). Obligate parasites are more likely to be affected by changes in host population density than facultative parasites and integration with control measures that reduce nematode pest infestations may reduce the performance of the agent, unless inundative applications are used. Such applications may be impractical as broadcast soil treatments in large scale agriculture. Although density dependence has been demonstrated for both obligate and facultative parasites of nematodes (Jaffee, Phillips, Muldoon, & Mangel, 1992; Ciancio, 1995; Ciancio & Bourijate, 1995) there is little theoretical basis to underpin the development of strategies for the biological control of nematodes and much is assumed from experience in other disciplines, especially entomology.

All plant parasitic nematodes are obligate parasites and must feed on plants to complete their development. The time spent in the rhizosphere where they are exposed to a wide range of micro-organisms depends on the parasitic habit of the nematode species. Unlike insects and fungi, nematodes do not spread rapidly through a particular field during a growing season and management strategies can be individual field- or even infested patch- based.

Control measures aim to reduce nematode feeding and invasion of roots to reduce crop damage and/or to reduce the fecundity of adult females and decrease post-crop populations left in soil (Kerry & Hominick, 2002). Of course, plant parasitic nematodes do not exist alone in soil and they have complex interactions with other soil organisms, including bacterial and fungal feeding nematodes, and the abiotic factors that affect them. Generalist natural enemies may be affected by the relative abundance of the populations of free-living nematodes in soil. Indeed, the earliest experiments to manipulate the fungal parasites of nematodes used the application of organic matter to soil to increase microbial abundance and the populations of free-living nematodes, which in turn supported increases in activity of the nematode trapping fungi able also to kill any plant parasitic species present (Linford, Yap, & Oliveira, 1938). However, it was found that the relationship between the activity of nematode trapping fungi and the nature and type of the soil organic matter was more complex and there was no simple relationship with nematode population density (Cooke, 1962). The efficacy of trapping fungi and other facultative parasites of nematodes may not be directly related to their abundance (see Section 3.1).

Apart from in very intensive agricultural systems, growers have used integrated pest management (IPM) strategies against nematode pests, as single measures are often inadequate to control them (Kerry, 2000). Demands in many countries to reduce dependence on nematicides for nematode management and the need to provide other control measures in situations where nematicides have always been uneconomic or inappropriate, present a significant challenge for applied nematologists. Some biological control agents have shown much promise but there are still considerable doubts about their utility. In this chapter we focus on a discussion of control measures, which may increase the robustness of biological control agents and lead to sustainable methods of nematode management. At the same time we are very aware that methods of pest control that require careful management will be very difficult to exploit in countries where growers are not adequately supported by extension advisors. Even in developed agricultural systems the uptake of IPM has often been slow (Van Emden & Peakall, 1996).

2. METHODS TO REDUCE NEMATODE POPULATIONS

A general overview of some nematode management methods which reduce nematode populations in soil, such as crop rotation, antagonistic crops, resistant cultivars, soil solarization, biofumigation and nematicides is provided with especial reference to those that could be used in appropriate combination with biological control in an integrated nematode pest management strategy to improve the effectiviness of biological agents. Excellent books have been published that have been devoted to integrated nematode management (e.g. Barker, Pederson, & Windham, 1998; Whitehead, 1998; Luc, Sikora, & Bridge, 2005; Perry & Moens, 2006), and should be consulted for guidelines in the structuring of integrated management programmes.

2.1. Crop Rotation

Seasonal rotations of susceptible crops with non-host or poor-host crops on the same area of land remain one of the most important techniques used for nematode management worldwide. The occurrence of nematode communities containing multiple pest or polyphagous species with wide host ranges, such as some species of *Meloidogyne*, limits the potential of using acceptable non-host crops for rotation (Viaene, Coyne, & Kerry, 2006). Hence, it is necessary to determine the host status of individual crop cultivars for local nematode populations before a rotation scheme is recommended for a particular field. Rotations using poor hosts or tolerant crops together with highly susceptible vegetable crops have been used for control of root-knot nematodes in tropical condition (Stefanova & Fernández, 1995; Gómez & Rodríguez, 2005). However, crop rotations have economic costs for the grower. In the past 20 years in the UK, the number of farmers producing potato crops has declined by 80% to around 5,500 individuals but the cropped area has remained relatively unchanged. Those specialist growers remaining have invested heavily in chilled storage facilities and machinery and must grow potatoes intensively to obtain a return on their investment. As a consequence, potatoes are grown on average every 6 years instead of the 9 year rotation recommended and potato cyst nematodes continue to spread despite the use of nematicides. Devine, Dunne, O'Gara, and Jones (1999) first recorded the effects of microbes on the decline of potato cyst nematode populations between potato crops and estimated it at only 10% with most egg loss resulting from their spontaneous hatch.

Use of witchgrass in a peanut rotation has beneficial effects on soil, reducing parasitic nematode populations and increasing numbers of free-living nematodes, and also causing shifts in rhizosphere microbial ecology (Kokalis-Burelle, Mahaffee, Rodríguez-Kabana, Kloepper, & Bowen, 2002). Some bacteria and fungi that affect the development of nematodes are dependent on specific plants to support their endophytic development or growth in th rhizosphere and so can only be used in certain crop rotations. Similarly, rotation crops, such as beans, maize and cabbage that support extensive growth of the nematophagous fungus, *Pochonia chlamydosporia* in their rhizospheres but support only limited reproduction of root-knot nematodes, are used to maintain the abundance of the fungus in soil (Table 1) whilst suppressing populations of the nematode (Puertas & Hidalgo-Díaz, 2007). Hence, growing an approved crop in the rotation to maintain populations of natural enemies on roots is another alternative to improve the efficacy of nematode management programmes based on crop rotations (Fig. 1). For obligate parasites such as the bacterium *Pasteuria penetrans*, it is essential that it is introduced into the soil with a nematode susceptible crop, which will provide developing nematodes on which the bacterium will multiply (Oostendorp, Dickson, & Mitchell, 1991). Timper et al. (2001) demonstrated in rotations of peanuts with 2 years of bahiagrass, cotton or corn, in a field naturally infested with *M. arenaria* and *P. penetrans* that the abundance of the bacterium was related to the population densities of the nematode and were greatest under continuous peanut cropping and next most abundant under the bahiagrass-peanut rotation.

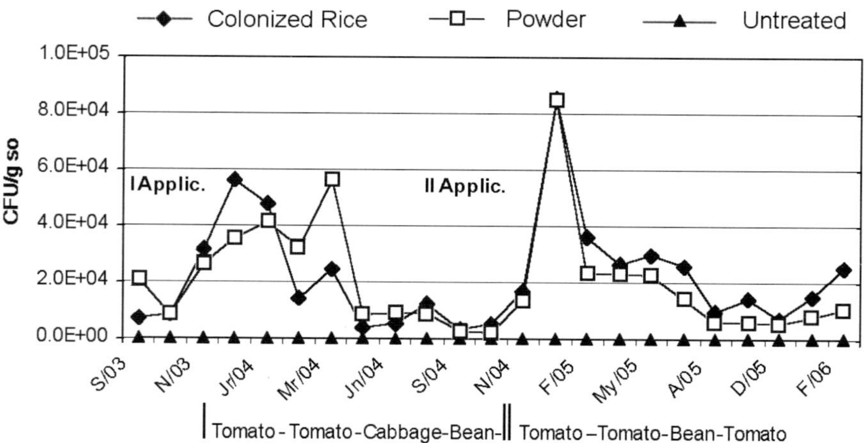

Figure 1. Changes in abundance of Pochonia chlamydosporia *in soil from September, 2003 until February, 2006 under different vegetable crops treated with two applications of the fungus in a field trial in Cuba. The fungus was applied on colonised rice or as a suspension of chlamydospores at a rate of 5,000 chlamydospores* g^{-1} *soil.*

2.2. Antagonistic Crops

Plants antagonistic to nematodes are those that are considered to produce toxic substances, usually, while the crops are growing or after incorporation into the soil. In practical nematode management strategies the use of this approach relies on pre-plant cover crops, intercropping or green manures.

Marigold, neem, sunn hemp, castorbean, partridge pea, asparagus, rape seed and sesame have been extensively studied and used as antagonistic crops for nematode control. Sunn hemp (*Crotalaria* spp.) is often cultivated as a cover crop for direct seeding, intercrops or soil amendment and is considered an antagonistic crop for most plant parasitic nematodes, especially root-knot nematodes (Wang, Sipes, & Schmitt, 2002). Population densities of *M. incognita* were affected by previous cover crops of *C. juncea* in north Florida (Wang, Mc Sorley, & Gallaher, 2004). Germani and Plenchette (2004), recommend the use of some *Crotalaria* spp. from Senegal as pre-crops for providing green manure while at the same time decreasing the level of root-knot nematode and increasing the level of beneficial mycorrhizal fungi.

Marigolds (*Tagetes* spp.) have been shown to suppress plant parasitic nematodes, such as root-lesion and root knot nematodes. Kimpinski, Arsenault, Gallant, and Sanderson (2000) demonstrated consistent reduction of *Pratylenchus penetrans* populations when marigolds were used as a cover crop followed by potato crops, with a significantly higher average yield. In Japan, where the continuous cropping of vegetables has led to nematodes (*P. coffeae* and *M. incognita*) becoming a major problem, a practical method using marigold has been developed, which requires only one season to incorporate these plants with only minor changes in the cropping system (Yamada, 2001). Biofumigation using fresh marigold as an amendment is used effectively in root knot management in the protected cultivation of vegetables in Morocco (Sikora, Bridge, & Starr, 2005).

Most antagonistic plants cultivated as pre-plant cover crops may be followed by soil incorporation of the biomass with a subsequent reduction of plant-parasitic nematode numbers and the enhancement of nematode antagonists (see Section 3.1). However, it should be noted that grower acceptance of new strategies using antagonistic plants are based on economic and logistical considerations, as well as efficacy. Too often the large amounts of biomass required restrict the use of the approach to cheap sources of local species/waste products. The value of these products may be enhanced by using them as media on which to culture nematophagous microbial agents either prior to or after their addition to soil. Although some empirical tests have been made, the combined use of antagonistic plants and biological control agents has been little studied.

2.3. Resistant Cultivars

Host plant resistance is currently the most effective and environmentally safe tactic for nematode management (Koenning, Barker, & Bowman, 2001; Castagnone-Sereno, 2002). When it is available in a high-yielding cultivar, it should be the foundation upon which other management measures build (Sikora et al., 2005), because resistance is highly specific, being effective against only a single species or even only one race of a species, it will not control other potential pests in the nematode community. This can be a major limitation to the use of resistance, except where the crop or soil is infested with only one pest species.

*Table 1. Main vegetable crops cultivated in rotations*** *in organoponic systems in Cuba and their ability to support* Pochonia chlamydosporia *colonisation of their rhizospheres.*

Common Name	Scientific Name	Botanic Family	Cultivar	Host	Status for
				M. incognita	*P. chlamydosporia**
Tomato	*Lycopersicon esculentum* Mill.	Solanaceae	Amalia	Host	Good
Sweet Pepper	*Capsicum annuum* L.	Solanaceae	Español	Host	Good
Eggplant	*Solanum melongena* L.	Solanaceae	FHB-1	Host	Good
Pak-choi	*Brassica rapa* L. subsp. *chinensis* (L.) Manelt.	Brassicaceae	Pak-Choi Canton	Non-host	Good
Broccoli	*Brassica oleracea* L. var. *italica* Plenek	Brassicaceae	Tropical F-8	Non-host	Good
Cabbage	*Brassica oleracea* L.	Brassicaceae	Hércules	Non-host	Good
Cauliflower	*Brassica oleracea* var. *botrytis* L.	Brassicaceae	Verano-6	Non-host	Moderate
Green bean	*Vigna unguiculata* (L.) Walp.	Fabaceae	Lina	Host	Moderate
Cucumber	*Cucumis sativus* L.	Cucurbitaceae	Tropical SS-5	Host	Moderate
Okra	*Abelmoschus esculentus* (L.) Moench	Malvaceae	Tropical C-17	Host	Moderate
Spinach	*Talinum triangulare* (Jacq.) Willd.	Portulacaceae	Baracoa	Host	Poor
Celery	*Apium graveolens* L.	Apiaceae	UTA	Host	Poor
Parsley	*Petroselinum crispum* (Mill.) Fuss.	Apiaceae	KD-77	Host	Poor

* The host status defines the ability to grow in the rhizosphere: good host (> 200 CFU cm^{-2} of root), moderate host (100-200 CFU cm^{-2} of root) and poor host (< 100 CFU cm^{-2} of root), see Kerry (2001).

** On the basis of selecting good hosts for the fungus and poor hosts for *Meloidogyne* spp. the recommended one year crop rotation:is: tomato/sweet pepper-cabbage/pak-choi/cauliflower-green bean (3 crops in one year).

Resistance is currently available to one or more nematode species in a limited number of food crops (see Cook & Starr, 2006) but it is widely used for cyst nematodes in potato crops in Europe and soybean crops in the USA, Brazil and Argentina. Cotton cultivars with moderate resistance to *M. incognita* are recommended in the USA as a valuable management approach to be used in rotation or with nematicides (Koenning et al., 2001; Davis & May, 2003). Resistance to *Meloidogyne* species in tomato is widely used in California and in crops under protected cultivation in the Mediterranean region of Europe, but not in many other regions especially in the tropics because the resistance gene breaks down at soil temperatures above 28ºC. Despite this limitation, Sorribas, Ornat, Verdejo-Lucas, Galeano, and Valero (2005) documented the economic value of using three successive resistant crops to *M. javanica* compared with three crops of a susceptible cultivar, in the production of tomato in glasshouses in Spain. Apparently, even if the *Mi* resistance gene is effective only during the first few weeks of the growing season before higher temperatures reduce its effectiveness, this period of resistance could be useful if it is combined with other management tactics, such as the use of biological control agents that provide longer term protection.

Resistant root-stocks in perennial crops, such as peach and citrus, have been used successfully for several decades. More recently, the grafting of resistant root-stocks to susceptible scions has been used for management of root-knot nematodes on annual crops. This practice is being widely used on cucumber, melon, pepper and aubergine in South East Asia and the Mediterranean regions of Europe, and is being introduced into Central American countries as part of the international programme to reduce the use of methyl bromide in large-scale melon cultivation.

The use of biological control agents may provide an environmentally friendly tactic that could be more effective in combination with resistant or partially resistant cultivars that reduce nematode reproduction enough to affect the residual nematode population density in a field. Cook and Starr (2006), suggest that the natural decline of cereal cyst nematodes, in monocultures of cereal crops in Western Europe, associated with fungal parasites of the nematode females and eggs may be assisted by the unwitting use of partial resistance. The combined use of a biological control agent that reduced the fecundity of females with a partially resistant cultivar could slow the selection of virulent species and pathotypes of nematodes. Timper and Brodie (1994) observed that the combined use of the fungus *Hirsutella rhossiliensis* and a potato cultivar resistant to *Pratylenchus penetrans* caused greater control than if either treatment was applied alone and this interaction was synergistic.

2.4. Soil Solarization

Soil solarization with plastic mulches leads to lethal temperatures which kill plant parasitic nematodes (around 45ºC) and is being used mainly in regions where high levels of solar energy are available for long periods of time (Whitehead, 1998). The effect of this approach is reduced with depth, but solarization for at least 4–6 weeks will increase soil temperatures to about 35–50ºC to depths of up to 30 cm and, depending on soil type, soil moisture content and prior tillage, will reduce nematode infestations significantly (Viaene et al., 2006).

In Japan and other East Asian countries, several farmers growing successive crops, such as tomato and melon susceptible to root-knot nematodes use solarization in plastic tunnels for 30 days in summer as an alternative to methyl bromide fumigation (Sano, 2002).

In Cuba, root knot nematode infestations are reduced, in peri-urban and small organic farm production, using solarization under sub-optimum conditions (Fernández & Labrada, 1995) but for subsistence agriculture, the cost of plastic sheeting may be limiting. The length of time required for effective solarization is a great limitation too, but it could be reduced when it is used with biofumigation.

Infection of *M. javanica* by *P. penetrans* was increased in naturally infested soils in a S. Australian vineyard treated by solarisation and decreased in soils treated with the nematicides oxamyl or phenamiphos but the bacterium did not significantly reduce nematode populations (Walker & Wachtel, 1988). Similarly, in a cucumber crop in a glasshouse trial the use of solarisation and *P. penetrans* had an additive detrimental effect on *M. javanica* populations (Tzortzakakis & Goewn, 1994).

2.5. Biofumigation

The term biofumigation is used when volatile substances are produced through microbial degradation of organic amendments that result in significant toxic activity towards nematodes or diseases (Bello, González, & Tello, 1997). Generally, biofumigation is more effective when there is an optimum combination of organic matter, high soil temperature and adequate moisture to promote microbial activity.

In Spain, biofumigation has been largely applied successfully as an alternative to methyl bromide in several crops (Bello, López-Pérez, Díaz-Viruliche, & Tello, 2001). Soil amended with fresh or dry cruciferous residues reduce significantly root-knot nematode infestations due, principally, to isothiocyanates released in soil when glucosinolates present in these crop residues are hydrolysed (Staplenton & Duncan, 1998; Ploeg & Staplenton, 2001; Díaz-Viruliche, 2000; D'Addabbo, De Mastro, Sasanelli, & Di Stefano, 2005). However, the practical application of this approach is limited due to the large amount of organic matter to be transported to the field or the cost of cover crops to be incorporated into the soil, together with the plastic mulch and drip irrigation system often necessary to improve the effectiveness of biofumigation. Also, the provision of large amounts of nutrients to soils may affect the activity of facultative parasites of nematodes (see Section 3.1).

2.6. Nematicides

Nematicides are commonly used in developed cropping systems and may directly kill nematodes or are effective by paralysing the nematodes for a variable period of time (nematostatic). Nematicides may be fumigants and non-fumigants and are classified according to their mode of action. Fumigant nematicides consist of compounds based on halogenated hydrocarbons (1,3-D and methyl bromide) and those which release methyl isothiocyanate (metham sodium and dazomet). They are mostly used pre-planting, and most are liquids which enter the soil water solution from a gas phase. In most cases the fumigants are broad-spectrum contact nematicides effective against adults, juveniles and eggs as well as other pests, diseases and weeds and have significant effects on non-target organisms, including the natural enemies of nematodes.

Non-fumigant nematicides are organophosphate (e.g. fenamiphos, ethoprophos and fosthiazate) and carbamate (e.g. aldicarb, carbofuran and oxamyl), which are applied to the soil, at planting time, as granular or liquid formulations that are water soluble. They have either contact or nematostatic effects and often some plant systematic activity against nematodes and insects. At low concentrations, they disrupt chemoreception and the ability of nematodes to locate their host roots; at higher concentrations, they disrupt nematode hatch and movement, but do not kill eggs. At field rates, the biochemical effect is reversible. Hence, to improve the effectiveness, nematicide concentration and time of exposure must be maximized by correct timing of application and incorporation in the target zone of the soil. They, mainly, protect the plant during the highly sensitive seedling or post-transplant stage of plant development.

Nematicides still continue to be a main nematode management approach, whether used as part of an integrated management programme or as the sole control component. The global market for nematicides is about 250,000 t of active ingredient each year, with USA and Western Europe as the main consumers; vegetable crops accounting for the greatest proportion of nematicide use and *Meloidogyne* spp. as the target for approximately half of this usage (Haydock, Woods, Grove, & Hare, 2006). However, in the last years some nematicides have been phased out, such as methyl bromide and restrictions in the use of others are increasing due to public and governmental concern about their detrimental impact on human health and the environment.

Several nematophagous fungi including trapping fungi, *P. lilacinus* and *P. chlamydosporia* have been grown in the presence of a range of pesticides and often shown to be little affected by standard dosages applied to soil (Kerry, 1987). It is therefore possible that these agents could be applied with nematostats to prolong and increase nematode control. Oxamyl increased the efficacy of *P. penetrans* in trials against *M. javanica* infection of tomato and cucumber crops and the effects on nematode control were additive (Tzortzakakis & Goewn, 1994). Aldicarb and ethoprop applications to soil infested with *M. arenaria* had no detrimental effects on the number of nematode juveniles parasitized by *P. penetrans* (Timper, 1999; Timper et al., 2001). Little work has been done on the combined use of nematicides and fungal biological control agents. However, Taba, Moromizato, Takaesu, Ooshiru, and Nasu (2006) combined the nematicide, fosthiazate with the nematode-trapping fungus, *Monacrosporium ellipsosporum* in a granular application, which effectively controlled *M. incognita* on tomato plants and established the fungus in the soil.

Nematode management in the future will never again be able to rely on one type of measure, as it has in the past. Management will require the logical use of effective control methodologies in combinations that are economically acceptable to the grower (Sikora et al., 2005). We should also recognize that effective use of nematode management tactics into IPM programmes demands educational input at the grower level. The success of several IMP programmes in Cuba have been built upon close interaction between farmers and researchers in successful extension advisor programmes (Fig. 2).

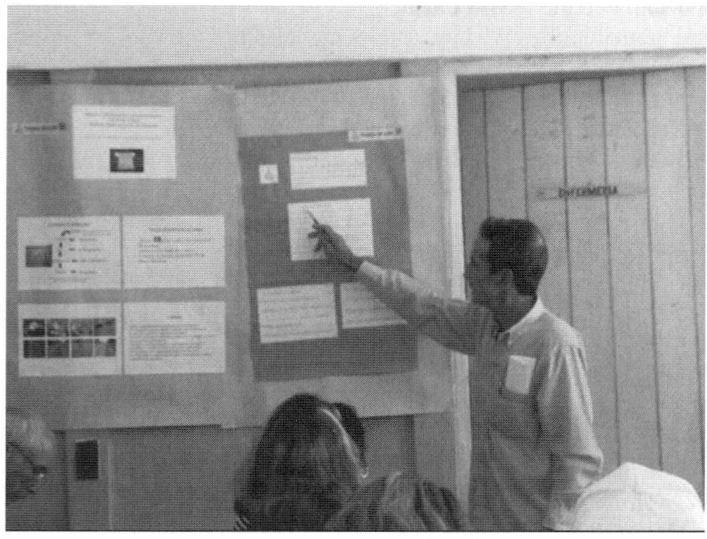

Figure 2. Essential training for extension workers: learning how to manage biomanagement strategies for nematode pests.

3. METHODS TO INCREASE MICROBIAL ABUNDANCE AND/OR ACTIVITY

A range of treatments have been applied to soil to increase its organic matter status and the associated increase in the diversity and activity of the microbial community has in turn been suggested as a cause for any detrimental effects on populations of plant parasitic nematodes (Akhtar & Malik, 2000). However, the effects of organic amendments in soil are complex and effects on nematodes may be due to the nematicidal action of breakdown products, direct and indirect increases in the activity of natural enemies, and indirect effects mediated through increases in the activity of the resident soil microbial community that is stimulated to produce nematicidal metabolites at active concentrations.

Chitin applied to soil at 1% (w/w) controlled *M. incognita* on cotton and there were significant changes in the microflora in amended soil and in the rhizosphere and within roots, including an increase in the chitinolytic bacteria (Hallmann, Rodriguez-Kabana, & Kloepper, 1999). Although the latter mechanism has often been suggested (Rodriguez-Kabana, Morgan-Jones, & Chet, 1987), active (μM) concentrations of enzymes such as the chitinases, which degrade nematode eggshells, have not been demonstrated in the rhizosphere and, if present, might increase the hatch of mature eggs. There is a need for critical research to determine the major modes of action to account for the effects of many organic amendments, especially as such research may enable rates of application to be reduced to amounts that would increase their practicality in a range of soils.

3.1. Organic Amendments, Green Manures and Companion Crops

Organic nutrients may be added to soil as composted or fresh plant material or as the root exudates from growing plants. All have been shown to affect the growth of microbial natural enemies and their activity against nematodes and may offer opportunities for their exploitation in management systems. Although, organic amendments may be expected to influence the activities of facultative parasites during their saprotrophic phase more than obligate parasites that have limited growth in soil, it is clear that there are a range of indirect effects. Hence, empirical studies examining the effects of organic amendments on the applications of organic matter to soil inoculated with root-knot nematodes encumbered with spores of *P. penetrans*, improved plant growth and multiplication of both nematode pest and bacterium (Gomes, De Freitas, Ferraz, Oliveira, & Da Silva, 2002).

Applications of organic matter (lucerne meal) to soil increased the abundance of the endoparasite, *Drechmeria coniospora* indirectly by increasing populations of bacterivorous nematodes, which were parasitized by the fungus (Van den Boogert, Velvis, Ettema, & Bouwman, 1994). As is the case with *P. penetrans*, this fungus survives in soil as infective spores that adhere to passing nematodes; soil factors including organic amendments that may affect the abundance and activity of nematodes would increase the chances of contact between parasite and host. However, as *D. coniospora* is a relatively weak parasite of plant parasitic nematodes (Jansson, Dackman, & Zuckerman, 1987), it has limited potential as a biological control agent.

Parasitism of nematode hosts by *Hirsutella rhossiliensis*, was not enhanced by large applications of chicken manure, wheat straw or composted cow manure

(Jaffee, Ferris, Stapleton, Norton, & Muldoon, 1994). Populations of this weakly competitive saprotroph may have succumbed to competition from the much enhanced populations of the resident soil microflora.

The effects of organic amendments on the interactions between facultative fungal parasites of nematodes and their hosts are also difficult to interpret and a range of different mechanisms are probably involved. Research on nematode trapping fungi has demonstrated that the enhancement of trapping activity resulting from the application of organic matter to soil is dependent on the fungal species and the type and amount of organic material added (Jaffee, Ferris, & Scow, 1998; Jaffee, 2004). Population density and activity were correlated for *Dactylellina haptotyla* but not for *Arthrobotrys oligospora* in these experiments conducted in microcosms.

Two theories have been proposed to explain the effects of organic amendments on trapping fungi (Jaffee, 2004). The numerical theory assumes that the fungi are obligate parasites in nature and organic amendments that stimulate microbial activity and the abundance of bacterivorous nematodes will increase populations of trapping fungi. The supplemental nitrogen model assumes that the fungi are facultative parasites that obtain their nitrogen from nematodes which allows them to compete for other nutrients in nitrogen-depleted organic matter in soil. Presumably, different species of trapping fungi may conform to either model (Jansson & Nordbring-Hertz, 1980).

Although there have been considerable advances in our knowledge of the ecology of trapping fungi in soil (Jaffee, 2002, 2003, 2004), key questions remain concerning the relationship between nutrition and trapping activity. Their role in the biological control of nematode pests will rely on this further understanding and the ability to promote trapping during the periods of nematode activity in the soil and rhizosphere.

Similarly, the parasitism of nematode eggs by opportunistic fungi, such as *P. chlamydosporia* and *Paecilomyces lilacinus* is also not necessarily related to the abundance of these fungi in the rhizosphere. Although organic soils may support many more propagules of *P. chlamydosporia* than mineral soils, the numbers of eggs of *Meloidogyne* spp. parasitized by the fungus were similar in both types of soils (Leij de, Kerry, & Dennehy, 1993). The availability of easily metabolised nutrient sources may sustain the fungus in its saprotrophic phase and prevent the switch to parasitism. Circumstantial evidence for such an hypothesis is provided by laboratory studies in which the secretion of a serine proteinase enzyme designated VCP1 involved in the degradation of the outer vitelline membrane of the eggshell and the early stages of infection, was repressed by the presence of glucose and simple nitrogen sources and induced by transfer to minimal media and the presence of nematode egg masses (Segers, 1996).

Pochonia chlamydosporia proliferates in the rhizosphere of a range of crop species and is more abundant on the surface of galls during the period of egg laying of *Meloidogyne* spp. than on healthy roots (Bourne, Kerry, & De Leij, 1996). The successful use of this fungus for control of root-knot nematodes in organic vegetable production systems depends on its use in rotations that include crops, which are poor hosts for the nematodes but support substantial populations of the fungus on their roots. Such rotations maintain effective levels of the fungus without excessive build up of root-knot nematode infestations and provide a practical method of nematode management in intensive horticulture (Atkins et al., 2003).

Applications of chlamydospores, with limited nutrient reserves, or as colonised rice grains to soil infested with *Meloidogyne* species enabled the fungus to establish in the rhizosphere of tomato plants and parasitise similar numbers of nematode eggs (Peteira et al., 2005). Presumably, readily metabolised nutrients in the rice were removed by the fungus and the resident soil microflora before the nematode egg masses were produced on roots and exposed to parasitism. Addition of neem (*Azadirachta indica*) leaves to soil but not those of calotropis (*Calotropis procera*) caused small increases in the abundance of *P. chlamydosporia* in the rhizosphere of tomato plants and increases in the proportion of eggs of *M. incognita* parasitized in pots (Reddy, Rao, & Nagesh, 1999).

Green manures incorporated in soil have been used to increase the activity of natural enemies of nematode pests. Applications of *Bacillus megaterium* reduced *M. chitwoodi* populations to a greater extent if oil radish or rapeseed green manures had been added to soil than if no manures had been used (Al-Rehiayani, Hafez, Thornton, & Sundararaj, 1999). Green manures have also been used with limited success in pot experiments to increase the activity of trapping fungi against *Heterodera schachtii* (Hoffmann-Hergarten & Sikora, 1993) whereas Pyrowolakis, Schuster, and Sikora (1999) were able to increase the parasitism of eggs of *H. schachtii* by <50% when chopped oil radish tops had been mixed in soil in pots. The activity of egg-parasitc fungi has also been increased in the field by the incorporation of oil radish as a green manure (Schlang, Steudel, & Miller, 1988). It is clear from the literature that the benefits of a combined green manure and a microbial agent depend on the soil, the type of green manure and the species of agent.

The rhizosphere of some plants antagonist to plant parasitic nematodes have distinct microfloras that have physiological traits, which indicate that at least part of the antagonism may be due to the bacterial and fungal community on roots (Kloepper, Rodriguez-Kabana, McInroy, & Collins, 1991; Insunza, Alstrom, & Eriksson, 2002). Although such associations have been found and provide a method for managing nematode populations, it has not been demonstrated that potential antagonistic microorganisms produce toxins or enzymes in the rhizosphere in sufficient concentrations to affect nematodes. However, the use of plants to manipulate the rhizosphere microbial community to the detriment of nematode pests is an attractive concept worthy of more research. Indeed, the use of *P. chlamydosporia* for the control of root-knot nematodes in intensive vegetable production is dependent on the use of crops in the rotation that are poor host for the nematode but support high densities of the fungus in their rhizospheres (Kerry, 1995; Atkins et al., 2003; Kerry & Hidalgo-Díaz, 2004; Puertas & Hidalgo-Díaz, 2007). However, the efficacy of *P. lilacinus* was not related to the host crop and its rhizosphere competence was not essential for effective nematode control and so unlike *P. chlamydosporia* this fungus may not be so restricted to particular rotations (Rumbos & Kiewnick, 2006).

4. THE COMBINED USE OF BIOLOGICAL CONTROL AGENTS

The above discussion has concentrated on approaches to combine biological control agents with other measures to increase the overall levels of control achieved. Another widely discussed approach to improve control has been to increase the diversity of the natural enemy community to which a specific pest is exposed. However, there is little direct evidence to suggest that several agents in soil provide better control than one agent present at the same total population density. Frequently, in the literature empirical studies have compared, for example, the control achieved by agent A applied at x propagules g soil^{-1} and agent B applied at the same rate with either agent applied at the same rate alone; rarely has the benefit achieved with the combined agent been compared with a single agent applied at twice ($2\times$) the rate. Also synergy is frequently reported when the data reveal only additive effects. There is therefore a need for more critical experimentation to demonstrate whether combined applications of agents compete or act additively or synergistically to improve the control achieved through the addition of a single agent.

Despite these concerns, it is clear that many potential biological control agents are compatible and, if considered appropriate, could be used in combined applications. Thus, *P. penetrans* has been used in combined applications with *P. lilacinus* and other soil inoculants such as *B. subtilis* and *Talaromyces flavus* used to control soil borne diseases (Zaki & Maqbool, 1991) and with *P. chlamydosporia* (Leij de, Davies, & Kerry, 1992). It has been suggested "helper bacteria" in the rhizosphere increase attachment of the endospores of *P. penetrans* (Duponnois, Netscher, & Mateille , 1997) and the bacterium is compatible with mycorrihzae (Talavera, Itou, & Mizukubo, 2002). It seems reasonable to expect a more diverse natural enemy community to be more resilient to changes in the soil environment and provide more consistent nematode control.

It is clear that in some suppressive soils there is much diversity within an individual agent such as *P. chlamydosporia* and the use of molecular diagnostic methods is beginning to reveal key differences between isolates of the fungus that may affect their performance as biological control agents (Mauchline, Kerry, & Hirsch, 2004). Also in the bacterium *P. penetrans* there is considerable variation in the range of attachment of the infective spores to different nematode populations, even if those spores have been derived from a single infected female (Davies, Redden, & Pearson, 1994). It may be that this variation within the natural enemy population reflects the nematode's ability to rapidly alter its surface coat as a defence mechanism in an evolutionary arms race (Davies et al., 2001).

5. FUTURE APPROACHES

5.1. Use of Genes from Natural Enemies

Although much of the discussion above concerns improvements in the use of biological control agents through their application and integration with other control measures, there remains a problem of producing sufficient inoculum for economic use against nematodes on broad-acre crops. An alternative approach, which may be more likely to generate consistent and economic control, is to use natural enemies as a source of novel bioactive compounds that could be used as nematicides or delivered through the genetic transformation of plants.

A chitinase gene from *Trichoderma harzianum* was first used to improve plant resistance to a range of fungal pathogens and suggested that biocontrol fungi were a rich source of genes that could be used to control diseases in plants (Lorito et al., 1998). For example, the gene from Pseudomonad spp., which produce toxins that kill the eggs of *Mesocriconema xenoplax* has been cloned and has potential for the control of this important nematode pest of peach trees (Kluepfel, Nyczepir, Lawrence, Wechter, & Leverentz, 2002).

Methods have been developed to transform nematode-trapping fungi (Ahman et al., 2002; Xu, Mo, Huang, & Zhang, 2005) and *A. oligospora* transformed to overproduce a subtilisin increased the virulence of the fungus and when the construct was used to transform *Aspergillus niger*, the transgenic fungus had nematoxic activity (Ahman et al., 2002). Similar subtilisin genes have been identified in *P. lilacinus* (Bonants et al., 1995) and *P. chlamydosporia* (Segers, Butt, Kerry, & Peberdy, 1994) and polymorphisms in the enzyme of the latter fungus suggest it may be a host range and virulence determinant (Morton, Hirsch, Peberdy, & Kerry, 2003).

The genome of *P. penetrans* is currently being sequenced and could be a source of novel anti-nematode genes. However, much research has to be done to identify key genes involved in antagonism or the infection processes of natural enemies of nematodes but this approach has considerable potential for the development of new control measures. The release of genetically-modified microorganisms will present very significant registration issues. For example Shaukat and Siddiqui, (2003) demonstrated that those mutant strains of *Pseudomonas fluorescens*, which over- or under-produced an antibiotic had significant effects on the diversity of rhizosphere fungi.

5.2. Improved Formulations and Application Methods

In scaling up the use of microbial biocontrol agents there is a need to optimise the amount of inoculum applied. The application of rhizosphere bacteria as seed treatments (Oostendorp & Sikora, 1989) and endophytic fungi as bare root dips (Pinochet, Camprubi, Calvet, Fernandez, & Kabana-Rodriquez, 1998) or in tissue cultured plantlets (Sikora, 2001) provide an opportunity for the large scale use of biological control.

Economic applications to soil, even as in-row treatments, are much more demanding. However, relatively little research on the improvement of inoculum quality, production methods and formulations of nematophagous microorganisms has been reported in the public domain. However, as a number of products have been marketed there is sufficient knowledge within commerce (Powell & Faull, 1989). Similarly, some empirical tests have been done on different media and the production of some potential biological control agents for nematodes have been optimised but little critical information is available on the impacts of different production methods to optimise competence (Jenkins & Grzywacz, 2000). Future research should determine the relationships between pest population densities and the performance of biological control agents, which would be required for their possible patch application.

In Honduras, melon producers are evaluating the use of *P. chlamydosporia* applications in addition to antagonistic crops in areas where root-knot nematode populations are moderate or low and the use of fumigants to reduce large infestations to levels that may be managed with more environmentally benign methods (B.R. Kerry & L. Hidalgo-Diaz, personal communication).

Formulations that are compatible with the delivery of microbial agents through drip irrigation systems may also enable precise application and reductions in inoculum rates. Procedures have been defined for risk assessments of biological control agents released into the environment (Kiewnick, Rumbos, & Sikora, 2004) and some studies have been done on the impact of releases on the rhizosphere microbial community (O'Flaherty, Hirsch, & Kerry, 2003) but more research is required.

In practice, improvements in the development of biological control agents either through improved selection procedures or through better production methods and the formulation of inoculum are still likely to require support from other control measures for the sustainable management of most nematode pests. Biological control will not be a replacement for nematicides and will require careful integration with other management practices. The practical challenge of such an approach is that growers may need the support of an expert extension service, often absent in many parts of the world, to exploit biological control agents. However, advances in the genomics of microbial natural enemies will provide new opportunities for chemical and genetic interventions through the identification of gene products with novel bioactivity, which may be easier to deploy than classical biological control. Whatever the approach, research on the natural enemies of nematodes remains an exciting and productive topic of endeavour.

REFERENCES

Ahman, J., Johansson, T., Olsson, M., Punt, P. J., van den Hondel, C. A., & Tunlid, A. (2002). Improving the pathogenicity of a nematode-trapping fungus by genetic engineering of a subtilisin with nematoxic activity. *Applied and Environmental Microbiology, 68,* 3408–3415.

Akhtar, M., & Malik, A. (2000). Roles of organic soil amendments and soil organisms in the biological control of plant-parasitic nematodes: A review. *Biosource Technology, 74,* 35–47.

Al-Rehiayani, S., Hafez, S. L., Thornton, M., & Sundararaj, P. (1999). Effects of *Pratylenchus neglectus, Bacillus megaterium,* and oil radish or rapeseed green manure on reproductive potential of *Meloidogyne chitwoodi* on potato. *Nematropica, 29,* 37–49.

Atkins, S. D., Hidaldo-Diaz, L., Kalisz, H., Mauchline, T. H., Hirsch, P. R., & Kerry, B. R. (2003). Development of a new management strategy for the control of root-knot nematodes (*Meloidogyne* spp.) in organic vegetable production. *Pest Management Science, 59,* 183–189.

Barker, K. R., Pederson, G. A., & Windham, G. L. (Eds.). (1998). *Plant and Nematode Interactions.* Madison, WI: American Society of Agronomy, Inc. 771 pp.

Bello, A., González, J. A., & Tello, J. C. (1997). La biofumigación como alternativa a la desinfección del suelo. *Horticultura Internacional, 17,* 41–43.

Bello, A., López-Pérez, J. A., Díaz-Viruliche, L., & Tello, J. (2001). Alternatives to methyl bromide for soil fumigation in Spain. In R. Labrada & L. Fornasari (Eds.), *Global report on validated alternatives to the use of methyl bromide for soil fumigation* (Chapter III, pp. 31–42). FAO and UNEP, Rome, Italy.

Bonants, P. J. M., Fitters, P. F. L., Thijs, H., Den Belder, E., Waalwijk, C., & Henfling, J. W. D. M. (1995). A basic serine protease from *Paecilomyces lilacinus* with biological activity against *Meloidogyne hapla* eggs. *Microbiology, 141,* 775–784.

Bourne, J. M., Kerry, B. R., & De Leij, F. A. A. M. (1996). The importance of the host plant on the interaction between root-knot nematodes (*Meloidogyne* spp.) and the nematophagous fungus, *Verticillium chlamydosporium* Goddard. *Biocontrol Science and Technology, 6,* 539–548.

Castagnone-Sereno, P. (2002). Genetic variability in parthenogenesis root-knot nematodes, *Meloidogyne* spp., and their ability to overcome plant resistance genes. *Nematology, 4,* 605–608.

Ciancio, A. (1995). Density-dependent parasitism of *Xiphinema diversicaudatum* by *Pasteuria penetrans* in a naturally infested field. *Phytopathology, 85,* 144–149.

Ciancio, A., & Bourijate, M. (1995). Relationship between *Pasteuria penetrans* infection levels and density of *Meloidogyne javanica*. *Nematologia Mediterranea, 23*, 43–49.

Cook, R., & Starr, J. L. (2006). Resistant cultivars. In R. Perry & M. Moens (Eds.), *Plant nematology* (pp. 370–389). Wallingford, UK: CABI Publishing.

Cooke, R. C. (1962). The ecology of nematode-trapping fungi during decomposition of organic matter in soil. *Annals of Applied Biology, 50,* 507–513.

D'Addabbo, T., De Mastro, G., Sasanelli, N., & Di Stefano, Y. (2005). Contro i nematodi: Azione biocida di differenti specie di *Brassica* spp. sul nematode galligeno *Meloidogyne incognita* su pomodoro. *Colture Protette, 12,* 55.

Davies, K. A., Fargette, M., Balla, G., Daudi, A., Duponnois, R., Gowen, S. R., et al. (2001). Cuticle heterogeneity as exhibited by *Pasteuria* spore attachment is not linked to the phylogeny of parthenogenetic root-knot nematodes (*Meloidogyne* spp.). *Parasitology, 122*, 111–120.

Davis, R. F., & May, O. L. (2003). Relationships between tolerance and resistance to *Meloidogyne incognita* in cotton. *Journal of Nematology, 35,* 411–416.

Davies, K. G., Redden, M., & Pearson, T. K. (1994). Endospore heterogeneity in *Pasteuria penetrans* related to adhesion to plant-parasitic nematodes. *Letters in Applied Microbiology, 19*, 370–373.

Devine, K. J., Dunne, C., O'Gara, F., & Jones, P. W. (1999). The influence of in-egg mortality and spontaneous hatching on the decline of *Globodera rostochiensis* during crop rotation in the absence of the host potato crop in the field. *Nematology, 1*, 637–645.

Díaz-Viruliche, L. (2000). Interés fitotécnico de la biofumigación en los suelos cultivados. Tesis en opción al grado de Doctor en Ciencias Agrícolas. Universidad Politécnica de Madrid, Escuela Técnica Superior de Ingenieros Agrónomos, 591 pp.

Duponnois, R., Netscher, C., & Mateille, T. (1997). Effect of the rhizosphere microflora on *Pasteuria penetrans* parasitizing *Meloidogyne graminicola*. *Nematologia Mediterranea, 25*, 99–103.

Fernández, E., & Labrada, R. (1995, 18–21 September). Experiencias en el uso de la solarización en Cuba. In *Memorias del Taller Solarización del Suelo* (pp. 5–6) Escuela Agrícola Panamericana El Zamorano Honduras.

Gair, R., Mathias, P. L., & Harvey, P. N. (1969). Studies of cereal nematode populations and cereal yields under continuous or intensive culture. *Annals of Applied Biology, 63,* 503–512.

Germani, G., & Plenchette, C. (2004). Potential of *Crotalaria* species as green manure crops for the management of pathogenic nematodes and beneficial mycorrhizal fungi. *Plant and Soil, 266,* 333–342.

Gomes, C. B., De Freitas, L. G., Ferraz, S., Oliveira, R. D. D. L., & Da Silva, R. V. (2002). Influence of cattle manure content in the substrate on the multiplication of *Pasteuria penetrans* in tomato. *Nematologia Brasileira, 26*, 59–65.

Gómez, L., & Rodríguez, M. (2005). Evaluación de un esquema de rotación de cultivos para el manejo de *Meloidogyne* spp. en sistemas de cultivos protegidos. *Revista da Protección Vegetal, 20*, 67–69.

Hallmann, J., Rodriguez-Kabana, R., & Kloepper, J. W. (1999). Chitin-mediated changes in bacterial communities of the soil, rhizosphere and within roots of cotton in relation to nematode control. *Soil Biology and Biochemistry, 31*, 551–560.

Haydock, P. P. J., Woods, S. R., Grove, I. G., & Hare, M. C. (2006). Chemical control of nematode. In R. Perry & M. Moens (Eds.), *Plant nematology* (pp. 392–408). Wallingford, UK: CABI Publishing.

Hoffmann-Hergarten, S., & Sikora, R. A. (1993). Enhancing the biological control efficacy of nematode-trapping fungi towards *Heterodera schachtii* with green manure. *Zeitschrift fur Pflanzenkrankheiten und Pflanzenschutz – Journal of Plant Diseases and Protection, 100,* 170–175.

Insunza, V., Alstrom, S., & Eriksson, K. B. (2002). Root bacteria from nematicidal plants and their biocontrol potential against trichodorid nematodes in potato. *Plant and Soil, 241*, 271–278.

Jaffee, B. A. (2002). Soil cages for studying how organic amendments affect nematode-trapping fungi. *Applied Soil Ecology, 21*, 1–9.

Jaffee, B. A. (2003). Correlations between most probable number and activity of nematode-trapping fungi. *Phytopathology, 93*, 1599–1605.

Jaffee, B. A. (2004). Do organic amendments enhance the nematode-trapping fungi *Dactylellina haptotyla* and *Arthrobotrys oligospora*? *Journal of Nematology, 36*, 267–275.

Jaffee, B. A., Ferris, H., & Scow, K. M. (1998). Nematode-trapping fungi in organic and conventional cropping systems. *Phytopathology, 88*, 344–350.

Jaffee, B. A., Ferris, H., Stapleton, J. J., Norton, M. V. K., & Muldoon, A. E. (1994). Parasitism of nematodes by the fungus *Hirsutella rhossiliensis* as affected by certain organic amendments. *Journal of Nematology, 26*, 152–161.

Jaffee, B., Phillips, R., Muldoon, A., & Mangel, M. (1992). Density-dependent host-pathogen dynamics in soil microcosms. *Ecology, 73*, 495–506.

Jansson, H.-B., Dackman, C., & Zuckerman, B. M. (1987). Adhesion and infection of plant parasitic nematodes by the fungus *Drechmeria coniospora*. *Nematologica, 33*, 480–487.

Jansson, H.-B., & Nordbring-Hertz, B. (1980). Interactions between nematophagous fungi and plant parasitic nematodes: attraction, induction of trap formation and capture. *Nematologica, 26*, 383–389.

Jenkins, N. E., & Grzywacz, D. (2000). Quality control of fungal and viral biocontrol agents – assurance and product performance. *Biocontrol Science and Technology, 10*, 753–777.

Kerry, B. R. (1987). Biological control. In R. H. Brown & B. R. Kerry (Eds.), *Principles and practice of nematode control in crops* (pp 233–263). Sydney, Australia: Academic Press.

Kerry, B. R. (1995). Ecological considerations for the use of the nematophagous fungus *Verticillium chlamydosporium* to control plant parasitic nematodes. *Canadian Journal of Botany, 73*, (Suppl. 1), S65–S70.

Kerry, B. R. (2000). Rhizosphere interactions and the exploitation of microbial agents for the biological control of plant-parasitic nematodes. *Annual Review of Phytopathology, 38*, 423–441.

Kerry, B. R. (2001). Exploitation of the nematophagous fungus *Verticillium chlamydosporium* Goddard for the biological control of root-knot nematodes (*Meloidogyne* spp.). In T. M. Butt, C. Jackson, & N. Magan (Eds.). *Fungi as Biocontrol Agents: Progress, Problems and Potential* (pp 155–168). Wallingford, UK: CABI International

Kerry, B. R., & Crump, D. H. (1998). The dynamics of the decline of the cereal cyst nematode, *Heterodera avenae*, in four soils under intensive cereal production. *Fundamental and Applied Nematology, 21*, 617–625.

Kerry, B. R., & Hidalgo-Díaz, L. (2004). Application of *Pochonia chlamydosporia* in the integrated control of root-knot nematodes on organically grown vegetable crops in Cuba. In R. Sikora, S. Gowen, R. Hauschild, & S. Kiewinick (Eds.), *Multitrophic interactions in soil and integrated control. IOBC/WPRS Bulletin 27*, 123–127.

Kerry, B. R., & Hominick, W. M. (2002). Biological control. In D. Lee (Ed.), *The biology of nematodes* (pp. 483–509). Taylor & Francis, London – New York.

Khan, Z., & Kim, Y. H. (2007). A review of the role of predatory nematodes in the biological control of plant parasitic nematodes. *Applied Soil Ecology, 35*, 370–379.

Kiewnick, S., Rumbos, C., & Sikora, R. A. (2004). Risk assessment of fungal biocontrol agents. *IOBC/WPRS Bulletin, 27*, 137–143.

Kimpinski, J., Arsenault, W. J., Gallant, C. E., & Sanderson, J. B. (2000). The effect of marigolds (*Tagete* spp.) and other cover crops on *Pratylenchus penetrans* and on following potato crops. *Supplement to the Journal of Nematology, 32* (4S), 531–536.

Kloepper, J. W., Rodriguez-Kabana, R., McInroy, J. A., & Collins, D. J. (1991). Analysis of populations and physiological characterization of microorganisms in rhizospheres of plants with antagonistic properties to phytopathogenic nematodes. *Plant and Soil, 136*, 95–102.

Kluepfel, D. A., Nyczepir, A., Lawrence, J. E., Wechter, W. P., & Leverentz, B. (2002). Biological control of the phytoparasitic nematode *Mesocriconema xenoplax* on peach trees. *Journal of Nematology, 34*, 120–123.

Koenning, S. R., Barker, K. R., & Bowman, D. T. (2001). Resistance as tactic for management of *Meloidogyne incognita* on cotton in North Carolina. *Journal of Nematology, 33*, 126–131.

Kokalis-Burelle, N., Mahaffee, W. F., Rodríguez-Kabana, J., Kloepper, W., & Bowen, K. L. (2002). Effects of switchgrass (*Panicum virgatum*) rotations with peanut (*Arachis hypogaea* L.) on nematode populations and soil microflora. *Journal of Nematology, 34*, 98–105.

Leij de, F. A. A. M., Davies, K. G., & Kerry, B. R. (1992). The use of *Verticillium chlamydosporium* Goddard and *Pasteuria penetrans* (Thorne) Sayre & Starr alone and in combination to control *Meloidogyne incognita* on tomato plants. *Fundamental and Applied Nematology, 15*, 235–242.

Leij de, F. A. A. M., Kerry, B. R., & Dennehy, J. A. (1993). *Verticillium chlamydosporium* as a biological control agent for *Meloidogyne incognita* and *M. hapla* in pot and micro-plot tests. *Nematologica, 39*, 115–126.

Linford, M. B., Yap, F., & Oliveira, J. M. (1938). Reduction of soil populations of the root-knot nematode during decomposition of organic matter. *Soil Science, 45*, 127–141.

Lorito, M., Woo, S. L., Garcia Fernandez, I., Colucci, G., Harman, G. E., Pintor-Toro, J. A., et al. (1998). Genes from mycoparasitic fungi as a source for improving plant resistance to fungal pathogens. *Proceedings of the National. Academy of Science, USA, 95*, 7860–7865.

Luc, M., Sikora, R. A., & Bridge, J. (Eds.). (2005). Plant parasitic nematode in subtropical and tropical agriculture (871 pp). Wallingford, UK: CABI Publishing.

Mauchline, T. H., Kerry, B. R., & Hirsch, P. R. (2004). The biocontrol fungus *Pochonia chlamydosporia* shows nematode host preference at the infraspecific level. *Mycological Research, 108*, 161–169.

Morton, C. O., Hirsch, P. R., Peberdy, J. P., & Kerry, B. R. (2003). Cloning of and genetic variation in protease VCP1 from the nematophagous fungus *Pochonia chlamydosporia. Mycological Research, 107*, 38–46.

O'Flaherty, S., Hirsch, P. R., & Kerry, B. R. (2003). The influence of the root-knot nematode *Meloidogyne incognita*, the nematicide aldicarb and the nematophagous fungus *Pochonia chlamydosporia* on heterotrophic bacteria in soil and the rhizosphere. *European Journal of Soil Science, 54*, 759–766.

Oostendorp, M., Dickson, D. W., & Mitchell, D. J. (1991). Population development of *Pasteuria penetrans* on *Meloidogyne arenaria. Journal of Nematology, 23*, 58–64

Oostendorp, M., & Sikora, R. A. (1989). Seed treatment with antagonistic rhizobacteria for the suppression of *Heterodera schachtii* early root infection of sugar beet. *Revue de Nematologie, 12*, 77–83.

Perry, R., & Moens, M. (2006). Plant nematology (447 pp.). Wallingford, UK: CABI Publishing.

Peteira, B., Puertas, A., Hidalgo-Díaz, L., Hirsch, P. R., Kerry, B. R., & Atkins, S. D. (2005). Real-time PCR to monitor and assess the efficacy of the nematophagous fungus *Pochonia chlamydosporia* var. *catenulata* against root-knot nematode populations in the field. *Biotecnología Aplicada, 22*, 261–266.

Pinochet, J., Camprubi, A., Calvet, C., Fernandez, C., & Kabana-Rodriquez, R. (1998). Inducing tolerance to the root-lesion nematode *Pratylenchus vulnus* by early mycorrhizal inoculation of micropropagated myrobalan 20C plum rootstock. *Journal of the American Society for Horticultural Science, 1223*, 342–347.

Ploeg, A. T., &. Staplenton, J. J. (2001). Glasshouse studies on the effects of time, temperature and amendment of soil with broccoli plant residues on the infestation of melon plants by *Meloidogyne incognita* and *M. javanica*. *Nematology, 3*, 855–861.

Powell, K. A., & Faull, J. L. (1989). Commercial approaches to the use of biological control agents. In J. M. Whipps & R. D. Lumsden (Eds.), *Biotechnology of fungi for improving plant growth* (pp. 259–275). Cambridge, UK: Cambridge University Press.

Puertas, A., & Hidalgo-Díaz, L. (2007). Influencia de la planta hospedante y su interacción con *Meloidogyne incognita* sobre la efectividad de *Pochonia chlamydosporia* var. *catenulata* como agente de control biológico. *Revista de Protección Vegetal, 22*, (*in press*).

Pyrowolakis, A., Schuster, R.-P., & Sikora, R. A. (1999). Effect of cropping pattern and green manure on the antagonistic potential and the diversity of egg pathogenic fungi in fields with *Heterodera schachtii* infection. *Nematology, 1*, 165–171.

Reddy, P. P., Rao, M. S., & Nagesh, M. (1999). Eco-friendly management of *Meloidogyne incognita* on tomato by integration of *Verticillium chlamydosporium* with neem and calotropis leaves. *Zeitschrift fur Pflanzenkrankheiten und Pflanzenschutz, 106*, 530–533.

Rodriguez-Kabana, R., Morgan-Jones, G., & Chet, I. (1987). Biological control of nematodes: soil amendments and microbial antagonists. *Plant and Soil, 100*, 237–248.

Rumbos, C. I., & Kiewnick, S. (2006). Effect of plant species on persistence of *Paecilomyces lilacinus* strain 251 in soil and on root colonization by the fungus. *Plant and Soil, 283*, 25–51.

Sano, Z. (2002). Nematode management strategies in East Asian countries. *Nematology, 4*, 129–130.

Schlang, J., Steudel, W., & Miller, J. (1988). Influence of green manure crops on the population dynamics of *Heterodera schachtii* and its fungal egg parasites. [Abstract]. *Nematologica, 34*, 293.

Segers, R. (1996). *The nematophagous fungus* Verticillium chlamydosporium: *Aspects of pathogenicity* (222 p.). PhD thesis, University of Nottingham.

Segers, R., Butt, T. M., Kerry, B. R., & Peberdy, J. F. (1994). The nematophagous fungus *Verticillium chlamydosporium* produces a chymoelastase-like protease which hydrolyses host nematode proteins in situ. *Microbiology, 140*, 2715–2723.

Shaukat, S. S., & Siddiqui, I. A. (2003). Impact of biocontrol agents *Pseudomonas fluorescens* CHA0 and its genetically modified derivatives on the diversity of culturable fungi in the rhizosphere of mungbean. *Journal of Applied Microbiology, 95*, 1039–1048.

Sikora, R. A. (2001). Use of mutualistic fungal endophytes for biological enhancement of tissue culture derived planting material for the control of fungal wilt and plant parasitic nematodes on banana. Joint Meeting of the American Phytopathological Society, the Mycological Society of America and the Society of Nematologists, Salt Lake City, Utah, USA, *Phytopathology,* S82.

Sikora, R. A., Bridge, J., & Starr, J. L. (2005). Management practice: an overview of integrated nematode management technologies. In M. Luc, R. A. Sikora, & J. Bridge (Eds.), *Plant parasitic nematodes in subtropical and tropical agriculture* (pp. 793–827). Wallingford, UK: CABI Publishing.

Sorribas, F. J., Ornat, C., Verdejo-Lucas, S., Galeano, M., & Valero, J. (2005). Effectiveness and profitability of the Mi-resistance gene in tomato over three consecutive growing seasons. *European Journal of Plant Pathology, 11*, 29–38.

Staplenton, J. J., & Duncan, R. A. (1998). Soil desinfestation with cruciferous amendments and sub-lethal heating effect on *Meloidogyne incognita*, *Sclerotium rolfsii* and *Pythium ultimum. Plant Pathology, 47*, 737–742.

Stefanova, M., & Fernández, E. (1995). Principales Patógenos del Suelo en las Hortalizas y su Control. In R. Labrada (Ed.), *Producción Intensiva de Hortalizas en los Trópicos Húmedos (*pp. 111–120). *División de Producción y Protección Vegetal*, FAO, Roma.

Stirling, G. R. (1991). *Biological control of plant parasitic nematodes: Progress, Problems and Prospects* (282 pp.). Wallingford, UK: CABI International.

Talavera, M., Itou, K., & Mizukubo, T. (2002). Combined application of *Glomus* sp. and *Pasteuria penetrans* for reducing *Meloidogyne incognita* (Tylenchida: Meloidogynidae) populations and improving tomato growth. *Applied Entomology and Zoology, 37*, 61–67.

Taba, S., Moromizato, K., Takaesu, Z., Ooshiru, A., & Nasu, K. (2006). Control of the southern root-knot nematode, *Meloidogyne incognita* using granule formulations containing nematode-trapping fungus, *Monacrosporium ellipsosporum* and a nematicide. *Japanese Journal of Applied Entomolgy and Zoology, 50*, 115–122.

Timper, P. (1999). Effect of crop rotation and nematicide use on abundance of *Pasteuria penetrans. Journal of Nematology, 31*, 575 [Abstract].

Timper, P., & Brodie, B. B. (1994). Effect of host-plant resistance and a nematode pathogenic fungus on *Pratylenchus penetrans*. *Phytopathology, 84*, 1090 [Abstract].

Timper, P., Minton, N. A., Johnson, A. W., Brenneman, T. B., Culbreath, A. K., Burton, G. W., et al. (2001). Influence of cropping systems on stem rot (*Sclerotium rolfsii*), *Meloidogyne arenaria* and the nematode antagonist *Pasteuria penetrans* in peanut. *Plant Disease, 85*, 767–772.

Tzortzakakis, E. A., & Goewn, S. R. (1994). Evaluation of *Pasteuria penetrans* alone and in combination with oxamyl, plant resistance and solarization for control of *Meloidogyne* spp. On vegetables grown in greenhouses in Crete. *Crop Protection, 13*, 455–462.

Van den Boogert, P. H. J. F., Velvis, H., Ettema, C. H., & Bouwman, L. A. (1994). The role of organic matter in the population dynamics of the endoparasitic nematophagous fungus *Drechmeria coniospora* in microcosms. *Nematologica, 40*, 249–257.

Van Emden, H. F., & Peakall, D. B. (1996). The practice of pest management in developing countries. In H. F. van Emden & D. B. Peakall (Eds.), *Beyond silent spring* (pp. 167–222). London, UK: Chapman and Hall.

Viaene, N., Coyne, D. L., & Kerry, B. (2006). Biological and cultural management. In R. Perry & M. Moens (Eds.), *Plant nematology* (pp. 346–369). Wallingford, UK: CABI Publishing.

Walker, G. E., & Wachtel, M. F. (1988). The influence of soil solarization and non-fumigant nematicides on infection of *Meloidogyne javanica* by *Pasteuria penetrans*. *Nematologica, 34*, 477–483.

Wang, K.-H., Mc Sorley, H. R., & Gallaher, R. N. (2004). Effect of winter cover crops on nematode population levels in north Florida. *Journal of Nematology, 36*, 517–523.

Wang, K.-H., Sipes, B. S., & Schmitt, D. P. (2002). Crotalaria as a cover crop for nematode management: A review. *Nematropica, 32*, 35–37.

Whitehead, A. G. (1998). *Plant nematode control* (384 pp.). Wallinford, UK: CAB International.

Xu, J., Mo, M.-H., Huang, X.-W., & Zhang, K.-Q. (2005). Improvement on genetic transformation in the nematode-trapping fungus *Arthrobotrys oligospora* and its quantification on dung samples. *Mycopathologia, 159*, 533–538.

Yamada, M. (2001). Methods of control of injury associated with continuous vegetables cropping in Japan-crop rotation and several cultural practices. *Japan Agricultural Reasearch Quarterly, 35*, 39–45.

Zaki, M. J., & Maqbool, M. A. (1991). Combined efficacy of *Pasteuria penetrans* and other biocontrol agents on the control of root-knot nematode on okra. *Pakistan Journal of Nematology, 9*, 49–52.

3

L. V. LOPEZ-LLORCA, J. G. MACIÁ-VICENTE AND H.-B. JANSSON

MODE OF ACTION AND INTERACTIONS OF NEMATOPHAGOUS FUNGI

Laboratory of Plant Pathology, Multidisciplinary Institute for Environmental Studies (MIES) "Ramon Margalef", University of Alicante, 03080 Alicante, Spain

Abstract. Nematophagous fungi are potential candidates for biological control of plant-parasitic nematodes, and an important constituent in integrated pest management programs. In this chapter we describe various aspects on the biology of these fungi. Nematophagous species can be found in most fungal taxa, indicating that the nematophagous habit evolved independently in the different groups of nematophagous fungi. Regarding their mode of action we discuss recognition phenomena (e.g. chemotaxis and adhesion), signaling and differentiation, and penetration of the nematode cuticle/eggshell using mechanical, as well as enzymatic (protease and chitinase) means. The activities of nematophagous fungi in soil and rhizosphere is also discussed.

1. INTRODUCTION

The term "nematophagous fungi" is used to describe a diverse group of organisms with the ability to infect and parasitize nematodes for the benefit of nutrients. The first description of their nematophagous habit came in the late 1800's and has been followed by work of many scientists describing this fascinating group of fungi. Apart from infecting nematodes, nematophagous fungi also have the ability to colonize and parasitize other organisms, such as plants and even other fungi. Some of them are obligate parasites of nematodes, but the majority are facultative saprophytes.

Because of their capability to parasitize plant- and animal-parasitic nematodes they have a potential for development as biocontrol agents. In the current chapter we describe and discuss some of the research that has been performed on nematophagous fungi. We will focus on fundamental aspects such as their mode of action and interactions, especially regarding their behaviour in the rhizosphere and their endophytic behaviour within the scenario of a complex trophic web, with the soil and its biota as background. Our working hypothesis is that an adequate management of this ecosystem will lead to the establishment of long-term nematode suppression as it happens under natural conditions in a wide array of

A. Ciancio & K. G. Mukerji (eds.), Integrated Management and Biocontrol of Vegetable and Grain Crops Nematodes, 51–76.

soils worldwide. The plant host defences are triggered unspecifically by biotic and abiotic factors. Therefore, better knowledge about the mode of action of nematophagous fungi, especially regarding the host plant, may lead to control of other root pathogens such as fungi and may in turn improve plant growth.

2. NEMATOPHAGOUS FUNGI

2.1. Biology

Depending on their mode of attacking nematodes, the nematophagous fungi are divided into four groups: (*i*) nematode-trapping (formerly sometimes called predacious or predatory fungi), (*ii*) endoparasitic, (*iii*) egg- and female-parasitic and (*iv*) toxin-producing fungi (Jansson & Lopez-Llorca, 2001). Some of the characteristics of these grops are resumed and shown in Fig. 1.

The nematode-trapping fungi, as the name implies, capture nematodes with the aid of hyphal trapping devices of various shapes and sizes, e.g. adhesive three-dimensional nets, adhesive knobs, non-adhesive constricting rings. A few "nematode-trappers" capture nematodes without visible traps in an adhesive substance formed on their hyphae, e.g. *Stylopage* spp.

Endoparasitic fungi use their spores (conidia or zoospores) to infect nematodes. The propagules adhere to the nematode cuticle, and the spore contents is then injected into the nematode, or the spores are swallowed by the host. Most of these fungi are obligate parasites of nematodes and live their entire vegetative stages inside infected nematodes.

The egg- and female-parasitic fungi infect nematode females and the eggs they contain, using appressoria or zoospores. Finally, the toxin-producing fungi immobilize the nematodes by a toxin, prior to hyphal penetration through the nematode cuticle. In all four nematophagous groups, nematode parasitism results in a complete prey or egg digestion, activity which supplies the fungus with nutrients and energy for continued growth.

2.2. Taxonomy and Phylogeny

Nematophagous fungi are found in most fungal taxa: Ascomycetes (and their hyphomycete anamorphs), Basidiomycetes, Zygomycetes, Chytridiomycetes and Oomycetes (Fig. 2). It therefore appears that the nematophagous habit evolved independently in the different fungal taxonomic groups. Barron (1992) suggested that the nematophagous habit evolved from lignolytic and cellulolytic fungi, as an adaptation to overcome competition for nutrients in soil.

Recently, the egg-parasitic fungi previously placed within the genus *Verticillium* were transferred to the new genus *Pochonia*, in parallel with entomopathogenic species of *Verticillium*, which were transferred to the genus *Lecanicillium* based both on morphological and molecular characters (Zare & Gams, 2001; Zare, Gams, &

Evans, 2001). The teleomorphs of the *Pochonia* species are located within *Cordyceps*. The best known species of egg parasites are *P. chlamydosporia* and *P. rubescens*, but species of other genera such as *Paecilomyces lilacinus* and *Lecanicillium lecanii,* are also known to parasitize nematode eggs.

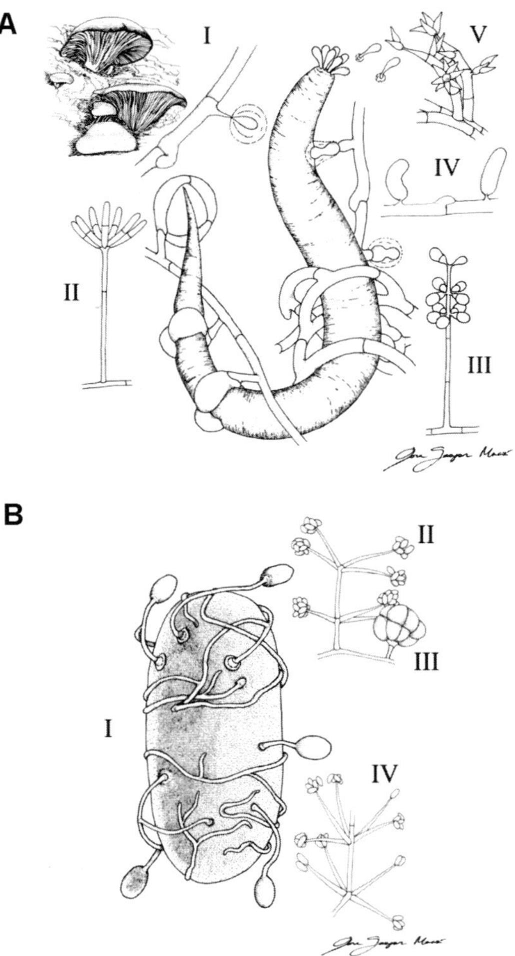

Figure 1. Biology of nematophagous fungi. Vermiform (motile) nematode (A) displaying infection structures: (I) toxin-producing fungus [Pleurotus *sp.*], *nematode-trapping fungi (II)* Drechslerella *sp. (III)* Arthrobotrys *sp., (IV)* Nematoctonus *sp. and (V) endoparasitic* Drechmeria *sp.. Nematode (sedentary) egg (B) (similar features can be found in egg masses, females and cysts) displaying infection structures: penetrating hyphae and appressoria of egg-parasitic fungi (I), conidia (II) and chlamydospores (III) of* Pochonia *sp., and conidia of* Lecanicillium *sp. (IV).*

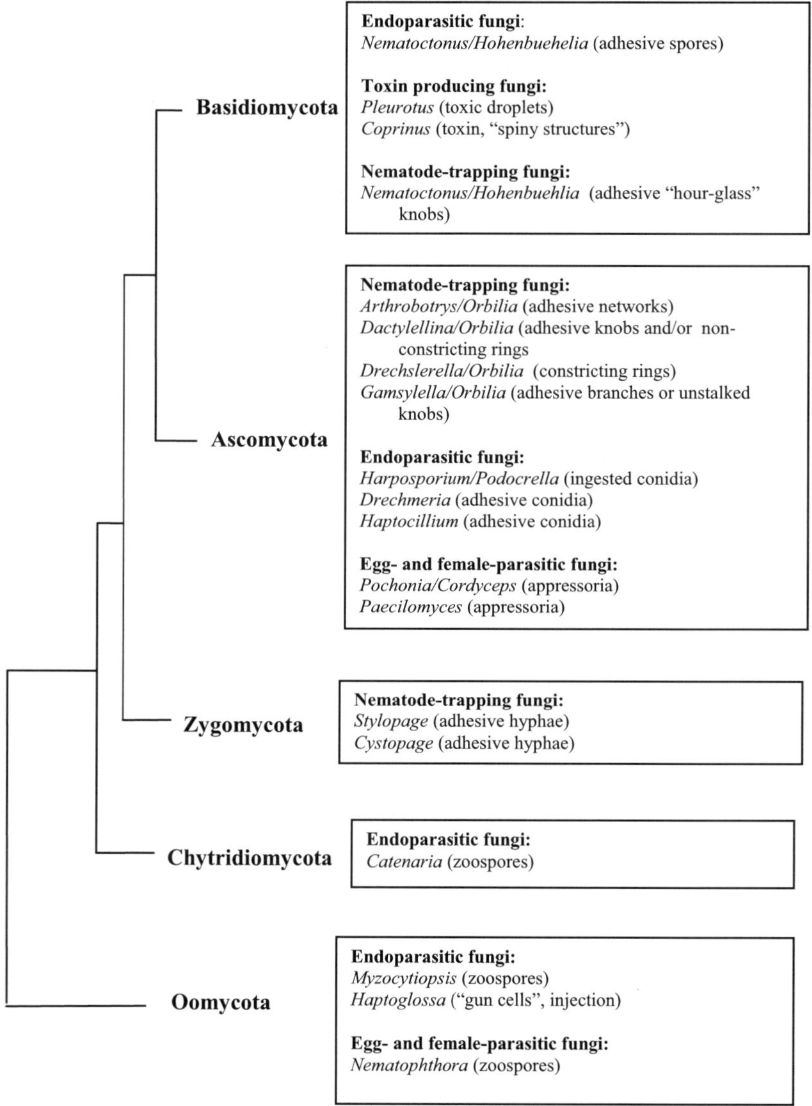

Figure 2. Taxonomic position of nematophagous fungi with examples of genera. The first genus names are anamorphs, and genus names after slashes indicate known teleomorphs. Infection structures are shown in parenthesis.

Most nematode-trapping species have a teleomorph within *Orbilia*, and their taxonomic positions have been arranged according to their type of trapping device (Ahrén, Ursing, & Tunlid, 1998). Scholler, Hagedorn, and Rubner (1999) suggested the following classification based on molecular data: *Arthrobotrys* (adhesive three-dimensional networks), *Dactylellina* (stalked adhesive knobs and/or non-constricting rings), *Drechslerella* (constricting rings) and *Gamsylella* (adhesive branches and unstalked knobs). This classification was questioned by Li et al. (2005) who suggested that the species in *Gamsylella* should be transferred to either *Arthrobotrys* or *Dactylellina* based on more and refined DNA sequencing. In this review we follow the taxonomy suggested by Scholler et al. (1999). Li et al. (2005) put forward a hypothesis of an evolutionary pathway of traps of the nematode-trapping Orbiliales. According to this hypothesis, two lines have evolved originating from adhesive knobs, in one line the adhesive was lost and evolved to form constricting rings, whereas the other evolutive line retained the adhesive and became three-dimensional networks.

Much less is known about the taxonomy/phylogeny of the endoparasitic fungi. Some of these are placed in the Chytridiomycetes, e.g. the zoosporic *Catenaria anguillulae*, others in *Haptocillium* (formerly *Verticillium*), *Harposporium* or *Drechmeria*. The teleomorph of *Harposporium* spp. has recently been transferred from *Atricordyceps* to *Podocrella* (Chaverri, Samuels, & Hodge, 2005). The basidiomycete genus *Hohenbuehelia* (anamorph: *Nematoctonus*) contains fungi that can be classified as both nematode-trapping and endoparasites (Thorn & Barron, 1986). The genus *Pleurotus* includes species, such as the oyster mushroom *P. ostreatus,* and constitutes the toxin-producing fungi. Recently, *Coprinus comatus* was shown to have similar capabilities (Luo, Mo, Huang, Li, & Zhang, 2004), suggesting that the nematophagous habit may be more widespread among Basidiomycetes than previously thought.

2.3. Fungal Parasites of Invertebrates

Entomopathogenic and nematophagous fungi are generally facultative parasites, usually implying a low host specificity and consequently a wide host range. They can also colonize a wide array of habitats and their main species can be found worldwide.

Entomopathogenic and nematophagous fungi bear multiple similarities. The most important species of both fungal groups have been described as soil inhabitants, where they spend most of the saprophytic growth phase. Soil is also the environment of nearly all plant-parasitic nematodes and of soil dwelling insects such as roots pests or other underground plant organs. For further details on these aspects see Lopez-Llorca and Jansson (2006).

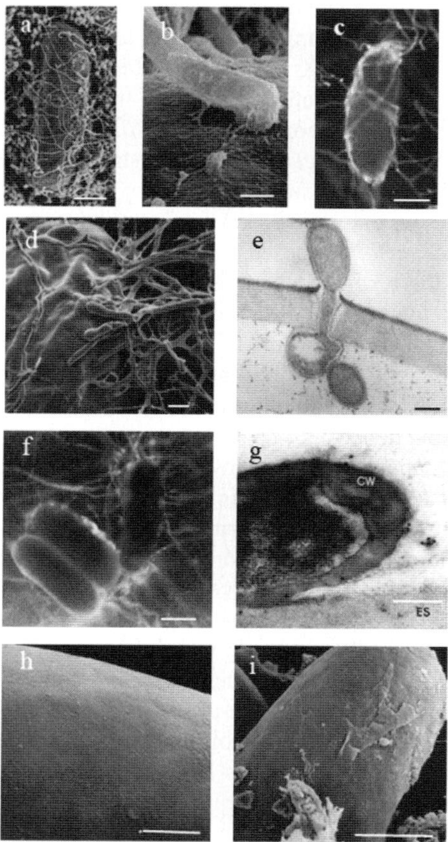

*Figure 3. Mode of action of fungal parasites of nematode eggs. (a) Field emission scanning electron microscopy (FESEM) of nematode (*Heterodera schachtii*) egg inoculated with conidia of* Pochonia rubescens *(Bar = 25 µm). (b) Detail of the fungus appressoria showing adhesive secretions on the eggshell (Bar = 2 µm). (c) Labelling of nematode-infected egg with Con A lectin fluorescently labelled (Bar = 25 µm). (d) Detail of advanced infection by* P. rubescens *showing fully developed appressoria on eggshell (Bar = 5µm). (e) Eggshell penetration by* P. rubescens *(Bar = 0.25 µm). (f) immunofluorescence detection of P32 protease produced by* P. rubescens *(Bar = 5 µm). (g) Immunogold detection of P32 (Bar = 1 µm) (Lopez-Llorca & Robertson, unpublished). (h) and (i) Effect of purified P32 on eggshell of* H. schachtii. *(h) control (Bar = 5 µm) and (i) P32-treated. (Bar = 10 µm). (FESEM, Lopez-Llorca & Claugher, unpubl.). (a) From Lopez-Llorca and Claugher, 1990, courtesy of Elsevier. (c) From Lopez-Llorca, Olivares-Bernabeu, Salinas, Jansson, and Kolattukudy, 2002b, courtesy of Elsevier. (d) and (e) adapted from Lopez-Llorca and Robertson, 1992b, courtesy of Springer. (f) From Lopez-Llorca and Robertson, 1992a, courtesy of Elsevier.*

3. MODE OF ACTION

The infection of nematodes and their eggs by various nematophagous fungi follows a similar, general pattern. This is illustrated here by infection of nematode eggs by *Pochonia rubescens* (Fig. 3) and also by the zoospores of *Catenaria anguillulae*, which infect vermiform nematodes (Fig. 4).

Penetration of nematode eggs by *P. rubescens* starts with contact of the hyphae with the egg (Fig. 3a) and subsequent formation of an appressorium (Figs. 3b, d). An extracellular material (ECM) or adhesive, is formed on the appressorium, and is revealed by labelling with the lectin Concanavalin A (Con A), indicating that it contains glucose/mannose residues (Fig. 3c). From the appressorium the fungus penetrates the nematode eggshell (Fig. 3e) by means of both mechanical and enzymatic components. The nematode eggshell contains mainly chitin and proteins (Bird & Bird, 1991) and therefore chitinases and proteases play an important role during eggshell penetration (Lopez-Llorca, 1990; Tikhonov, Lopez-Llorca, Salinas, & Jansson, 2002). The ECM contains the protease P32 that can be immunologically detected using both fluorescent stains (Fig. 3f) or colloidal gold (Fig. 3g). The proteolytic activity causes the degradation of eggshells (Fig. 3i).

The life cycle of *C. anguillulae* starts with uniflagellate zoospores which become attracted to natural orifices (mouth, anus, excretory pores, etc.) of nematodes (Figs. 4a, 4b). The flagellar movement is supported by the mitochondria at the base of the flagellum (Fig. 4c). Upon contact with the nematode cuticle the zoospores show an "amoeboid movement" before encystment takes place (Fig. 4d). During encystment a cell wall is formed covered by an adhesive, and the flagellum is withdrawn (Fig. 4e). The encysted zoospore forms an infection peg which penetrates the nematode cuticle (Fig. 4f). Within 24 hours the developing fungus invades and digests the nematode contents, and zoosporangia are formed (Fig. 4g) from which the zoospores are released (Fig. 4h) to infect new hosts. *Catenaria anguillulae* also has the ability to infect nematode eggs (Wyss et al., 1992).

3.1. Recognition: Chemotaxis and Adhesion

Nematodes infection starts with a recognition phase including attraction, host chemotaxis towards fungal hyphae or traps, or chemotaxis of zoospores towards the host's natural openings (Jansson & Nordbring-Hertz, 1979; Jansson & Thiman, 1992). The compounds involved in chemotactic events are not known (Jansson & Friman, 1999; Bordallo et al., 2002). The adhesive on the traps of *A. oligospora* switches from an amorphous to a fibrillar appearance after contact with a nematode, which is in contrast to the adhesive on conidia of *D. coniospora* which always appears fibrillar (Jansson & Nordbring-Hertz, 1988). The adhesive on the appressoria of *P. chlamydosporia* and *P. rubescens* can be labelled with the lectin Concanavalin A, suggesting a glycoprotein nature with mannose/glucose moieties (Lopez-Llorca et al., 2002b). Involvement of a Gal-NAc-specific lectin of *A. oligospora* (Nordbring-Hertz & Mattiasson, 1979) and a sialic acid-specific lectin of *D. coniospora* (Jansson & Nordbring-Hertz, 1984) in nematode recognition have been suggested. Infection events eventually lead to a signalling cascade necessary for penetration and colonisation of the nematode prey (Tunlid, Jansson, & Nordbring-Hertz, 1992).

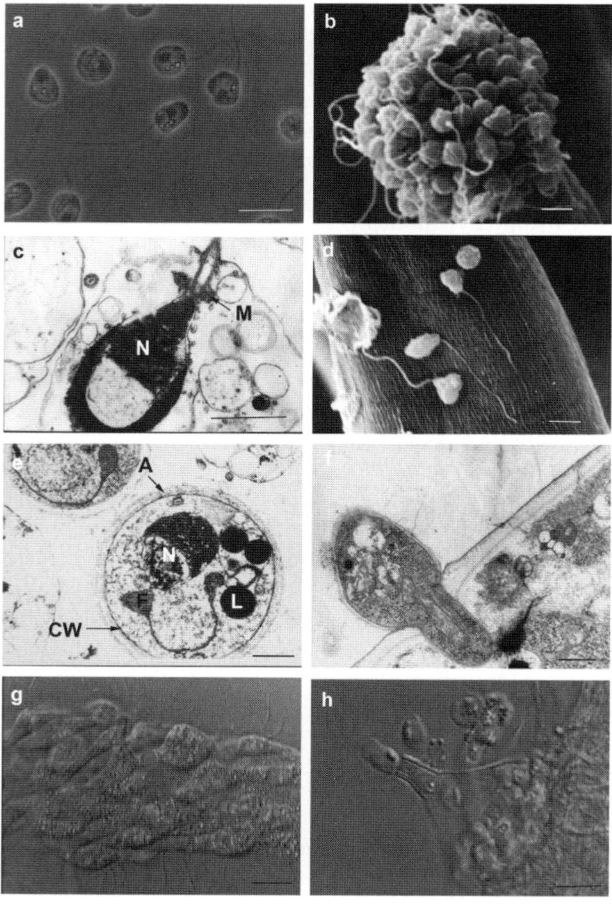

Figure 4. Infection of nematodes by the zoosporic fungus Catenaria anguillulae. *Monoflagellate zoospores (a). Zoospores (b) accumulated at the mouth of a nematode. Ultrastructure of a zoospore (c): N = nucleus and nuclear cap, M = mitochondrium at flagellar base. Zoospores show typical amoeboid movement prior to encystment (d). Encysted zoospore (e): A = adhesive, CW = cell wall, F = withdrawn flagellum, L = lipid droplet, N = nucleus and nuclear cap. Penetration of nematode cuticle (f) and development of zoosporangia (g) inside an infected nematode. The cycle is completed by the release of zoospores (h). Scale bars: a, b, d, h = 2 µm; c, e, f = 1 µm; g = 5 µm. (Figs. a, g, h) from Jansson et al., 1995, courtesy of IWF Wissen und Medien, Göttingen; (b) from Jansson & Thiman, 1992, courtesy of Mycological Society of America; (e) from Tunlid, Nivens, Jansson, and White, 1991b, courtesy of Experimental Mycology; (c, d and f) H–B. Jansson, unpublished.*

After contact, an extracellular material, or adhesive, is formed which keeps the fungus onto the nematode surface (Figs. 3b, 3c, 4e). Nematophagous fungi adhesives commonly contain proteins and/or carbohydrates (Tunlid, Johansson, & Nordbring-Hertz, 1991a; Tunlid et al., 1991b).

Carbohydrates present on the surface of nematodes are involved in the recognition step of lectin binding, but also appear to be involved in nematode chemotaxis (Zuckerman & Jansson, 1984; Jansson, 1987). The main nematode sensory organs, amphids and inner labial papillae, are located in the cephalic and labial region, around their mouth (Ward, Thomson, White, & Brenner, 1975).

A hypothesis of the involvement of carbohydrates in nematode chemoreception was put forward by Zuckerman (1983) and Zuckerman and Jansson (1984). The chemoreceptors, purportedly glycoproteins, could be blocked by lectins (Concanavalin A binding to mannose/glucose residues, and Limulin binding to sialic acid) resulting in loss of chemotactic behaviour of bacterial-feeding nematodes to bacterial exudates (Jeyaprakash, Jansson, Marban-Mendoza, & Zuckerman, 1985). Furthermore, treating nematodes with enzymes (mannosidase, sialidase) obliterating the terminal carbohydrates also decreased chemotactic behaviour (Jansson, Jeyaprakash, Damon, & Zuckerman, 1984), showing the role of carbohydrate moieties in nematode chemotaxis.

The endoparasitic nematophagous fungus *D. coniospora* infects nematodes with conidia which adhere to the host chemosensory organs (Jansson & Nordbring-Hertz, 1983). Conidial adhesion was suggested to involve a sialic acid-like carbohydrate since treatment of nematodes with the lectin Limulin, and treatment of spores with sialic acid, decreased adhesion (Jansson & Nordbring-Hertz, 1984). Furthermore, nematodes with newly adhered spores lost their ability to respond chemotactically to all attraction sources tested, i.e. conidia, hyphae and bacteria, indicating a connection between adhesion and chemotaxis through carbohydrates present on the nematode surface (Jansson & Nordbring-Hertz, 1983).

The conidia of *D. coniospora* adhere to the chemosensory organs of *Meloidogyne* spp., but do not penetrate and cannot infect the nematodes. Irrespective of the lack of infection, the fungus was capable of reducing root galling in tomato in a biocontrol experiment (Jansson, Jeyaprakash, & Zuckerman, 1985), again indicating the involvement of chemotactic interference.

Interfering with nematode chemotaxis, thereby inhibiting their host-finding behavior, may be a possible way of controlling plant-parasitic species. In a pot experiment using tomato as host plant and *Meloidogyne incognita* as parasitic nematode, addition of Concanavalin A and *Limax flavus* agglutinin (sialic acid specific lectin) resulted in decreased plant damage by the nematode compared to controls (Marban-Mendoza, Jeyaprakash, Jansson, & Zuckerman, 1987). Addition of lectins (or enzymes) on a field is not feasible, but the possibility to use, for instance, lectin-producing leguminous plants have been shown to reduce galling by root knot nematodes (Marban-Mendoza, Dicklow, & Zuckerman, 1992).

3.2. Signalling and Differentiation

Most pathogenic fungi differentiate appressoria (Fig. 3b) when sensing the host's surface or even artificial surfaces. Appressoria development has been studied in detail in plant pathogenic fungi infecting leaves (Lee, D'Souza, & Kronstad, 2003; Basse & Steinberg, 2004). A hypothesis of signalling events during appressorium formation of the insect pathogen *Metarhizium anisopliae* was put forward by St. Leger (1993), partly based on knowledge acquired on plant pathogenic fungi.

Nematophagous fungi, especially egg parasites, differentiate appressoria on their hosts (Lopez-Llorca & Claugher, 1990). Very little is known about the signalling pathways leading to nematodes infection by nematophagous fungi. Recently, using expressed sequence tag (EST) techniques, it was shown that genes involved in the formation of infection structures and in fungal morphogenesis were expressed during trap formation of the nematophagous fungus *Dactylellina haptotyla* (syn. *Monacrosporium haptotylum*) (Ahrén et al., 2005). Similar results have also been presented for the entomopathogen *M. anisopliae* (Wang & St. Leger, 2005).

Fungi infecting vermiform nematodes differentiate several trapping organs as a response to environmental stimuli, chemical as well as tactile. The constricting ring traps function through the inflation of the three ring cells which form the trapping device. When a nematode starts touching the inner ring wall, an unknown mechanism triggers its inflation and closure, a process which takes about 0.1 seconds. The cells of the ring can also be manipulated to close in the laboratory by mild heat, pressure or Ca^{2+}. Chen, Hsu, Tsai, Ho, and Lin (2001) investigated signalling taking place in ring closure of the constricting ring trap of *D. dactyloides* using activators and inhibitors of G-proteins, and suggested a model in which the nematode exerts a pressure on the ring which activates G-proteins. This leads to an increase in cytoplasmic Ca^{2+}, activation of calmodulin and finally to opening of water channels resulting in trap inflation and nematode capture.

3.3. Penetration of Nematode Cuticles and Eggshells

After firm attachment to the host surface, nematophagous fungi penetrate the nematode cuticle (Fig. 4f) or eggshell (Fig. 3e). As in many other instances of fungal penetration of host surfaces, nematophagous fungi appear to use both enzymatic and physical means. The nematode cuticle mainly contains proteins (Bird & Bird, 1991) and therefore the action of proteolytic enzymes (Table 1) may be important for penetration. A serine protease, PII, from *A. oligospora*, has been characterized, cloned and sequenced (Åhman, Ek, Rask, & Tunlid, 1996). The expression of PII is increased by the presence of proteins, including nematode cuticles (Åhman et al., 1996). PII belongs to the subtilisin family and has a molecular mass of 32 kDa.

Table 1. Serine proteases and chitinases isolated and characterized from different nematophagous fungi.

Nematophagous species	*Enzyme*	*kDa*	*pI*	*Optimum pH*	*References*
		Proteases Nematode-trapping fungi			
Arthrobotrys oligospora	PII	35	4.6	7–9	Tunlid, Rosén, Ek, and Rask (1995) Åhman et al. (1996)
A. oligospora	Aoz	38	4.9	6–8	Zhao, Mo, and Zhang (2004)
Arthrobotrys (Monacrosporium) microscaphoides	Mlx	39	6.8	9	M. Wang, Yang, and Zhang (2006)
Arthrobotrys (Dactylella) shizishanna	Ds1	35	-	10	R.B. Wang, Yang, Lin, Y. Zhang, and K.Q. Zhang (2006)
		Egg-parasitic fungi			
Pochonia rubescens	P32	32	6.2	8.5	Lopez-Llorca (1990) Olivares-Bernabéu (1999) Lopez-Llorca and Robertson (1992b)
Pochonia chlamydosporia	VCP1	33	10.2	-	Segers, Butt, Kerry, and Peberdy (1994); Segers, Butt, Keen, Kerry, and Peberdy (1995)
Paecilomyces lilacinus	PL	33.5	>10.2	10.3	Bonants et al. (1995)
Lecanicillium psalliotae	Ver112	32	-	10	Yang et al. (2005a, 2005b)
		Chitinases/chitosanases			
P. rubescens	CHI43	43	7.6	5.2–5.7	Tikhonov et al. (2002)
P. chlamydosporia	CHI43	43	7.9	5.2–5.7	Tikhonov et al. (2002)
P. lilacinus	-	23	8.3	6	Chen, Cheng, Huang, and Li (2005)

Another serine protease from *A. oligospora* (Aoz1), with a molecular mass of 38 kDa showing 97% homology with PII was recently described (Zhao et al., 2004). Other serine proteases have been isolated and characterized from the nematode-trapping fungi *Arthrobotrys* (syn. *Monacrosporium*) *microscaphoides* designated Mlx (M. Wang et al., 2006) and *Arthrobortys* (syn. *Dactylella*) *shizishanna* (Ds1) (R. B. Wang et al., 2006) both showing high homology with the *A. oligospora* serine proteases (M. Wang et al., 2006).

Nematode eggshells mostly contain protein and chitin (Clarke, Cox, & Shepherd, 1967) organized in a microfibrillar and amorphous structure (Wharton, 1980). Therefore, a search for extracellular enzymes degrading those polymers was carried out. A 32 kDa serine protease (P32) was first purified and characterized from the egg parasite *P. rubescens* (Lopez-Llorca, 1990). Involvement of the enzyme in pathogenesis was suggested by quick *in vitro* degradation (Fig. 3i) of *Globodera pallida* egg shell proteins (Lopez-Llorca, 1990), but most of all by its immunolocalization (Fig. 3f, 3g) in appressoria of the fungus infecting *Heterodera schachtii* eggs (Lopez-Llorca & Robertson, 1992b).

Although pathogenesis is a complex process involving many factors, inhibition of P32 with chemicals and polyclonal antibodies reduced egg infection and penetration (Lopez-Llorca et al., 2002b). The similar species *P. chlamydosporia* also produces an extracellular protease (VcP1) (Segers et al., 1994) which is immunologically related to P32 and similar enzymes from entomopathogenic fungi (Segers et al., 1995). VcP1-treated eggs were more easily infected than untreated eggs, suggesting a role of the enzyme in eggshell penetration by egg-parasitic fungi.

Recently a serine protease (Ver112) was isolated and characterized from *Lecanicillium psalliotae* showing similarities with the *Arthrobotrys* proteases (PII and Aoz1) of ca 40%, and ca 60% homology with serine proteases of egg-parasitic fungi (Yang et al., 2005a, 2005b).

Other proteases from nematophagous fungi have been partly characterized, e.g. a chymotrypsin-like protease from conidia of the endoparasite *D. coniospora* (Jansson & Friman, 1999), and a collagenase produced by the nematode-trapping *Arthrobotrys tortor* (Tosi, Annovazzi, Tosi, Iadarola, & Caretta, 2001). Non-nematophagous fungi such as the mycoparasites *Trichoderma harzianum* and *Clonostachys rosea* (syn. *Gliocladium roseum*) are also sources of serine proteases with nematicidal activity (Suarez, Rey, Castillo, Monte, & Llobell, 2004; Li, Yang, Huang, & Zhang, 2006).

Several chitinolytic enzymes of *Pochonia rubescens* and *P. chlamydosporia* have been detected. One of those accounting for most of the activity was a 43 kDa endochitinase (CHI43) (Tikhonov et al., 2002). When *G. pallida* eggs were treated with both P32 and CHI43 damage to eggshell was more extensive than with each enzyme alone, suggesting a cooperative effect of both enzymes to degrade egg shells (Tikhonov et al., 2002). Recently a chitosanase was isolated and characterized from the egg-parasitic fungus *P. lilacinus* (Chen et al., 2005).

3.4. Fungal Pathogen Genomics and Proteomics

In the era of genomics, fungal pathogens are suitable candidates for the analysis under this new paradigm in modern biology. In the dawn of fungal pathogen genomics under the Fungal Genome Initiative, important fungal pathogens have been or are being sequenced (Xu, Peng, Dickman, & Sharon, 2006). A direct bonus is the finding of unique fungal genes and characterization of genome structure and function. Available gene predictions in genomes of fungal plant pathogens indicate 30% of no homologues. This situation, which could be similar in nematophagous fungi, indicates that new fungal genes or gene products (e.g. proteins of unknown function) can soon be discovered.

The re-evaluation of the study of fungal pathogenicity-related genes with a genomic approach is underway. One example is appressorium development. This awaits to be applied in nematophagous fungi. Signalling/reception are other fields which will follow.

Proteomic approaches complement genomics. There are expression, localization and interactions, which are unique to this global strategy. Our preliminary results indicate that plant-host fungal invertebrate pathogen "cross-talk" can be approached this way

The assembly of the Fungal Tree of Life project (Spatafora, 2005; Kuramae, Robert, Snel, Weiss, & Boekhout, 2006) which is at a very advanced stage, could represent a useful tool for deciding on how to proceed to establish genomic approaches. EST approaches to understand the pathogenicity of nematophagous fungi are already being used (Ahrén et al., 2005).

4. SOIL AND RHIZOSPHERE ENVIRONMENT

4.1. Activities in Soil

Nematophagous fungi are generally regarded as soil organisms (Dackman, Jansson, & Nordbring-Hertz, 1992), although there are reports on their frequent occurrence also in aquatic environments, especially in shallow, unpolluted water (Hao, Mo, Su, & Zhang, 2005). Most nematophagous fungi can live saprophytically in soil, but in presence of a host they change from a saprophytic to a parasitic stage. The exact mechanism behind this is not known. Nematophagous fungi inhabit soil pores where infection structures are formed and nematodes are captured (Fig. 5). The zoosporic fungi are obviously dependent of soil water films for their function.

When nematophagous species have to be applied to manage plant parasitic nematodes they have to be delivered to soil. Several approaches for introducing them have been used (see Stirling, 1991), but very little efforts have been paid to follow the fate of nematophagous fungi in the soil/rhizosphere environment, after their release.

Nematophagous fungi grow in almost all types of soil, but are generally regarded as being more frequent in soils with high organic matter (Duddington, 1962). Generally, they have few nutritional and vitamin requirements for growth, and hence are ubiquitous. Additions of glucose (Cooke, 1962) and chopped organic matter, e.g. grass (Duddington, 1962) increased activity of nematode-trapping species. This effect was probably due to an increase in the numbers of microbivorous nematodes. *Arthrobotrys* spp. have a teleomorph in *Orbilia*, which are weak wood decomposers (Pfister, 1997), and the wood decomposing *Pleurotus* spp. suggests that decomposition of wood may be an important supply of carbon and energy for the fungi. Capturing nematodes may hence support the fungi with nitrogen (Barron, 1992). In Petri dishes and sterilized microcosms there is a heavy reduction of nematodes due to nematophagous fungi (Jansson, 1982b), and a density dependence relationship exists between nematodes and endoparasites (Jaffee, Gaspard, & Ferris, 1989).

In field soil, there is no clear correlation between nematophagous fungi and nematodes (Persmark, Banck, & Jansson, 1996a) and nematode-trapping fungi are known to be sensitive to soil mycostasis (Cooke & Satchuthananthavale, 1968), as well as to feeding by soil enchytraeids (Jaffee, 1999).

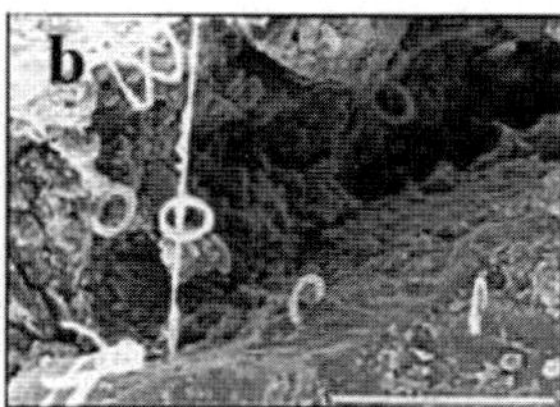

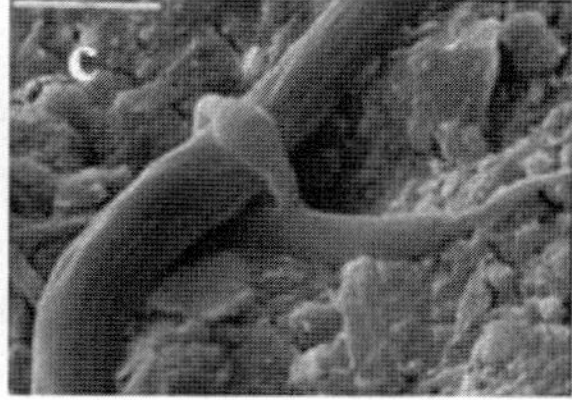

Figure 5. Low temperature scanning electron micrographs (LTSEM) of nematophagous fungi in soil. (a) Conidiophores with conidia of the nematode-trapping fungus Arthrobortys superba *(bar = 100 μm). (b) Constricting ring traps of* Drechslerella dactyloides *(bar = 50 μm). (c) Nematode captured in constricting ring of* D. dactyloides *(bar = 50 μm). From Jansson, Persson, and Odselius (2000), courtesy of Mycological Society of America.*

Introduction of nematophagous fungi, and most microbial biocontrol agents, to soil has been problematic due to both biotic and abiotic factors. Biocontrol experiments using the egg-parasite *P. chlamydosporia* showed low control efficiency against root-knot nematodes, and furthermore, the fungus was detected at very low rates, mainly in the rhizosphere of the test plants (Verdejo-Lucas, Sorribas, Ornat, & Galeano, 2003). One of the reasons for this may be that the soil was not receptive to the fungus.

We have used an *in vitro* assay to be able to easily study soil receptivity for nematophagous fungi (Monfort, Lopez-Llorca, Jansson, & Salinas, 2006). Using a soil-membrane technique 0, 25, 50, 75 and 100% sterilized soil was inoculated with several isolates of the nematophagous fungi *P. chlamydosporia* and *P. lilacinus*. After 4 weeks, colony radius was measured (expressed as relative growth) as well as hyphal density on the membrane placed on top of the soils.

When comparing two sandy soils (Spanish and Australian) with similar physico-chemical properties, large differences between the receptivity to the fungi were found, both regarding isolates as well as between soils. For instance, an Australian isolate of *P. chlamydosporia* was most inhibited in the Spanish soil, but the least inhibited in the Australian soil. The result suggests that a soil can be more receptive to indigenous isolates than to non-indigenous ones.

4.2. Nematophagous Fungi as Root Endophytes

Since nearly all plant-parasitic nematodes attack plant roots, the rhizosphere biology of nematophagous fungi is important from the point of view of a biological control strategy. Nematode-trapping fungi (Peterson & Katznelson, 1965; Gaspard & Mankau, 1986; Persmark & Jansson, 1997) and egg-parasitic fungi (Bourne, Kerry, & De Leij, 1996; Kerry, 2000) have been found to be more frequent in the rhizosphere than in the bulk soil.

External root colonisation varies between plant species. The pea rhizosphere harboured by far the highest frequency and diversity of nematode-trapping fungi compared to other plant species tested (Persmark & Jansson, 1997). In an investigation on chemotropic growth of nematophagous fungi towards roots of several plants, only isolates of *A. oligospora* were attracted (Bordallo et al., 2002). In a 3-month pot experiment, *Dactylellina ellipsospora* (syn. *Monacrosporium ellipsosporum*) and *D. dactyloides* were especially competent in colonising tomato roots (Persson & Jansson, 1999).

Several nematode-trapping fungi are able to form so-called conidial traps in response to roots and root exudates (Persmark & Nordbring-Hertz, 1997). The external root colonisation by the egg-parasite *Pochonia chlamydosporia* also varied with plant species and was increased when plants were infected with the root-knot nematode *Meloidogyne incognita* (Bourne et al., 1996). This effect is possibly due to increased leakage of root exudates after damage to the root surface by the nematodes.

In recent investigations we studied the endophytic root colonization of the four groups of nematophagous species. The nematode-trapping species *A. oligospora, D. dactyloides* (Figs. 6a, b), and *N. robustus* (Figs. 6b, c) were all capable of endophytic colonization of barley roots. Similar root colonization was also detected for the egg-parasite *P. chlamydosporia* (Figs. 6e, f) and the toxin-producing *P. djamor*. The only fungi which did not show root colonization were the endoparasitic fungi *H. rhossiliensis* and *N. pachysporus* (Lopez-Llorca, Bordallo, Salinas, Monfort, & Lopez-Serna., 2002a; Bordallo et al., 2002; Lopez-Llorca, Jansson, Macia Vicente, & Salinas, 2006). The fungi grew inter- and intracellularly, formed appressoria when penetrating plant cell walls of epidermis and cortex cells, but never entered vascular tissues (Lopez-Llorca et al., 2002a; Bordallo et al., 2002). In contrast to *Pochonia* spp., appressoria had never been observed previously in *A. oligospora*.

Using histochemical stains it was possible to reveal the plant defence reactions, e.g. papillae and other cell wall appositions induced by nematophagous fungi, but these never prevented root colonization. Nematophagous fungi grew extensively especially in monocotyledon plants producing abundant mycelia, conidia and chlamydospores. Necrotic areas of the roots were observed at initial stages of colonization by the nematode-trapping and toxin-producing fungi tested, but were never seen at later stages, even when the fungi proliferated in epidermal and cortical cells.

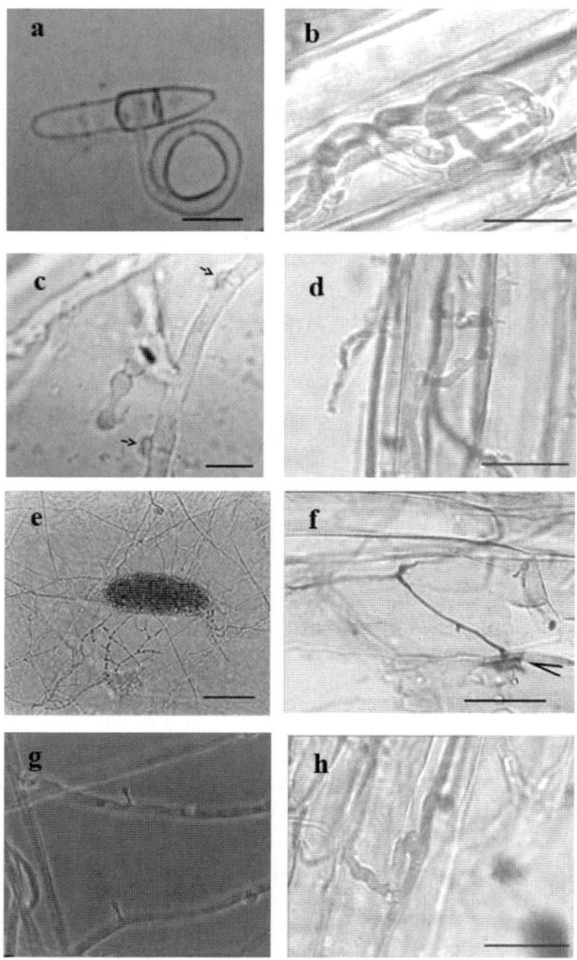

Figure 6. Parasitic (a, c, e, g) vs. endophytic (b, d, f, h) behaviour of nematophagous fungi. (a) Conidial trap of Drechslerella *sp. (c) Mycelia of a* Nematoctonus *sp. showing an "hour glass" trapping device and clamp connections (arrows). (e) Nematode egg infected by* Pochonia *sp. (g) Hyphae and toxin-producing organ of* Pleurotus *sp. (b, d, f, h). Display of endophytic colonisation of barley cortex cells by the nematophagous fungi displayed on the left hand side of each picture. Scale bars: a = 25 µm; b, d, h = 15 µm; c = 2 µm; e = 10 µm; f = 30 µm; g = 1 µm. (a and c: C. Olivares-Bernabéu, unpublished; b, d, h: from Lopez-Llorca* et al., *2006, courtesy of Springer; e: from Lopez-Llorca* et al., *2002b, courtesy of Elesevier; f: from Bordallo* et al., *2002, courtesy of the New Phytologist Trust; g: from Nordbring-Hertz* et al., *1995, courtesy of IWF Wissen und Medien, Göttingen).*

In cereal roots proceeding from soils naturally infested with the cereal cyst nematode *Heterodera avenae* and *Pochonia* spp., either the syncytia induced by the nematode and fungal hyphae could be detected inside the roots (Fig. 7c). Abundant sporulation of *Pochonia* spp. was also observed on the root surface (Fig. 7 a, b). The results at least indicate the possibility that nematode infection by the fungus may occurr inside roots, although so far this event has not been observed.

Actually, it is unknown whether endophytic colonization induces systemic resistance to nematodes and/or plant pathogens in plants. We have found that *P. chlamydosporia* could reduce growth of the plant pathogenic fungus *Gaeumannomyces graminis* var. *tritici* (take-all fungus, Ggt) in dual culture Petri dish and in growth tube experiments. In pot experiments *P. chlamydosporia* increased plant growth whether Ggt was present in the roots or not, suggesting a growth promoting effect by *P. chlamydosporia* (Monfort et al., 2005).

Endophytic rhizobacteria reducing plant-parasitic nematodes have been described (Hallmann, Quadt-Hallmann, Miller, Sikora, & Lindow, 2001), as well as the reduction of root knot nematodes by arbuscular mycorrhizal fungi (Waecke, Waudo, & Sikora, 2001). If this is true also in nematophagous fungi this will open up a new area of biocontrol using these fungi. The endophytic root colonization by egg-parasitic fungi, e.g. *Pochonia* spp., may provide them an opportunity to infect eggs of economically important endoparasitic nematodes (e.g. cyst and root-knot species) inside the roots and to reduce subsequent spread and roots infection by the second generation of juveniles.

Structures resembling trapping organs were observed in epidermal cells colonized by *A. oligospora*, and these may serve the purpose of trapping newly hatched juveniles escaping the roots. The ability to colonize plant roots may also be a survival strategy of these fungi and could explain soil suppressiveness to plant-parasitic nematodes in nature. The colonization of plant roots by nematophagous fungi is a new area of research that deserves in-depth investigations, not the least for biocontrol purposes and is presently underway in our laboratory.

4.3. Rhizosphere Dynamics and Biocontrol

The rhizosphere is a microecosystem in which roots release nutrients which in turn will affect microbes and their grazers. The former will modify these nutrients and could affect root and plant development. In this complex scenario, nematophagous fungi are both "hunters" and "hunted" since they predate on nematodes and can be affected, for instance, by myceliophagous species. It is tempting to use a combination of current non-destructive methods to analyse dynamics of the biotic component of the rhizosphere. Modification, or engineering, of the rhizosphere resource exchange could be vital for modifying the endophytic behaviour of nematophagous fungi. This may in turn affect their capability to control root diseases. Recently, microbiosensors, i.e. hybrids of soil sensors and

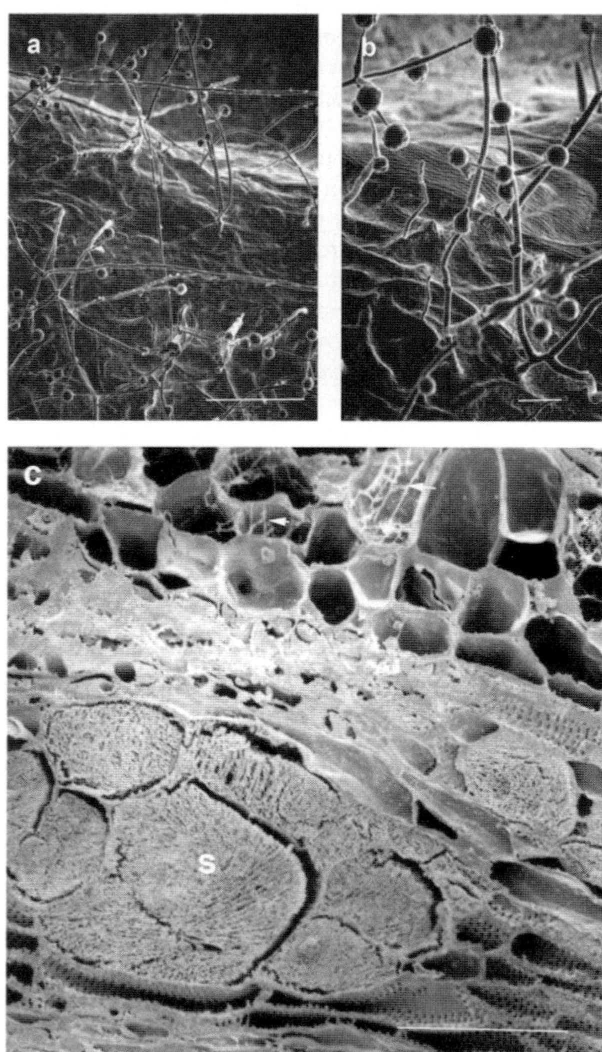

Figure 7. Rhizosphere colonization by fungal egg parasites in nematode suppressive soils. (a) Profuse hyphal growth and sporulation (LTSEM) in oat rhizosphere. (b) Close-up of phialides and slimy conidia of Pochonia *spp. (c) Field emission scanning electron microscopy (FESEM) of longitudinal section through a cereal root infected by the nematode* Heterodera avenae, *showing syncytia (S) and fungal colonization (arrowheads) in root cortex cells (a: Lopez-Llorca & Duncan, 1988; b,c: Lopez-Llorca & Claugher, unpublished).*

molecular methods for rhizosphere studies, have been devised (Cardon & Gage, 2006). These are genetically engineered bioreporter bacteria which join reporter genes, e.g. GFP and Lux, with promoters induced by several rhizosphere conditions (starvation, contaminants, quorum sensing). These are timely approaches for global studies on general rhizosphere function in ecosystems. Some of these bioreporters are biocontrol bacteria. Biocontrol fungi, e.g. nematophagous, are next on the list.

4.4. Root Exudates

To this point it is clear that the biocontrol scenario of plant-parasitic nematodes by nematophagous fungi relies on a multitrophic interaction in which plant roots play an important role. There is also abundant scientific evidence that roots produce compounds (exudates) which mediate plant-plant and plant-microbe interactions (Bais, Weir, Perry, Gilroy, & Vivanco, 2006). The latter would also include plant-nematode (and other micro- and meso-fauna) interactions.

Root exudates are very diverse structurally and chemically, and vary among plant species, but above all they may influence a wide array of processes relevant to the biocontrol action of nematophagous fungi. Leaving aside the effect of root exudates on nematode feeding and colonization, these compounds can influence nutrient availability in the rhizosphere (e.g. siderophores). Root exudates can also elicit release of compounds which could act in root defence or mediate signalling processes.

Root exudates also mediate plant-microbe interactions. The role of flavonoids on the specificity of rhizobia-*Leguminosae* interactions is well established (Perret, Staehelin, & Broughton, 2000). These root exudates induce the expression of rhizobia Nod genes, which are then involved in the synthesis of Nod factors (lipochitino-oligosaccharides with diverse chemical modifications) that are recognized by the appropriate host plant.

Closer to nematophagous fungi, arbuscular mycorrhizal fungi (AMF) recognize the presence of a compatible host plant through root exudates. A sesquiterpene has been identified as a branch-inducing factor for AMF in legumes (Akiyama, Matsuzaki, & Hayashi, 2005). Hyphal morphogenesis is vital for successful AMF-root colonization. This aspect may also be important in nematophagous fungi.

Root exudates affect nematodes, especially microbivorous species. On the other hand, plant-parasitic nematodes increase production of root exudates (rhizodeposition). The quality of root exudates is also changed. C/N-ratio in particular can alter the trophic stage of the fungus *Rhizoctonia solani* and turn it into a root pathogen (Van Gundy, Kirkpatrick, & Golden, 1977). These effects of root exudation on nematophagous fungi remain largely unknown, but are worth investigating.

There are new evidences that tri-trophic webs can be established in the rhizosphere leading to benefits for the plant host. Plant roots produce exudates which attract nematodes (Green, 1971). These can act as vectors of rhizobia that are thus transferred to roots (Horiuchi, Prithiviraj, Bais, Kimball, & Vivanco,

2005). It is also known that nematodes are attracted to nematophagous fungi to various extents (Jansson & Nordbring-Hertz, 1979; Jansson, 1982a). The role of non-parasitic nematodes as vectors to inoculate nematophagous fungi or root endophytes in nature has not yet been investigated.

4.5. Detection and Quantification

It is vital to be able to detect and quantify biocontrol agents, e.g. nematophagous fungi, in soil and rhizospheres, in the period following their addition. Many techniques for this purpose have been too unspecific or difficult to perform (Jansson, 2001). Antibodies have been tried with little success due to cross-reactions with other fungi (Eren & Pramer, 1966). Molecular markers such as the GUS gene have been transformed to *A. oligospora* (Persmark, Persson, & Jansson, 1996b; Tunlid, Åhman, & Oliver, 1999) and the GFP gene has been transformed to *P. chlamydosporia* (Atkins, Mauchline, Kerry, & Hirsch, 2004). In the former case it was not possible to quantify the growth of the fungus in soil at sufficiently low levels (Persmark et al., 1996b). The problem with *P. chlamydosporia* was to obtain stable transformants. A possible solution could be to try *Agrobacterium*-mediated transformation (Michielse, Arentshorst, Ram, & Van den Hondel, 2005).

Another promising approach is to use PCR-based techniques in combination with fluorogenic probes (e.g. scorpions and beacons). Such methods using real-time PCR and primers based on ITS sequences of *P. chlamydosporia* and *Paecilomyces lilacinus* have recently been presented (Ciancio, Loffredo, Paradies, Turturo, & Finetti Sialer, 2005; Atkins, Clark, Pande, Hirsch, & Kerry, 2005).

5. NEMATOPHAGOUS FUNGI AND BIOCONTROL

Nematophagous fungi have been tested for biological control of plant-parasitic nematodes for many years but, so far, met with little success, partly due to lack of knowledge on the ecology of these organisms (Stirling, 1991). One of these factors may be the soil receptivity to nematophagous fungi, which varies as discussed above. This receptivity will need to be part of a screening for possible biocontrol agents. Another important factor is the endophytic colonization of plant roots. This may protect the plants from nematode and fungal diseases through induced resistance or production of antibiotic secondary metabolites.

Nematophagous fungi (as endophytes or not) may also increase plant growth by participation in nutrient uptake, or by modification of plant growth regulators (hormones and related compounds). Therefore, in the search for nematophagous fungi as biocontrol agents, endophytic colonization also needs to be included.

The combination of several types of nematophagous fungi, e.g. egg-parasitic and nematode-trapping, which destroy nematodes at their different life stages may also be an important criterion. Interactions with other soil fungi, including both plant-parasitic and biocontrol agents, is also an important consideration when selecting the proper fungi for biological control of plant-parasitic nematodes.

6. CONCLUSIONS

Nematophagous fungi are ubiquitous organisms with the capacity to attack, infect and digest living nematodes at all stages, adults, juveniles and eggs. They may use trapping organs, spores and appressoria to initiate infection of their nematode hosts. The nematophagous fungi may not only infect nematodes, but may also infect other fungi as mycoparasites, and colonize plant roots endophytically. These various capabilities of nematophagous fungi, the latter in particular, may render them good candidates for biological control of plant root diseases.

ACKNOWLEDGEMENTS

In Memoriam: This chapter is dedicated to the memory of Mr. D. Claugher, a close friend and a fantastic microscopist at the Natural Science Museum in London, that we gratefully acknowledge and thank for his help in scanning "the secret life of nematophagous fungi". His contribution to the understanding of their mode of action is partly revealed in some of the wonderful images obtained with his Field Emission Scanning Electron Microscope. This work was financed by a grant (AGL2004-05808/AGR) from the Spanish Ministry of Education and Science. We also thank Dr. C. Olivares-Bernabéu and Mr W. Robertson for supplying unpublished images and data.

REFERENCES

Åhman, J., Ek, B., Rask, L., & Tunlid, A. (1996). Sequence analysis and regulation of a gene encoding a cuticle-degrading serine protease from the nematophagous fungus *Arthrobotrys oligospora. Microbiology, 142*, 1605–1616.

Ahrén, D., Ursing, B. M., & Tunlid, A. (1998). Phylogeny of nematode-trapping fungi based on 18S rDNA sequences. *FEMS Microbiology Letters, 158,* 179–184.

Ahrén, D., Tholander, M., Fekete, C., Rajashekar, B., Friman, E., Johansson, T., et al. (2005). Comparison of gene expression in trap cells and vegetative hyphae of the nematophagous fungus *Monacrosporium haptotylum. Microbiology, 151,* 789–803.

Akiyama, K., Matsuzaki, K. I., & Hayashi, H. (2005). Plant sequiterpenes induce hyphal branching in arbuscular mycorrhizal fungi. *Nature, 435,* 824–827.

Atkins, S. D., Mauchline,T. H., Kerry, B. R., & Hirsch, P. R. (2004). Development of a transformation system for the nematophagous fungus. *Pochonia chlamydospora. Mycological Research, 108,* 654–661.

Atkins, S. D., Clark, I. M., Pande, S., Hirsch, P. R., & Kerry, B. R. (2005). The use of real-time PCR and species-specific primers for the identification and monitoring of *Paecilomyces lilacinus. FEMS Microbiology Ecology, 51,* 257–264.

Bais, H. P., Weir, T. L., Perry, L. G., Gilroy, S., & Vivanco, J. M. (2006). The role of root exudates in rhizosphere interactions with plants and other organisms. *Annual Review of Plant Biology, 57,* 233–266.

Barron, G. L. (1992). Lignolytic and cellulolytic fungi as predators and parasites. In G. C. Carroll & D. T. Wicklow, (Eds.), *The fungal community, its organization and role in the ecosystems* (pp. 311–326). New York: Marcel Dekker.

Basse, C. W., & Steinberg, G. (2004). *Ustilago maydis*, model system for analysis of the molecular basis of fungal pathogenicity. *Molecular Plant Pathology, 5,* 83–92.

Bird, A. F., & Bird, J. (1991). *The structure of nematodes*. San Diego: Academic Press,.

Bonants, P. J. M., Fitters, P. F. L., Thijs, H., Den Belder, E., Waalwijk, C., Willem, J., & Henfling, D. M. (1995). A basic serine protease from *Paecilomyces lilacinus* with biological activity against *Meloidogyne hapla* eggs. *Microbiology, 141*, 775–784.

Bordillo, J. J., Lopez-Llorca, L. V., Jansson, H.-B., Salinas, J., Persmark, L., & Asensio, L. (2002). Colonization of plant roots by egg-parasitic and nematode-trapping fungi. *New Phytologist, 154*, 491–499.

Bourne, J. M., Kerry, B. R., & De Leij, F. A. A. M. (1996). The importance of the host plant on the interaction between root-knot nematodes (*Meloidogyne* spp.) and the nematophagous fungus, *Verticillium chlamydosporium* Goddard. *Biocontrol Science and Technology, 6*, 539–548.

Cardon, Z. G., & Gage, D. J. (2006). Resource exchange in the rhizosphere: Molecular tools and the microbial perspective. *Annual Review of Ecology, Evolution, and Systematics, 37*, 459–488.

Chaverri, P., Samuels, G. J., & Hodge, K. T. (2005). The genus *Podocrella* and its nematode-killing anamorph *Harposporium. Mycologia, 97*, 435–443.

Chen, Y. Y., Cheng, C. Y., Huang, T. L., & Li, Y. K. (2005). Chitosanase from *Paecilomyces lilacinus* with binding affinity for specific chitooligosaccharides. *Biotechnology and Applied Biochemistry, 41*, 145–150.

Chen, T. H., Hsu, C. S., Tsai, P. J., Ho, Y. F., & Lin, N. A. (2001). Heterotrimeric G-protein and signal transduction in the nematode-trapping fungus *Arthrobotrys Dactyloides. Planta, 212*, 858–863.

Ciancio, A., Loffredo, A., Paradies, F., Turturo, C., & Finetti Sialer, M. (2005). Detection of *Meloidogyne incognita* and *Pochonia chlamydosporia* by fluorogenic molecular probes. *OEPP/EPPO Bulletin, 35*, 157–164.

Clarke, A. J., Cox, P. M., & Shepherd, A. M. (1967). The chemical composition of the egg shells of the potato cyst-nematode, *Heterodera rostochiensis* Woll. *Biochemical Journal, 104*, 1056–1060.

Cooke, R. C. (1962). The ecology of nematode-trapping fungi in the soil. *Annals of Applied Biology, 50*, 507–513.

Cooke, R. C., & Satchuthananthavale, V. (1968). Sensitivity to mycostasis of nematode-trapping hyphomycetes. *Transactions of the British Mycological Society, 51*, 555–561.

Dackman, C., Jansson, H. B., & Nordbring-Hertz, B. (1992). Nematophagous fungi and their activities in soil. In G. Stotsky & J-M. Bollag (Eds.), *Soil biochemistry* (Vol. 7, pp. 95–130). New York: Marcel Dekker.

Duddington, C. L. (1962). Predaceous fungi and the control of eelworms. In C. L. Duddington & J. D. Carthy (Eds.), *Viewpoints in biology* Vol. 1. London: Butterworths.

Eren, J., & Pramer, D. (1966). Application of immunofluorescent staining to studies of the ecology of soil microorganisms. *Soil Science, 101*, 39–45.

Gaspard, J. T., & Mankau, R. (1986). Nematophagous fungi associated with *Tylenchulus semipenetrans* and the citrus rhizosphere. *Nematologica, 32*, 359–363.

Green, C. D. (1971). Mating and host finding behaviour of plant nematodes. In B. M. Zuckerman, W. F. Mai, & R. A. Rohde (Eds.), *Plant parasitic nematodes* (vol. II, pp. 247–266). New York: Academic Press.

Hallmann, J., Quadt-Hallmann, A., Miller, W. G., Sikora, R. A., & Lindow, S. E. (2001). Endophytic colonization of plants by the biocontrol agent *Rhizobium etli* G12 in relation to *Meloidogyne incognita* infection. *Phytopathology, 91*, 415–422.

Hao, Y., Mo, M., Su, H., & Zhang, K. (2005). Ecology of aquatic nematode-trapping hyphomycetes in southwestern China. *Aquatic Microbiology Ecology, 40*, 175–181.

Horiuchi, J. I., Prithiviraj, B., Bais, H. P., Kimball, B. A., & Vivanco, J. M. (2005). Soil nematodes mediate positive interactions between legume plants and rhizobium bacteria. *Planta, 222*, 848–857.

Jaffee, B. A. (1999). Enchytraeids and nematophagous fungi in tomato foelds and vineyards. *Phytopathology, 89*, 398–406.

Jaffee, B. A., Gaspard, J. T., & Ferris, H. (1989). Density-dependent parasitism of the soil-borne nematode *Criconemella xenoplax* by the nematophagous fungus *Hirsutella rhossiliensis. Microbial Ecology, 17*, 193–200.

Jansson, H.-B. (1982a). Attraction of nematodes to endoparasitic nematophagous fungi. *Transactions of the British Mycological Society, 79*, 25–29.

Jansson, H.-B. (1982b). Predacity of nematophagous fungi and its relation to the attraction of nematodes. *Microbial Ecology, 8*, 233–240.

Jansson, H.-B. (1987). Receptors and recognition in nematodes. In J. Veech & D. Dickson (Eds.), *Vistas on nematology* (pp. 153–158). Hyattsville, MD: Society of Nematologists.

Jansson, H.-B. (2001). Methods to monitor growth and activity of nematode-trapping fungi in soil. In R. Sikora (Ed.), *Tri-trophic interactions in the rhizosphere and root-health nematode-fungal-bacterial interrelationships. IOBC/WRPS Bulletin, 24*, 65–68.

Jansson, H.-B., & Friman, E. (1999). Infection-related surface proteins on conidia of the nematophagous fungus *Drechmeria coniospora. Mycological Research, 103*, 249–256.

Jansson, H.-B., Jeyaprakash, A., Damon, R. A., & Zuckerman, B. M. (1984). *Caenorhabditis elegans* and *Panagrellus redivivus*: enzyme-mediated modification of chemotaxis. *Experimental Parasitology, 58*, 270–277.

Jansson, H.-B., Jeyaprakash, A., & Zuckerman, B. M. (1985). Control of root knot nematodes on tomato by the endoparasitic fungus *Meria coniospora. Journal of Nematology, 17*, 327–330.

Jansson, H.-B., & Lopez-Llorca, L. V. (2001). Biology of nematophagous fungi. In J. D. Misrha & B. W. Horn (Eds.), *Trichomycetes and other fungal groups: Professor Robert W. Lichtwardt commemoration volume* (pp. 145–173). Enfield, NH: Science Publisher, Inc.

Jansson, H.-B., & Nordbring-Hertz, B. (1979). Attraction of nematodes to living mycelium of nematophagous fungi. *Journal of General Microbiology, 112*, 89–93.

Jansson, H.-B., & Nordbring-Hertz, B. (1983). The endoparasitic fungus *Meria coniospora* infects nematodes specifically at the chemosensory organs. *Journal of General Microbiology, 129,* 1121–1126.

Jansson, H.-B., & Nordbring-Hertz, B. (1984). Involvement of sialic acid in nematode chemotaxis and infection by an endoparasitic nematophagous fungus. *Journal of General Microbiology, 130,* 39–43.

Jansson, H.-B., & Nordbring-Hertz, B. (1988). Infection events in the fungus- nematode system. In G. O. Poinar & H. B. Jansson (Eds.), *Diseases of nematodes* (Vol. 2, pp. 59–72). Boca Raton: CRC Press.

Jansson, H.-B., Nordbring-Hertz, B., Wyss, U., Häusler, P., Hard, T., & Poloczek, E. (1995). Infection of nematodes by zoospores of *Catenaria anguillulae.* IWF Wissen und Medien, Göttingen, Germany. Film No. C 1868.

Jansson, H.-B., & Thiman, L. (1992). A preliminary study of chemotaxis of zoospores of the nematode-parasitic fungus *Catenaria anguillulae. Mycologia, 84,* 109–112.

Jansson, H.-B., Persson, C., & Odselius, R. (2000). Growth and capture activities of nematophagous fungi in soil visualized by low temperature scanning electron microscopy. *Mycologia, 92*, 10–15

Jeyaprakash, A., Jansson, H.-B., Marban-Mendoza, N., & Zuckerman, B. M. (1985). *Caenorhabditis elegans*: Lectin-mediated modification of chemotaxis. *Experimental Parasitology, 59*, 90–97.

Kerry, B. R. (2000). Rhizosphere interactions and the exploitation of microbial agents for the biological control of plant-parasitic nematodes. *Annual Review of Phytopathology, 38*, 423–441.

Kuramae, E. E., Robert, V., Snel, B., Weiss, M., & Boekhout, T. (2006). Phylogenomics reveal a robust fungal tree of life. *FEMS Yeast Research, 6*, 1213–1220.

Lee, N., D´Souza, C. A., & Kronstad, J. W. (2003). Of smuts, blasts, mildews, and blights: cAMP signaling in phytopathogenic fungi. *Annual Review of Phytopathology, 41,* 399–427.

Li, Y., Hyde, K. D., Jeewon, R., Cai, L., Vijaykrishna, D., & Zhang, K. (2005). Phylogenetics and evolution of nematode-trapping fungi (Orbiliales) estimated from nuclear and protein coding genes. *Mycologia, 97*, 1034–1046.

Li, J., Yang, J., Huang, X., & Zhang, K. Q. (2006). Purification and characterization of an extracellular serine protease from *Clonostachys rosea* and its potential as a pathogenic factor. *Process Biochemistry, 41*, 925–929.

Luo, H., Mo, M., Huang, X., Li, X., & Zhang, K. (2004). *Coprinus comatus*: A basidiomycete fungus forms novel spiny structures and infects nematodes. *Mycologia, 96*, 1218–1225.

Lopez-Llorca, L. V. (1990). Purification and properties of extracellular proteses produced by the nematophagous fungus *Verticillium suchlasporium. Canadian Journal of Microbiology, 36,* 530–537.

Lopez-Llorca, L. V., Bordallo, J. J., Salinas, J., Monfort, E., & Lopez-Serna, M. L. (2002a). Use of light and scanning electron microscopy to examine colonisation of barley rhizosphere by the nematophagous fungus *Verticillium chlamydosporium*. *Micron, 33*, 61–67.

Lopez-Llorca, L. V., & Claugher, D. (1990). Appressoria of the nematophagous fungus *Verticillium suchlasporium*. *Micron and Microscopica Acta, 21*, 125–130.

Lopez-Llorca, L. V., & Duncan, G. H. (1988). A scanning electron microscopy study of fungal endoparasitism of cereal cyst nematode (*Heterodera avenae*). *Canadian Journal of Microbiology, 34*, 613–619.

Lopez-Llorca, L. V., & Jansson, H.-B. (2006). Fungal parasites of invertebrates: multimodal biocontrol agents. In G. D. Robson, P. van West, & G. M. Gadd (Eds.), *Exploitation of fungi* (pp. 310–335). Cambridge, UK: Cambridge University Press.

Lopez-Llorca, L. V., Jansson, H.-B., Macia Vicente, J. G., & Salinas, J. (2006). Nematophagous fungi as root endophytes. In B. Schulz, C. Boyle, T. Sieber (Eds.), *Soil biology: Microbial root endophytes* (Vol 9, pp. 191–206) Berlin, Heidelberg: Springer-Verlag.

Lopez-Llorca, L. V., Olivares-Bernabeu, C., Salinas, J., Jansson, H. B., & Kolattukudy, P. E. (2002b). Prepenetration events in fungal parasitism of nematode eggs. *Mycological Research, 106*, 499–506.

Lopez-Llorca, L. V., & Robertson, W. M. (1992a). Ultrastructure of infection of cyst nematode eggs by the nematophagous fungus *Verticillium suchlasporium*. *Nematologica,* 39, 65–74.

Lopez-Llorca, L. V., & Robertson, W. M. (1992b). Immunocytochemical localization of a 32-kDa protease from the nematophagous fungus *Verticillium suchlasporium* in infected nematode eggs. *Experimental Mycology,* 16, 261–267.

Marbán-Mendoza, N., Dicklow, M. B., & Zuckerman, B. M. (1992). Control of *Meloidogyne incognita* on tomato by two leguminous plants. *Fundamental and Applied Nematology, 15*, 97–100.

Marbán-Mendoza, N., Jeyaprakash, A., Jansson, H.-B., & Zuckerman B. M. (1987). Control of root knot nematodes in tomato by lectins. *Journal of Nematology,* 19, 331–335.

Michielse, C. B., Arentshorst, M., Ram, A. F. J., & Van den Hondel, C. A. M. J. (2005). *Agrobacterium*-mediated transformation leads to improved gene replacement efficiency in *Aspergillus awamori. Fungal Genetics and Biology, 42*, 9–19.

Monfort, E., Lopez-Llorca, L. V., Jansson, H.-B., Salinas, J., Park, J. O., & Sivasithamparam, K. (2005). Colonisation of seminal roots of wheat and barley by egg-parasitic nematophagous fungi and their effects on *Gaeumannomyces graminis* var. *tritici* and development of root-rot. *Soil Biology and Biochemistry, 37*, 1229–1235.

Monfort, E., Lopez-Llorca, L. V., Jansson, H.-B., & Salinas, J. (2006). *In vitro* soil receptivity to egg-parasitic nematophagous fungi. *Mycological Progress, 5*, 18–23.

Nordbring-Hertz, B., Jansson, H.-B., Friman, E., Persson, Y., Dackman, C., Hard, T., et al. (1995). Nematophagous fungi. IWF Wissen und Medien, Göttingen, Germany. Film No. C 1851.

Nordbring-Hertz, B., & Mattiasson, B. (1979). Action of a nematode-trapping fungus shows lectin-mediated host-microorganism interaction. *Nature, 281*, 477–479.

Olivares-Bernabéu, C. (1999). *Caracterización biológica y molecular de hongos patógenos de huevos de nematodos*. Ph.D. thesis, University of Alicante, Alicante, Spain.

Perret, X., Staehelin, C., & Broughton, W. J. (2000). Molecular basis of symbiotic promiscuity. *Microbiology and Molecular Biology Reviews, 64*, 180–201.

Persmark, L., Banck, A., & Jansson, H.-B. (1996a). Population dynamics of nematophagous fungi and nematodes in an arable soil: vertical and seasonal fluctuations. *Soil Biology and Biochemistry, 28*, 1005–1014.

Persmark, L., & Jansson H.-B. (1997). Nematophagous fungi in the rhizosphere of agricultural crops. *FEMS Microbiology Ecology, 22*, 303–312.

Persmark L., & Nordbring-Hertz, B. (1997) Conidial trap formation of nematode-trapping fungi in soil and soil extracts. *FEMS Microbiology Ecology, 22,* 313–323

Persmark, L., Persson, Y., & Jansson, H. B. (1996b). Methods to quantify nematophagous fungi in soil: microscopy or GUS gene activity. In D. F. Jensen, H. B. Jansson, & A. Tronsmo (Eds.), *Monitoring antagonistic fungi deliberately released into the environment* (pp. 71–75). Dordrecht: Kluwer Academic Publishers,.

Persson, C., & Jansson, H.-B. (1999). Rhizosphere colonization and control of *Meloidogyne* spp. by nematode-trapping fungi. *Journal of Nematology, 31*, 164–171.

Peterson, E. A., & Katznelson, H. (1965). Studies on the relationships between nematodes and other soil microorganisms. IV. Incidence of nematode-trapping fungi in the vicinity of plant roots. *Canadian Journal of Microbiology, 11*, 491–495.

Pfister, D. H. (1997). Castor, Pollux and life histories of fungi. *Mycologia, 89*, 1–23.

Scholler, M., Hagedorn, G., & Rubner, A. (1999). A reevaluation of predatory orbiliaceous fungi. II. A new generic concept. *Sydowia, 51*, 89–113.

Segers, R., Butt, T. M., Keen, J. N., Kerry, B. R., & Peberdy, J. F. (1995). The subtilisins of the invertebrate mycopathogens *Verticillium chlamydosporium* and *Metarhizium anisopliae* are serologically and functionally related. *FEMS Microbiology Letters, 126*, 227–232.

Segers, R., Butt, T. M., Kerry, B. R., & Peberdy, F. (1994). The nematophagous fungus *Verticillium chlamydosporium* Goddard produces a chymoelastase-like protease which hydrolyses host nematode proteins in situ. *Microbiology, 140*, 2715–2723.

Spatafora, J. W. (2005). Assembling the fungal tree of life (AFTOL). *Mycological Research, 109*, 755–756.

Stirling, G. R. (1991). *Biological control of plant parasitic nematodes: Progress, problems and prospects*. Wallingford: CAB International.

St. Leger, R. J. (1993). Biology and mechanism of insect-cuticle invasion by Deuteromycete fungal pathogens. In N. E. Beckage, S. N. Thompson, & B. A. Federici (Eds.), *Parasites and pathogens of insects: Pathogens* (Vol 2, pp. 211–229). San Diego: Academic Press.

Suarez, B., Rey, M., Castillo, P., Monte, E., & Llobell, A. (2004). Isolation and characterization of PRA1, a trypsin-like protease from the biocontrol agent *Trichoderma harzianum* CECT 2413 displaying nematicidal activity. *Applied Microbiology and Biotechnology, 65*, 46–55.

Thorn, R. G., & Barron, G. L. (1986). Nematoctonus and the Tribe resupinateae in Ontario, Canada. *Mycotaxon, 25*, 321–453.

Tikhonov, V. E., Lopez-Llorca, L. V., Salinas, J., & Jansson, H. B. (2002). Purification and characterization of chitinases from the nematophagous fungi *Verticillium chlamydosporium* and *V. suchlasporium*. *Fungal Genetics and Biology, 35*, 67–78.

Tosi, S., Annovazzi, L., Tosi, I., Iadarola, P., & Caretta, G. (2001). Collagenase production in an Antarctic strain of *Arthrobotrys tortor* Jarowaja. *Mycopathologia, 153*, 157–162.

Tunlid, A., Åhman, J., & Oliver, R. P. (1999). Transformation of the nematode-trapping fungus *Arthrobotrys oligospora*. *FEMS Microbiol Letters, 173*, 111–116.

Tunlid, A., Jansson, H.-B., & Nordbring-Hertz, B. (1992). Fungal attachment to nematodes. *Mycological Research, 96*, 401–412.

Tunlid, A., Johansson, T., & Nordbring-Hertz, B. (1991a). Surface polymers of the nematode-trapping fungus *Arthrobotrys oligospora*. *Journal of General Microbiology, 137*, 1231–1240.

Tunlid, A., Nivens, D. E., Jansson, H. B., & White, D. C. (1991b). Infrared monitoring of the adhesion of *Catenaria anguillulae* zoospores to solid surfaces. *Experimental Mycology, 15*, 206–214.

Tunlid, A., Rosén, S., Ek, B., & Rask, L. (1995). Purification and characterization of an extracellular serine protease from the nematode-trapping fungus *Arthrobotrys oligospora*. *Microbiology,* 140, 1687–1695.

Van Gundy, S. D., Kirkpatrick, J. D., & Golden, J. (1977). The nature and role of metabolic leakage from root-knot nematode galls and infection by *Rhizoctonia solani*. *Journal of Nematology, 9*, 113–121.

Verdejo-Lucas, S., Sorribas, F. J., Ornat, C., & Galeano, M. (2003). Evaluating *Pochonia chlamydosporia* in a double-cropping system of lettuce and tomato in plastic houses infested with *Meloidogyne javanica*. *Plant Pathology, 52*, 521–528

Waecke, J. W., Waudo, S. W., & Sikora, R. (2001). Suppression of *Meloidogyne hapla* by arbuscular mycorrhiza fungi (AMF) on pyrethrum in Kenya. *International Journal of Pest Management, 47*, 135–140.

Wang, C., & St. Leger, R. J. (2005). Developmental and transcriptional responses to host and nonhost cuticles by the specific locust pathogen *Metarhizium anisopliae* var. *acridum*. *Eukaryotic Cell, 4*, 932–947.

Wang, M., Yang, J., & Zhang, K. Q. (2006). Characterization of an extracellular protease and its cDNA from the nematode-trapping fungus *Monacrosporium microscaphoides*. *Canadian Journal of Microbiology, 52*, 130–139.

Wang, R. B., Yang, J. K., Lin, C., Zhang, Y., & Zhang, K. Q. (2006). Purification and characterization of an extracellular serine protease from the nematode-trapping fungus *Dactylella shizishanna. Letters in Applied Microbiology, 42*, 589–594.

Ward, S., Thomson, N., White, J. G., & Brenner, S. (1975). Electron microscopical reconstruction of the anterior sensory anatomy of the nematode *Caenorhabditis elegans*. *Journal of Comparative Neurology, 160*, 313–337.

Wharton, D. A. (1980). Nematode egg-shells. *Parasitology, 81,* 447–463.

Wyss, U., Voss, B., & Jansson, H.-B. (1992). *In vitro* observations on the infection of *Meloidogyne incognita* eggs by the zoosporic fungus *Catenaria anguilulae* Sorokin. *Fundamental and Applied Nematology, 15*, 133–139.

Yang, J., Huang, X., Tian, B., Sun, H., Duan, J., Wu, W., & Zhang, M. (2005a). Characterization of an extracellular serine protease gene from the nematophagous fungus *Lecanicillium psalliotae. Biotechnology Letters*, *27*, 1329–2334.

Yang, J., Huang, X., Tian, B., Wang, M., Niu, Q., & Zhang, K. (2005b). Isolation and characterization of a serine protease from the nematophagous fungus *Lecanicillium psalliotae,* displaying nematicidal activity. *Biotechnology Letters*, *27*, 1123–1128.

Xu, J. R., Peng, Y. L., Dickman, M. B., & Sharon, A. (2006). The dawn of fungal pathogen genomics. *Annual Review of Phytopathology*, *44*, 337–366.

Zare, R., & Gams, W. (2001). A revision of *Verticillium* section *Prostrata*. IV. The genera *Lecanicillium* and *Simplicillium* gen. nov. *Nova Hedwigia*, *73*, 1–50.

Zare, R., Gams, W., & Evans, H. C. (2001). A revision of *Verticillium* section *Prostrata*. V. The genus *Pochonia*, with notes on *Rotiferophthora. Nova Hedwigia, 73*, 51–86.

Zhao, M., Mo, M., & Zhang, Z. (2004). Characterization of a serine protease and its full-length cDNA from the nematode-trapping fungus *Arthrobotrys oligospora*. *Mycologia*, *96*, 16–22.

Zuckerman, B. M. (1983). Hypothesis and possibilities of intervention in nematode chemoreceptors. *Journal of Nematology*, *15*, 173–182.

Zuckerman, B. M., & Jansson, H.-B. (1984). Nematode chemotaxis and mechanisms of host/prey recognition. *Annual Review of Phytopathology*, *22*, 95–113.

Section 2

Crops Ecology and Control

4

T. MATEILLE, P. CADET AND M. FARGETTE

CONTROL AND MANAGEMENT OF PLANT PARASITIC NEMATODE COMMUNITIES IN A SOIL CONSERVATION APPROACH

IRD, UMR CBGP, Campus de Baillarguet, CS30016, 34988 Montferrier-sur-Lez Cedex, France

Abstract. The nematodes specificities and their interactions with plants are reviewed, considering host plants quality and compatibility. The potentials of nematode resistance and diversity of antagonists and parasitism are discussed, in relation to host specificity and obligate multitrophic relationships. The ecology and management of nematode communities are also reviewed, focusing on soil health approaches and new paradigms for plant protection.

1. INTRODUCTION

Farmers in West Africa, whatever their technological level and social status, are aware of the fact that root galls are caused by plant parasitic nematodes present in soil. In their native wolof language, Senegal farmers say *"nematod bi la"* (meaning *"here are nematodes"*) and use the wolof word *"krous",* which refers to their pray chain, to indicate the series of galls, formed by the roots histological alterations. Unfortunately, farmers belive that methods developed to control plant-parasitic nematodes, even if largely available, suffer unsatisfactory applications, and that their prayers represent the last resort to face the expansion of these parasites.

1.1. Soil Fertility and Plant-parasitic Nematodes

Low crop production in tropical and subtropical countries is mainly due to soil erosion and low natural fertility, especially because of their low content in organic matter. The example of sub-Saharan Africa is significant: in the area extending from Mauritania to Nigeria, fertile soils represent only 10.5 million hectares on a total of 434 (Mensah, 1989). Any increase in productivity implies, hence, a direct fertility improvement. In order to compensate the difficulty to increase soil fertility levels, farmers practise itinerant farming, which have a detrimental effect on ecosystems

A. Ciancio & K. G. Mukerji (eds.), Integrated Management and Biocontrol of Vegetable and Grain Crops Nematodes, 79–97.

sustainability. Sedentary farming systems, such as vegetable crops, compel an excessive use of soil resources, and this depletion is amplified by the demographic expansion.

Crop losses due to parasites and predators are boosted in these agronomic situations. In the world, more than 8,000 fungal species, 250 virus species, 60 bacterial species and 1,500 nematode species cause crop damages (Bachelier, 1978). Yield losses in basic food crops due to plant parasitic nematodes reached 10.7% in the world and 12.6% in developing countries (Sasser, 1989). For root-knot nematodes only, yield losses reach 60% for vegetables or rice, 25% for potato, 50% for groundnut, 20% for tobacco. Therefore, economic impacts of plant-parasitic nematodes are significant.

1.2. Nematode Specificities

The global pathogenicity of nematode communities, commonly considered as more diversified under tropical than temperate climates (Luc, Bridge, & Sikora, 2005), is seldom taken into account. Nematode control is focused on some taxa (Evans, Trudgill, & Webster, 1993), mostly according to their reputation, whereas nematode communities in tropics gather many species with diverse pathogenic effects: 17 species attack sugar-cane in Burkina Faso (Cadet, 1987), whereas 21 species parasitize banana in Ivory Coast (Fargette & Quénéhervé, 1988), and 25 are reported from vegetables in Senegal (Netscher, 1970).

Climatic conditions are also essential, since both high temperatures and bimodal seasons in the Tropics increase the multiplication rate of several nematode populations (mainly through increased fecundity and life cycles, per year). For example, reproduction of *Heterodera* species are 25 times faster (Merny, 1966), whereas *Meloidogyne* females are 5 times more fertile (De Guiran & Netscher, 1970) in warm than in temperate countries.

The greatest differences come from farming systems. In the Tropics, most of the cropped surfaces are managed with low inputs, and crops cover a broad range of plant species, frequently cultivated in complex agroecosystems (associations, rotations, etc.). These practices usually maintain species diversity and thus damages cannot be allocated to only a single nematode species.

Chemical control is still widely under-used in developing countries. Moreover, the application of large quantities of nematicides and the very small range of molecules available induce pollution and, possibly, their soil degradation by microbes. Physical methods are not as secure as expected and crop practices (fallowing, crop rotations, etc.) or resistant varieties are not easy to implement, especially by wide or small scale farmers. Eventually, management of plant-parasitic nematodes with biocontrol agents is also poorly developed because of lack of informations and strategies and/or because of costs and economic constrains.

Crop protection is now subjected to quality rules related to environmental protection and human/animal health. Integrated pest management, mainly based on non chemical practices, appears, for public opinion, as one of the most attractive alternatives. However, one can wonder on both knowledge and relevance of processes led by research and development programs. That is the case for nematode control exclusively focused on plant-nematode and on nematode-predator relationships (binary approaches) at the expense of more integrative approaches including nematode ethology in communities and mesological relationships. Some examples will point up limits of such approaches.

2. PLANT-NEMATODE INTERACTIONS

2.1. Host Plants Quality

Natural practices as fallowing or flooding (no host-plant condition), if water is available, were developed in banana plantations as an alternative to chemical control (Sarah, Lassoudière, & Guérout, 1983). But fallows and flooding (Fig. 1) controlled only *Radopholus similis*, the most dangerous species on banana (Mateille, Foncelle, & Ferrer, 1988).

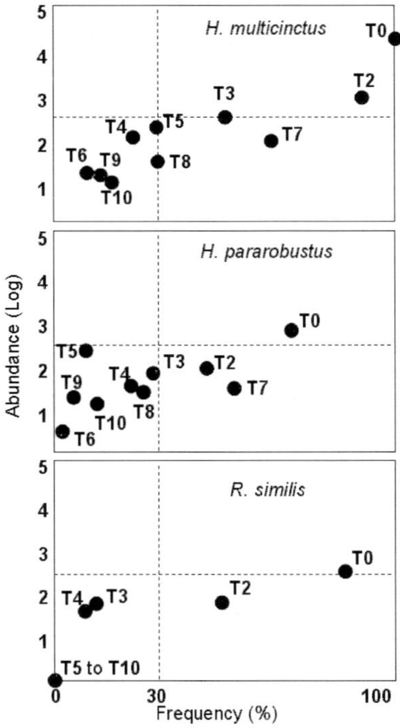

Figure 1. Abundance and frequency of Radopholus similis*,* Helicotylenchus multicinctus *and* Hoplolaimus pararobustus *in soil of flooded banana plots during 10 weeks (from Mateille* et al., *1988).*

Comparing infestation processes of banana *vitro*-plants (nematode free material) and of corms and suckers (infested material), the transplantation of infested corms and suckers restored the initial nematode community whereas *vitro*-plants prevented soil infestation with *R. similis* and did not keep the development of other pathogenic species such as *Helicotylenchus multicinctus* or *Hoplolaimus pararobustus* (Mateille, Quénéhervé, & Hugon, 1994) (Fig. 2). Nematode investigations carried out in vegetable cropping systems in Senegal showed that population levels were not only related to the susceptibility of the vegetable crops, but also either to plant rotations or to cropping systems (flooded or dry areas) (Sawadogo, Diop, Thio, Konate, & Mateille, 2000).

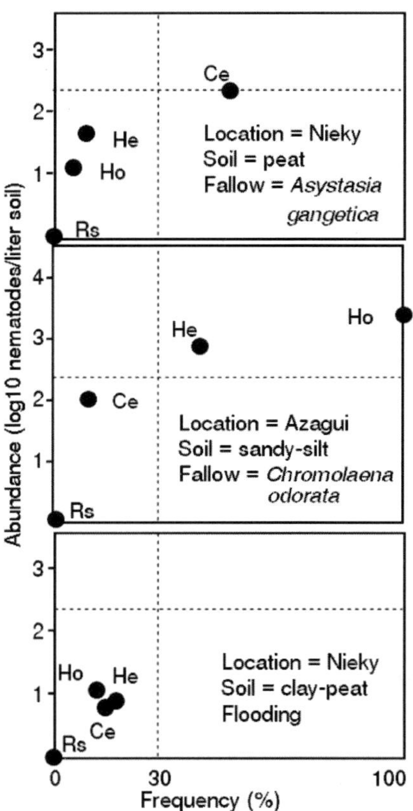

Figure 2. Abundance and frequency of Helicotylenchus multicinctus (He), Hoplolaimus pararobustus (Ho), Radopholus similis (Rs) *and* Cephalenchus emarginatus (Ce) *populations in peat or sandy-silt soils after fallowing, and in a clay-peat soil after flooding in banana plots (from Mateille* et al., *1994).*

In Sahelian regions, vegetable cropping systems are based on bimodal rotations: irrigated vegetables during cold and dry seasons – cereals or groundnut during hot and wet seasons. Reduction of root-knot nematode populations was confirmed on non-susceptible groundnut crops, but other pathogenic species such as *Scutellonema cavenessi*, whose populations increase on groundnut or sorghum (Fig. 3), did not decline on vegetables (Diop, Ndiaye, Mounport, & Mateille, 2000).

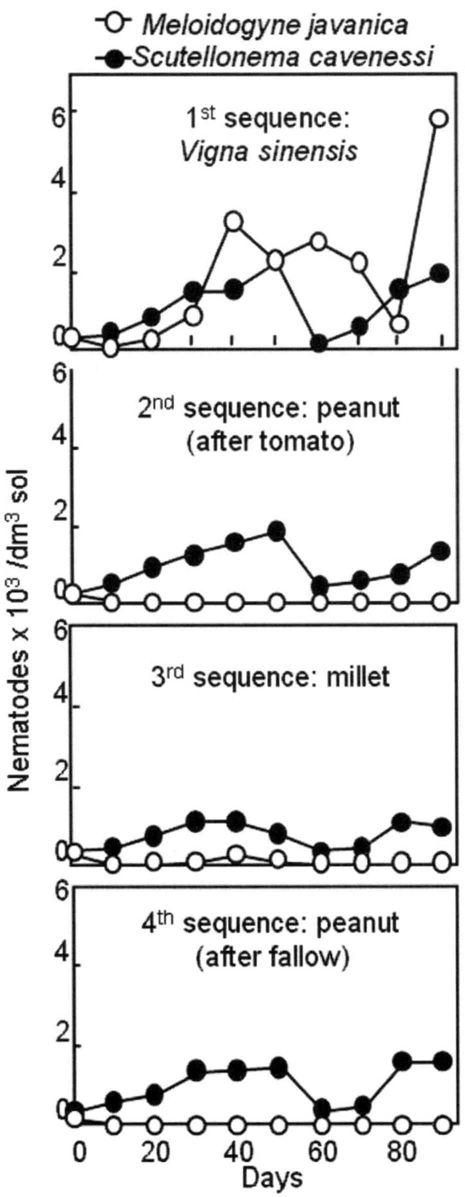

Figure 3. Decade evolution of Meloidogyne javanica *and* Scutellonema cavenessi *population densities on crop rotation in a vegetable plot (from Diop* et al., *2000).*

Plant quality has also a significant effect on species demography (Yeates, 1987). On banana, multiplication of nematode populations depends on the species, according to the physiological activity of the suckers they parasitize (Mateille, Cadet, & Quénéhervé, 1984): *H. multicinctus* and *H. pararobustus* populations develop indifferently on growing (fruit-bearing) suckers or on dying plant material (mother plant or cut suckers), and *R. similis* develops only on growing suckers.

2.2. Plant-nematode Compatibility

Studies conducted with *R. similis*, *H. multicinctus* and *H. pararobustus* on banana cultivars showed that (*i*) penetration and multiplication of nematodes, demographic structure and distribution in roots (Mateille, 1992), (*ii*) effects of nematodes on plant growth and on mineral/organic absorption and photosynthesis (Mateille, 1993), (*iii*) histological and physiopathological root disturbances (Mateille, 1994a; 1994b) were specific to the nematode species. Plant defence mechanisms commonly resulted in stimulating their secondary metabolism. In that way, peroxidase/polyphenoloxidase balances appear to be highly requested: peroxidase activities are highly involved in banana cultivars resistant to *R. similis* only.

In the same way, significant differences in alkaloid composition and content of plant tissues with nematostatic effects, such as *Crotalaria* species, used as green manure or cover crops, can explain their different efficiences against a number of nematode populations representing four *Meloidogyne* species (Jourand, Rapior, Fargette, & Mateille, 2004) (Fig. 4).

2.3. Resistance to Nematodes

Tomato yied losses due to root-knot nematodes are well identified all over the world. *Meloidogyne* species are very polyphagous and ubiquitous. Most of them, especially tropical species, reproduce through parthenogenesis.

In developing countries, chemical control and grafted resistant varieties (Beaufort®, KingKong®, etc.) are widely used by vegetable producing companies. Small scale farmers mainly use resistant varieties. Four *Meloidogyne* species were detected in Senegal: *M. arenaria* (6.8%), *M. incognita* (28.5%), *M. javanica* (69.7%) and *M. mayaguensis* (31.3%), and 27% of the samples shelter *Meloidogyne* communities. More than 30% of *M. incognita* and *M. javanica* populations were virulent (resistance breaking) and *M. mayaguensis* is able to parasitize all the plants resistant to other *Meloidogyne* species (Fig. 5).

In Senegal, tomato crops cover 32% of the vegetable producing areas and resistant varieties account for 95% of the seeds used. Nevertheless, 89% of these resistant plants are highly infested by root-knot nematodes, with more than 5 000 nematodes per gram of roots (Trudgill et al., 2000).

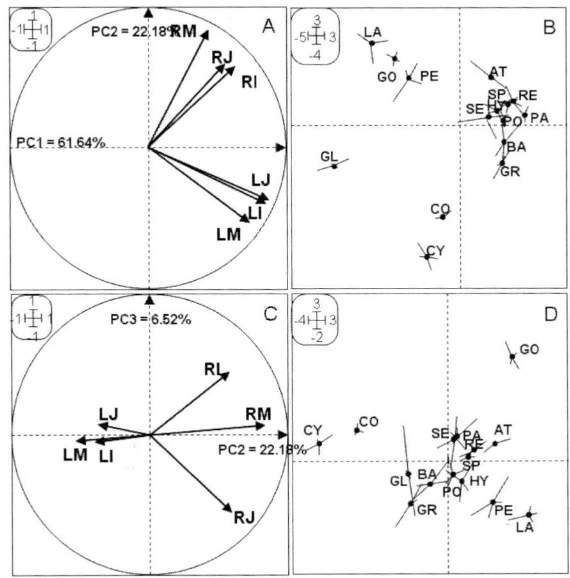

Figure 4. Nematostatic activity of leaf (L) and root (R) extracts from 15 West African Crotalaria *spp. on second stage juveniles of* Meloidogyne incognita *(I),* M. javanica *(J) and* M. mayaguensis *(M). PCA loading plots (A, C) and score plots for* Crotalaria *species (B, D) (adapted from Jourand* et al., *2004).*

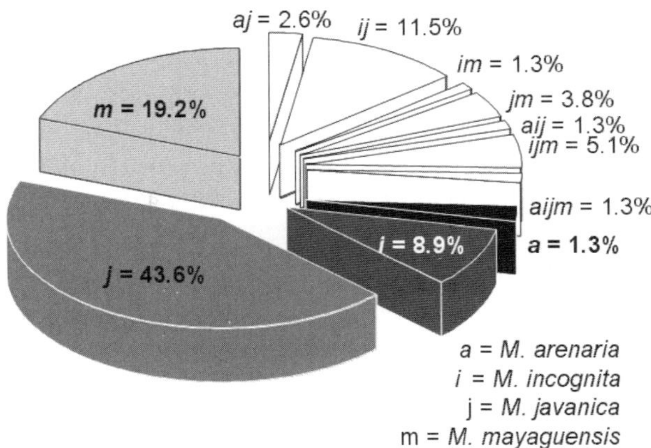

Figure 5. Distribution of Meloidogyne *spp. in single populations and communities in vegetable producing areas in Senegal (adapted from Trudgill* et al., *2000).*

3. PREY-PREDATOR INTERACTIONS

3.1. Diversity of Antagonists and Parasitism

Most of nematode-parasitic prokaryotes include unicellular rickettsiae, with binary fission, viruses (whose pathogenic effects on nematodes remain unclear), and bacteria, among which only one group is currently recognized for its parasitic action: *Pasteuria penetrans sensu lato* (Sayre & Starr, 1988). Among eucaryotes, only Hyphomycete and Zygomycete fungi are nematode parasites. They were commonly classified according to their predation activity: endoparasite fungi of juvenile and adult stages, such as *Catenaria anguillulae*, or trapping fungi, such as *Arthrobotrys oligospora* (Gray, 1988), and ovicide fungi, such as *Pochonia chlamydosporia* (Morgan-Jones & Rodriguez-Kabana, 1988). Nematophagous fungi can be obligate parasites of nematodes (*Catenaria anguillulae*), or can be opportunists as trapping fungi. The growth of opportunist fungi in soil can be influenced by plants (root exudates), and traps can be stimulated by the presence of nematodes. Consequently, dependence of fungi on their environment is very wide.

The Gram+ bacterium *P. penetrans* has three dependence levels with respect to its environment. Present in the soil (first level), this organism is an obligate parasite of nematodes (second level). *P. penetrans* was never detected on other organisms than on soil nematodes. Moreover, its parasitoïd behavior makes it absolutely dependent on its nematode hosts. Spore germination is induced when the nematode begin to feed on the plant (third level). *P. penetrans* presents three phases in its life cycle: a free step in the soil, as sporanges, commonly called spores (survival form); an attachment step when the spores attach to the nematodes (aggression form) and a penetration/development step in the nematode (parasitic form).

Unfortunalely, systematics of both *P. penetrans* and nematophagous fungi remain poor, and their diversity and specificity to their nematode hosts are insufficiently evaluated. Up to now, research mainly focused on "nematode-predator" relationships and environment as factors mainly restricted to soil temperature, moisture and acidity. Concerning nematophagous fungi, the impact of organic matter and minerals on their growth was often highlighted. But, these studies focused on biological requirements for growth and predation with the aim of biopesticide production.

P. penetrans was detected on soilborne nematodes only, on more than 90 genera and 200 species. *P. penetrans* was initially classified in the genus *Duboscqia* (Cobb, 1906), then in the genus *Bacillus* (Mankau, 1975), and finally in the genus *Pasteuria* (Sayre & Starr, 1985). However, the characterization of this bacterium by description of its vegetative and reproduction structures only never made it possible to classify in the Bacillaceae family or in the Actinomycetaleae one. Data on ribosomal genes, however, clearly indicated its affinity with Bacillaceae (Charles et al., 2005). Six groups of *Pasteuria* were set up at species rank: *P. ramosa*, parasite of daphnia (Metchnikoff, 1888); *P. penetrans*, parasite of *Meloidogyne* nematodes; *P. thornei*, parasite of *Pratylenchus* nematodes (Starr & Sayre, 1988); *P. nishizawae*, parasite of *Heterodera* and *Globodera* nematodes (Nishizawa, 1986), and two other species, one parasite of *Heterodera goettingiana* (Sturhan, Winkelheide, Sayre, & Wergin, 1994) and one parasite of *Belonolaimus longicaudatus* (Gibkin-Davis, Williams, Hewlett, & Dickson, 1995). However, pure culture of this bacterium and more informations proceeding from biochemical and genomic taxonomy (Anderson, Preston, & Dickson, 1999; Atibalentja, Noel, & Domier, 2000; Bird, Opperman, & Davies, 2003) are required to understand its evolutive speciation and to characterise this bacterium.

3.2. Nematode-antagonists Specificity

Most of the known nematophagous fungi are able to trap and parasitize very different nematode species as bacteriophagous, plant-parasitic and insect-parasitic species (Rosenzweig, Premachandran, & Pramer, 1985). Some specificities were, however, observed in the composition of the fatty acids produced by fungi in culture, depending on the generic affiliation of the predaceous species and its predation mechanisms (Radzhabova, Gasanova, Mekhtieva, & Bekhtereva, 1987). A wide investigation reported specificities in morphological adaptations, host preference and prey recognition by nematophagous fungi (Nordbring-Hertz, 1988). Antagonistic effects of several *Arthrobotrys oligospora* and *A. conoides* strains isolated in Burkina faso and Senegal on three *Meloidogyne* species (*M. mayaguensis*, *M. incognita*, *M. javanica*) appeared very specific (Duponnois, Mateille, Sene, Sawadogo, & Fargette, 1996) (Table 1). This observation suggests that efficacy of biocontrol of major nematode pest communities requires mixed inocula, including more than a single fungus isolate/species.

Table 1. Fraction (%) of Meloidogyne *spp. juveniles trapped* in vitro *by different* Arthrobotrys *spp. isolates (adapted from Duponnois* et al., *1996)*[1].

Species	Isolate	*M. mayaguensis**	*M. incognita**	*M. javanica**
Arthrobotrys sp.	ORS 18690 S2	11 c	0	0
A. oligospora	ORS 18690 S5	26 b	3 b	0
A. oligospora	ORS 18692 S7	74 a	0	0
A. oligospora	S 30	78 a	65 a	0
A. oligospora	S 31	82 a	70 a	20
A. conoides	S 42	82 a	60 a	0
Arthrobotrys sp.	BF 10	10 c	4 b	0
Arthrobotrys sp.	BF 74	9 c	2 b	0
Arthrobotrys sp.	BF 80	14 c	2 b	0
Arthrobotrys sp.	SOSU 2	8 c	16 b	0

[1] Data followed by the same letter do not differ according to the one way analysis of variance (P>0.05)

Specificity studies were carried out mainly on *P. penetrans*. The endospores of this bacterium are passively intercepted in soil by moving hosts. They were detected also on *Meloidogyne* males and, rarely, were also reported as adhering to females (Carneiro, Randig, Freitas, & Dickson, 1999; Davies & Williamson, 2006). The low frequency of these observations suggest that the juveniles are the main target required by the parasite for the cycle initiation and its subsequent development.

Among the 80 *Meloidogyne* species identified, only 15 species were detected as hosts of *P. penetrans*. Isolates of *P. penetrans* detected to date appear very specific to nematode hosts (Stirling, 1985; Davies, Kerry, & Flynn, 1988a). However, isolates maintained in contact with various *Meloidogyne* species showed a host range wider than that of isolates maintained on the same nematode host during several generations. As a consequence, multiplication of these isolates must preferentially be carried out on the *Meloidogyne* species they were isolated from, in order to prevent modification or changes in specificity (Channer & Gowen, 1992).

Host specificity is usually measured by the percentage of infested *Meloidogyne* juveniles encumbered with endospores and by the number of adhering propagules per infested juvenile (Davies, Flynn, & Kerry, 1988b). Host specificity was observed among *Meloidogyne* species (interspecificity) (Table 2), and also among populations belonging to the same species (intraspecificity) (Fargette, Davies, Robinson, & Trudgill, 1994). However, it is very difficult to infer whether *P. penetrans* propagules present in soil represent different bacterial communities with changing levels of host specificitiy, or whether they proceed from populations of the same pathotype.

Table 2. Attachment of endospores from three populations of Pasteuria penetrans *to different lines and species of* Meloidogyne *tested (since then pVI lines belong to* M. mayaguensis*) (from Fargette* et al., *1994).*

Nematode	*Pasteuria penetrans* populations		
	PNG	*PCal*	*PP1*
M. incognita			
race 1	4*	4	4
race 2	4	4	4
race 3	1	2	3
race 4	1	2	3
Ivory Coast pVI lines	1	3	1
1	0	0	3
2	0	1	1
3	0	0	0
5	1	0	3
7	0	1	0
8	1	1	1
M. mayaguensis	1	1	1
M. arenaria	3	1	3

*Class of adhering endospores: 0=0; 1=1–5; 2=6–10; 3=11–50; 4=>50

Davies, Redden, & Pearson (1994) observed a bacterial surface heterogeneity in *P. penetrans* isolates. The nature of the specificity appears to depend on the biochemical interactions occurring during the contact between the bacterial parasporal fibers surrounding the endospores and some epitopes present on the nematode cuticle surface. Immunological studies on root-knot nematodes cuticle showed that some proteins are involved in the process of host recognition (Davies, Robinson, & Laird, 1992). Either the quality and amounts of these proteins differed among the nematodes species tested, suggesting that they may be responsible for the variability expressed in some specificity studies (Davies & Danks, 1992). However, as several aspects of the mechanism of endospore adhesion remain still undeciphered, the exact determinism of the host specificity process still remains to be completed (Davies et al., 2001).

3.3. Obligate Multitrophic Relationships

Pasteuria penetrans either reduces juvenile penetration in roots because of their strong spore encumberment (Stirling, Sharma, & Perry, 1990), or decreases the density of new generation juveniles in the soil (O' Brien, 1980; Sayre, 1980; Raj & Mani, 1988; Sayre & Starr, 1988). In this case, nematodes produce more *P. penetrans* spores than eggs (Sayre & Starr, 1985). This density-dependence relationship was suspected by Spaull (1981) on sugarcane and Verdejo-Lucas (1992) observed similar seasonal fluctuations between populations of *M. incognita*, *M. arenaria*, *M. hapla* and populations of *P. penetrans*.

Ciancio (1995, 1996) was the first to model density-dependence relationships between plant-parasitic nematodes and *P. penetrans*. But, as plants, nematodes and *P. penetrans* correspond to an obligate tritrophic system, plants can indirectly act on the prey-predator relationship too.

Nematode and *P. penetrans* surveys carried out in various vegetable producing areas revealed correspondences between vegetable species and the abundance of *Meloidogyne* juveniles infested by *P. penetrans* (Diop, Mateille, N'Diaye, Dabiré, & Duponnois, 1996; Hewlett, Cox, Dickson, & Dunn, 1994; Ko, Bernard, Schmitt, & Sipes, 1995; Tzortzakakis, Channer, & Gowen, 1995; Mateille, Duponnois, & Diop, 1995; Giannakou & Gowen, 1996). Most of them observed that either plant susceptibility to nematodes or crop practices (sequences of different plant susceptibilities) influence the proportions of infested juveniles, since development of bacterial populations depends on the abundance of nematode juveniles in the soil.

4. ECOLOGY AND MANAGEMENT OF NEMATODE COMMUNITIES

4.1. Soil Health Approach

Plant production is directly related to soil quality, defined by its functional capacity within an ecosystem with three concerns: biological productivity, environment quality, and plant/animal health (Doran, Sarrantonio, & Liebig, 1996). Soil quality includes three basic components: physical, chemical and biological soil properties.

Biological properties relate to four fields (Chaussod, 2002): fertility, health, environmental impact and resilience. Soil health relates more to ecological characteristics (Doran & Zeiss, 2000) and deals with agronomy (Doran & Safley, 2002). A good health situation is usually correlated with fast nutrient cycles, a strong stability (high resistance or resilience) and a wide biodiversity. Low biodiversities are commonly observed in very high value and high input agrosystems (Anderson, 1994).

In both temperate (Evans et al., 1993) and tropical regions (Luc et al., 2005), strong anthropized agrosystems are mostly characterized by low diversified nematode communities (less than 10 plant-parasitic species). Among these species, some of them can be very frequent and abundant. That is, for example, the case for root-knot nematodes in intensive vegetable producing areas. In these agrosystems, conditions favor nematodes as root-knot nematodes which display high rates of multiplication and cause important plant damages and yield losses. Nematode control methods necessarily imply substantial inputs and lead to both economic and ecological dead ends.

In low anthropized agrosystems, which relate to organic or extensive agriculture, mostly based on fallowing, crop associations or rotations, nematode diversity is usually higher. For example, studies conducted on fallowing (Cadet & Floret, 1999) showed that nematode diversity in communities increases with age of fallows and that damages caused by such communities to the next cereal crop were lower.

The species diversity in plant-parasitic nematode communities can be very high in ecosystems: for example, about 20 species were detected in Atlantic and Mediterranean French coastal sand dunes (Maher et al., 2004) and approximately 30 species in the French Landes forest (Baujard, Comps, & Scotto La Massèse, 1979), apparently without damage in these areas.

4.2. New Paradigms for Nematode Management

High-grade results achieved by crop protection researches carried out on crop practices, plant resistance to nematodes or biocontrol, are very diverse. But, all these control practices, included or not in integrated pest management strategies, seem not to be sufficient: they all target some nematode species (population approach), and then induce changes in nematode communities but do not necessarely decrease their overall pathogenicity. Also, drastic control methods induce biotic imbalances by killing parasites but also their antagonists (direct effect or indirect through host population depletion). So, binary researches focused on plant-nematode or nematode-parasite relationships should be extended to ecological investigations on nematode communities.

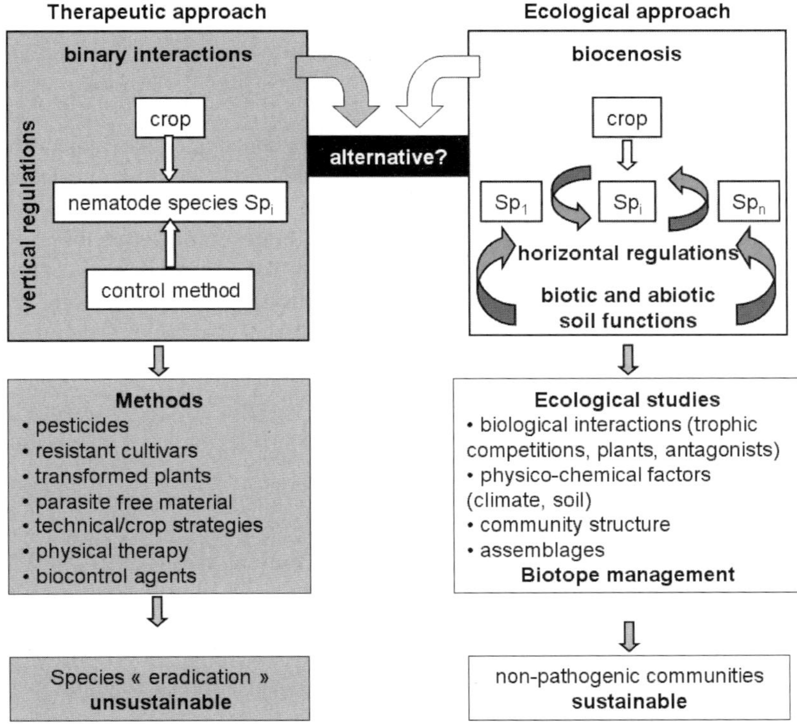

Figure 6. Nematode management, from therapeutic to ecological approaches.

An alternative consists in eco-epidemiological approaches through interactions existing within communities (interspecific competitions, biological and edaphic constrains) which focus on the management of the parasite biodiversity (Fig. 6).

Specific and functional evolution, and pullulation of plant-parasite populations, especially of plant-parasitic nematodes, are enhanced by agriculture intensification and by environment anthropisation. In fact, plant-parasitic nematodes as "predators" belong to a food web as parts of soil factors. Up to now, all control strategies developed in agriculture focus on the eradication of target species. This induces biotic gaps, community rearragements, insurgence of virulent races, increased aggressivity of minor species, etc. and this, the "soil cleaning" strategy, appears to be not sustainable.

The development of sustainable management strategies should move from such "therapeutic approach" (much in favour in research program strategies carried out in the world) to some more "ecological approach". This approach would seek for information and knowledge about biotic trade-offs in ecosystems, in order to introduce them in agrosystems (resilience). This strategy would question: why seeking plant-parasitic nematode eradication? Can agronomic problems be solved by agronomic strategies only? Can nematode diversity in communities be considered as an auxiliary for nematode management?

4.3. Proposed Approaches

The nematode diversity in communities would represent the central object to focus on, and species diversity as well as population levels in communities would be tested for their indication capacity. They could be related to or account for inform on environment disturbances and capacity to facilitate or not epidemics, for soil resistance and resilience. Comparative studies of environments displaying contrasted characteristics or different anthropisation levels should provide understanding of interactions and clues for management, strategies for more or less intensively run agrosystems or endangered environments.

Three types of contexts can be studied:

- Ecosystems: these systems are particularly appropriate to study plant-nematode tradeoffs. They involve "horizontal" biotic regulations defined by the interspecific competitions in the communities (Putten, Vet, Harvey, & Wäckers, 2001): competitions for habitat occupation and for food resources. Specific biological characteristics (life traits, rate of multiplication) will account for such interaction and it is essential to understand the different species fit with each other within a community. They also involve "vertical" biotic regulations related to crop and soil (microbial antagonists) constraints on the species within communities. Obviously, as plant-parasitic nematodes are obligate parasites, the plant plays a major role in the nematode community structure; this depends on both plant susceptibility to different species and on species pathogenicity. Because of their specificities, microbial antagonists also have a marked impact on community structures (De Rooij-van der Goes, Van der Putten, & Van Dijk, 1995). Eventually, abiotic regulations (soil physicochemical factors and functions) also affect the space-time structure of nematode communities (Cadet, Thioulouse, & Albrecht, 1994; Cadet & Thioulouse, 1998).
- Organic agriculture: in organic agriculture, the management of plant-parasitic nematodes implies crop diversification, rotations with non-host or poor host plants, amendments with green manures, biofumigation methods. All these methods enhance biodiversity in soils, as a source of significant biological competitions. Organic agriculture makes it possible to analyze consequences of methods specifically targeting "major" species on the whole nematode communities, without skews induced by chemical treatments.
- Land use changes: these situations induce shifts in community structures, the determinant of which should provide clues for processes involved in community structuring. The original nematode structure, before changing the land use, followed along a time course, should provide elements for understanding new interactions and patterns.

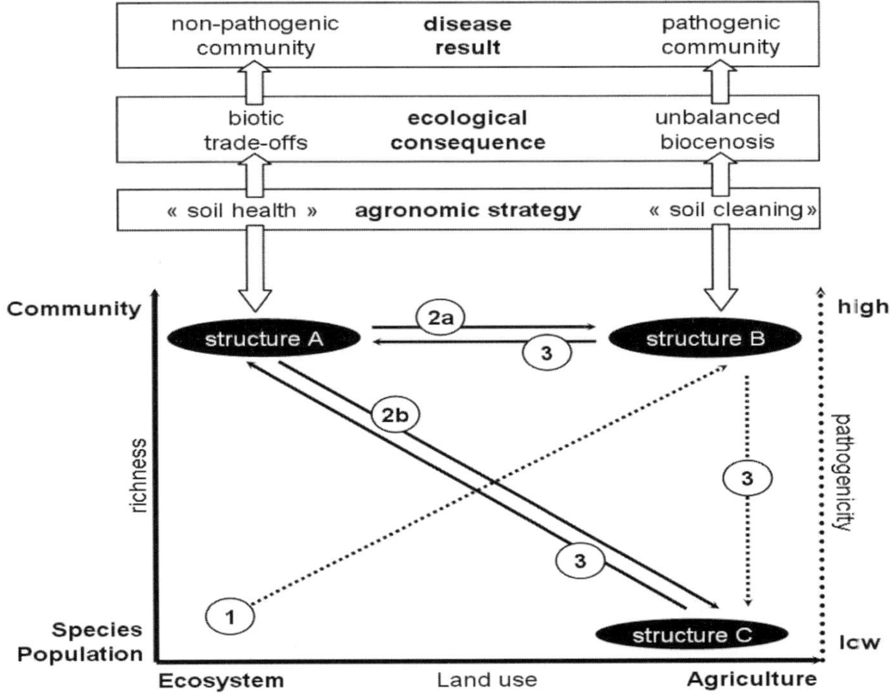

Figure 7. The soil health approach. The pathogenicity of a nematode community is higher in agrosystems than in ecosystems (1). Its increase in agrosystems depends on the structure of the community: same species but different proportions (2a), or lower richness (2b). Conservation strategies for resilience (3).

5. CONCLUSIONS

Significant conceptual progress took place in population biology sciences during the last 30 years with the emergence of the population genetics. This progress could answer questions about how populations function, preservation of polymorphisms and selection. However, the population approach shows limits since a population is not isolated from the other species, and because communities are not only random assemblages (Law, 1999).

Nematode community structures are described by descriptive ecology, but the subjacent rules and mechanisms are poorly known. It is thus necessary to support a cognitive approach to study ecology of nematode communities which will aim at understanding assemblage processes, the knowledge of how populations interact in communities (life trait evolution, adaptation) may provide clues and keys for population management and sustainable use/preservation of agro-eco-systems.

Crop practices favour monocultures, which reduce soil biodiversity. That supports proliferation of parasites. Imbalances induced by these situations strengthened by large inputs, may solve economic questions and on the short run only. Conversely, taking into account the whole nematode diversity in a more ecological approach open new ways for sustainable management of these parasites in agriculture, by testing the possible regulation of the global pathogenic effect of the nematode communities (Fig. 7). It is consequently critical to reconsider control practices inducing ecosystem fragility. Strategies implemented in order to eradicate plant parasites are prone to unsustainability, whereas biological interactions together with environment should be integrated into more environmental friendly strategies to "control" and "manage" plant-parasitic nematodes.

REFERENCES

Anderson, J. M. (1994). Functional attributes of biodiversity in land use systems. In D. J. Greenland & I. Szabolcs (Eds.), *Soil resilience and sustainable use* (pp. 267–290). Wallingford, UK: Cab International.

Anderson, J. M., Preston, J. F., & Dickson, D. W. (1999). Phylogenetic analysis of *Pasteuria penetrans* by 16S rRNA gene cloning and sequencing. *Journal of Nematology, 31*, 319–325.

Atibalentja, N., Noel, G. R., & Domier, L. L. (2000). Phylogenetic position of the North American isolate of *Pasteuria* that parasitizes the soybean cyst nematode, *Heterodera glycines*, as inferred from 16S rDNA sequence analysis. *International Journal of Systematic and Evolutionary Microbiology, 50*, 605–613.

Bachelier, G. (1978). *La faune du sol. Son écologie et son action*. Editions ORSTOM, Paris, 391.

Baujard, P., Comps, B., & Scotto La Massèse, C. (1979). Introduction à l'étude écologique de la nématofaune tellurique du massif landais. *Revue d'Ecologie et Biologie du Sol, 16*, 61–78.

Bird, D. M., Opperman, C. H., & Davies, K. G. (2003). Interactions between bacteria and plant-parasitic nematodes: now and then. *International Journal of Parasitology, 33*, 1269–1276.

Cadet, P. (1987). Etude comparative des peuplements naturels de nématodes parasites associés à la canne à sucre. *Nematologica, 33*, 97–105.

Cadet, P., & Floret, J. (1999). Effect of plant parasitic nematodes on the sustainability of a natural fallow cultural system in the Sudano-Sahelian area in Senegal. *European Journal of Soil Biology, 35*, 91–97.

Cadet, P., & Thioulouse, J. (1998). Identification of soil factors that relate to plant parasitic nematode communities on tomato and yam in the French West Indies. *Applied Soil Ecology, 8*, 35–49.

Cadet, P., Thioulouse, J., & Albrecht, A. (1994). Relationships between ferrisol properties and the structure of plant parasitic nematode communities on sugarcane in Martinique (French West Indies). *Acta Oecologica, 15*, 767–780.

Carneiro, R. M. D. G., Randig, O., Freitas, L. G., & Dickson, D. W. (1999). Attachment of endospores of *Pasteuria penetrans* to males and juveniles of *Meloidogyne* spp. *Nematology, 1*, 267–271.

Channer, A. G., & Gowen, S. R. (1992). Selection for increased host resistance and increased pathogen specificity in *Meloidogyne – Pasteuria penetrans* interaction. *Fundamental and Applied Nematology, 15*, 331–339.

Charles, L., Carbone, I., Davies, K. G., Bird, D., Burke, M., Kerry, B. R., & Opperman, C. H. (2005). Phylogenetic analysis of *Pasteuria penetrans* by use of multiple genetic loci. *Journal of Bacteriology, 187*, 5700–5708

Chaussod, R. (2002). La qualité biologique des sols: des concepts aux applications. *Comptes Rendus de l'Academie Agricole Française, 88*, 61–68.

Ciancio, A. (1996). Time delayed parasitism and density-dependence in *Pasteuria* spp. and host nematode dynamics. *Nematropica, 26*, 251.

Ciancio, A. (1995). Density-dependent parasitism of *Xiphinema diversicaudatum* by *Pasteuria penetrans* in a naturally infested field. *Phytopathology, 85*, 144–149.

Cobb, N. A. (1906). Fungus maladies of the sugar cane with notes on associated insects and nematodes. *Hawaiian Sugar Plant Association, Experiment Station, Division of Pathology and Physiology Bulletin, 2nd Edition, 5*, 163–195.

Davies, K. G., & Danks, C. (1992). Interspecific differences in the nematode surface coat between *Meloidogyne incognita* and M. *arenaria* related to the adhesion of the bacterium *Pasteuria penetrans*. *Parasitology, 105*, 475–480.

Davies, K. G., Fargette, M., Balla, G., Daudi, A., Duponnois, R., Gowen, S. R., et al. (2001). Cuticle heterogeneity as exhibited by *Pasteuria* spore attachment is not linked to the phylogeny of parthenocarpic root-knot nematodes *Meloidogyne* spp. *Parasitology, 122*, 111–120.

Davies, K. G., Kerry, B. R., & Flynn, C. A. (1988a). Observations on the pathogenicity of *Pasteuria penetrans*, a parasite of root-knot nematodes. *Annals of Applied Biology, 112*, 491–501.

Davies, K. G., Flynn, C. A., & Kerry, B. R. (1988b). The life cycle and pathology of the root-knot nematode parasite *Pasteuria penetrans*. *Brighton Crop Protection Conference – Pests and Diseases-*9C-19, 1221–1226.

Davies, K. G., Redden, M., & Pearson, T. K. (1994). Endospore heterogeneity in *Pasteuria penetrans* related to adhesion to plant-parasitic nematodes. *Letters in Applied Microbiology, 19*, 370–373.

Davies, K. G., Robinson, M. P., & Laird, V. (1992). Proteins involved in the attachment of a hyperparasite, *Pasteuria penetrans* to its plant parasitic nematode host, *Meloidogyne incognita*. *Journal of Invertebrate Pathoogy, 59i* 18–23.

Davies, K. G., & Williamson, V. M. (2006). Host specificity exhibited by populations of endospores of *Pasteuria penetrans* to the juvenile and male cuticle of *Meloidogyne hapla*. *Nematology, 8*, 475–476.

De Guiran, G., & Netscher, C. (1970). Les nématodes du genre *Meloidogyne*, parasites des cultures maraîchères au Sénégal. *Cahiers ORSTOM, Série Biologie, 11*, 151–158.

De Rooij-van der Goes, P. C. E. M., Van der Putten, W. H., & Van Dijk, C. (1995). Analysis of nematodes and soil-borne fungi from *Ammophila arenaria* (Marram grass) in Dutch coastal foredunes by multivariate techniques. *European Journal of Plant Pathology, 101*, 149–162.

Diop, M.T., Mateille, T., N'Diaye, S., Dabiré, K. R., & Duponnois, R. (1996). Effects of crop rotations on populations of *Meloidogyne javanica* and *Pasteuria penetrans*. 3rd International Nematology Congress, Pointe-à-Pitre, Guadeloupe, 7–12 Juillet 1996. Nematropica, 26: 256 [Abst.].

Diop, M. T., Ndiaye, S., Mounport, D., & Mateille, T. (2000). Développement des populations de *Meloidogyne javanica* et de *Scutellonema cavenessi* dans les systèmes de culture maraîchère au Sénégal. *Nematology, 2*, 535–540.

Doran, J. W., & Safley, M. (2002). Defining and assessing soil health and sustainable productivity. In C. E. Pankhurst, B. M. Doube, & V. V. S. R. Gupta (Eds.), *Biological indicators of soil health* (pp. 1–28). Wallingford, UK: CAB International.

Doran, J. W., Sarrantonio, M., & Liebig, M. A. (1996). Soil health and sustainability. *Advances in Agronomy, 56*, 1–54.

Doran, J. W., & Zeiss, M. R. (2000). Soil health and sustainability: managing the biotic component of soil quality. *Applied Soil Ecology, 15*, 3–11.

Duponnois, R., Mateille, T., Sene, V., Sawadogo, A., & Fargette, M. (1996). Effect of different West African species and strains of *Arthrobotrys* nematophagous fungi on *Meloidogyne* species. *Entomophaga, 41*, 475–483.

Evans, K., Trudgill, D. L., & Webster, J. M. (1993). *Plant parasitic nematodes in temperate agriculture* (666 pp). New York: Oxford University Press.

Fargette, M., Davies, K. G., Robinson, M. P., & Trudgill, D. L. (1994). Characterization of resistance breaking *Meloidogyne incognita* like populations using lectins, monoclonal antibodies and spore of *Pasteuria penetrans*. *Fundamental and Applied Nematology, 17*, 537–542.

Fargette, M., & Quénéhervé, P. (1988). Populations of nematodes in soils under banana, cv Poyo, in the Ivory Coast. 1. The nematofauna occuring in the banana producing areas. *Revue de Nématologie, 11*, 239–244.

Giannakou, O., & Gowen, S. R. (1996). The development of *Pasteuria penetrans* as affected by different plant hosts. *Brighton Crop Protection Conference – Pests and Diseases, 1*, 393–398.

Gibkin-Davis, R. M., Williams, D., Hewlett, T. E., & Dickson, D. W. (1995). Development and host attachment studies using *Pasteuria* from *Belonolaimus longicaudatus* from Florida. *Journal of Nematology, 27,* 500 [Abstract].

Gray, N. F. (1988). Fungi attacking vermiform nematodes. In G. O. Poinar & H. B. Jansson (Eds.), *Diseases of nematodes,* (Vol. II, pp. 3–38). Boca Raton, Florida: CRC Press, Inc.

Hewlett, T. E., Cox, R., Dickson, D. W., & Dunn, R. A. (1994). Occurence of *Pasteuria* spp. in Florida. *Journal of Nematology, 26*, 616–619.

Jourand, P., Rapior, S., Fargette, M., & Mateille, T. (2004). Nematostatic activity of aqueous extracts of West African *Crotalaria* species. *Nematology, 6*, 765–771.

Ko, M. P., Bernard, E. C., Schmitt, D. P., & Sipes, B. S. (1995). Occurence of *Pasteuria*-like organisms on selected plant-parasitic nematodes of pineapple in the Hawaiian Islands. *Journal of Nematology, 27*, 395–408.

Law, R. (1999). Theoretical aspects of community assembly. In Mac Glade (Ed.), *Advanced ecological theory* (pp. 143–171). Oxford: Blackwell Science.

Luc, M., Bridge, J., & Sikora, R. A. (2005). Reflections on nematology in subtropical and tropical agriculture. In M. Luc, R.A. Sikora, & J. Bridge (Eds.), *Plant parasitic nematodes in subtropical and tropical agriculture* (pp. 1–10). Wallingford, UK: CAB International.

Maher, N., Bouamer, S., Duyts' H., Van der Putten, W., Fargette, M., & Mateille, T. (2004). *A Europe-wide survey of nematode taxa occurring in coast sand dunes.* XXVII ESN Inernational Symposium, Rome.

Mankau, R. (1975). *Bacillus penetrans* n. comb. causing a virulent disease of plant-parasitic nematodes. *Journal of Invertebrate Pathology, 26*, 333–339.

Mateille, T. (1992). Comparative development of three banana-parasitic nematodes on *Musa acuminata* (AAA group) cv. Poyo and Gros Michel vitro-plants. *Nematologica, 38*, 203–214.

Mateille, T. (1993). Effects of banana-parasitic nematodes on *Musa acuminata* (AAA group) cvs. Poyo and Gros Michel vitro-plants. *Tropical Agriculture (Trinidad), 70*, 325–331.

Mateille, T. (1994a). Réactions biochimiques provoquées par trois nématodes phytoparasites dans les racines de *Musa acuminata* (groupe AAA) variétés Poyo et Gros Michel. *Fundamental and Applied Nematology, 17*, 283–290.

Mateille, T. (1994b). Comparative host tissue reactions of *Musa acuminata* (AAA group) cvs Poyo and Gros Michel roots to three banana-parasitic nematodes. *Annals of Applied Biology, 124*, 65–73.

Mateille, T., Cadet, P., & Quénéhervé, P. (1984). Influence du recépage du bananier Poyo sur le développement des populations de *Radopholus similis* et d'*Helicotylenchus multicinctus*. *Revue de Nématologie, 7*, 355–361.

Mateille, T., Duponnois, R., & Diop, M.T. (1995). Influence des facteurs telluriques abiotiques et de la plante hôte sur l'infection des nématodes phytoparasites du genre *Meloidogyne* par l'actinomycète parasitoïde *Pasteuria penetrans*. *Agronomie, 15*, 581–591.

Mateille, T., Foncelle, B., & Ferrer, H. (1988). Lutte contre les nématodes du bananier par submersion du sol. *Revue de Nématologie, 11*, 235–238.

Mateille, T., Quénéhervé, P., & Hugon, R. (1994). The development of plant-parasitic nematode infestations on micro-propagated banana plants following field control measures in Côte d'Ivoire. *Annals of Applied Biology, 125*, 147–159.

Mensah, M. C. (1989). The challenge of sustainable agriculture in Africa. In J. S. Yaninek & H. R. Herren (Eds.), *Biological control: a sustainable solution to crop pest problems in Africa.* Proceedings of the Inaugural Conference and Workshop, IITA Biological control program center for Africa (5–9 Dec. 1988), 8–17.

Merny, G. (1966). Biologie d'*Heterodera oryzae* Luc & Berdon, 1961. II. Rôle des masses d'œufs dans la dynamique des populations et la conservation de l'espèce. *Annales des Epiphytes, 17*, 445–449.

Metchnikoff, M. E. (1888). *Pasteuria ramosa. Annales de l' Institut Pasteur, 2*, 165–171.

Morgan-Jones, G., & Rodriguez-Kabana, R. (1988). Fungi colonizing cysts and eggs. In G. O. Poinar & H. B. Jansson (Eds.), *Diseases of nematodes* (Vol. II, pp. 39–58). Boca Raton, Florida: CRC Press, Inc.

Netscher, C. (1970). Les nématodes parasites des cultures maraichères du Sénégal. *Cahiers ORSTOM, Série Biologie, 11*, 209–229.

Nishizawa, T. (1986). On a strain of *Pasteuria penetrans* parasitic to cyst nematodes. *Revue de Nématologie, 9*, 303–304.

Nordbring-Hertz, B. (1988). Ecology and recognition in the nematode – nematophagous fungus system. In K. C. Marshall (Ed.), *Advances in microbial ecology*, (Vol. 10, pp. 81–114). New York: Plenum Press.

O' Brien, P C. (1980). Studies on parasitism of *Meloidogyne javanica* by *Bacillus penetrans*. *Journal of Nematology, 12*, 234 [Abstract].

Page, S. L. J., & Bridge, J. (1985). Observations on *Pasteuria penetrans* as a parasite of *Meloidogyne acronea*. *Nematologica, 31*, 238–240.

Putten, W. H. van der, Vet, L. E. M., Harvey, J. A., & Wäckers, F. L. (2001). Linking above- and belowground multitrophic interactions of plants, herbivores, pathogens, and their antagonists. *Trends in Ecology & Evolution, 16*, 547–554.

Radzhabova, A. A., Gasanova, S. G., Mekhtieva, N. A., & Bekhtereva, M. N. (1987). Comparative lipid composition of some species of predaceous fungi. *Mikrobiologiia, 56*, 179–184.

Raj, M. A. J., & Mani, A. (1988). Biocontrol of *Meloidogyne javanica* with the bacterial spore parasite *Pasteuria penetrans*. *International Nematology Network Newsletter, 5*, 3–4.

Rosenzweig, W. D., Premachandran, D., & Pramer, D. (1985). Role of trap lectins in the specificity of nematode capture of fungi. *Canadian Journal of Microbiology, 31*, 693–695.

Sarah, J. L., Lassoudière, A., & Guérout, R. (1983). La jachère nue et l'immersion du sol: deux méthodes intéressantes de lutte intégrée contre *Radopholus similis*. *Fruits, 38*, 35–43.

Sasser, J. N. (1989). *Plant parasitic nematodes: the farmer's hidden enemy*. Department of Plant Pathology and the Consortium for International Crop Protection, Raleigh, USA.

Sawadogo, A., Diop, M. T., Thio, B., Konate, Y. A., & Mateille, T. (2000). Incidence de quelques facteurs agronomiques sur les populations de *Meloidogyne* spp. et leurs principaux organismes parasites en culture maraîchère sahélienne. *Nematology, 2*, 895–906.

Sayre, R. M. (1980). Biocontrol: *Bacillus penetrans* and related parasites of nematodes. *Journal of Nematology, 12*, 260–270.

Sayre, R. M., & Starr, M. P. (1985). *Pasteuria penetrans* ex Thorne, 1940 nom. rev., comb. n., sp. n., a mycelial and endospore-forming bacterium parasitic in plant-parasitic nematodes. *Proceedings of the Helminthological Society of Washington, 52*, 149–165.

Sayre, R. M., & Starr, M. P. (1988). Bacterial diseases and antagonisms of nematodes. In G. O. Poinar & H. B. Jansson (Eds.), *Diseases of nematodes, vol. 1*. (pp. 69–101). CRC Press, Inc, Boca Raton, Florida,.

Spaull, V. W. (1981). *Bacillus penetrans* in South African plant-parasitic nematodes. *Nematologica, 27*, 244–245.

Starr, M. P., & Sayre, R. M. (1988). *Pasteuria thornei* sp. nov. and *Pasteuria penetrans sensu stricto* emend., mycelial and endospore-forming bacteria parasitic respectively, on plant-parasitic nematodes of the genera *Pratylenchus* and *Meloidogyne*. *Annales de l'Institut Pasteur/Microbiologie, 139*, 11–31.

Stirling, G. R. (1985). Host specificity of *Pasteuria penetrans* within the genus *Meloidogyne*. *Nematologica, 31*, 203–209.

Stirling, G. R., Sharma, R. D., & Perry, J. (1990). Attachment of *Pasteuria penetrans* spores to the root knot nematode *Meloidogyne javanica* in soil and its effects on infectivity. *Nematologica, 36*, 246–252.

Sturhan, D., Winkelheide, R., Sayre, R. M., & Wergin, W. P. (1994). Light and electron microscopical studies of the life cycle and developement stages of a *Pasteuria* isolate parasitizing the pea cyst nematode, *Heterodera goettingiana*. *Fundamental and Applied Nematology, 17*, 29–42.

Trudgill, D. L., Bala, G., Block, V. C., Daudi, A., Davies, K. G., Gowen, S. R., et al. (2000). The importance of tropical root-knot nematodes *Meloidogyne* spp. and factors affecting the utility of *Pasteuria penetrans* as a biocontrol agent. *Nematology, 8*, 823–845.

Tzortzakakis, E. A., Channer, A. G., & Gowen, S. R. (1995). Preliminary studies on the effect of the host plant on the susceptibility of *Meloidogyne* nematodes to spores attachment by the obligate parasite *Pasteuria penetrans*. *Russian Journal of Nematology, 3*, 23–26.

Verdejo-Lucas, S. (1992). Seasonal population fluctuations of *Meloidogyne* spp. and the *Pasteuria penetrans* group in kiwi orchards. *Plant Disease, 76*, 1275–1279.

Yeates, G. W. (1987). How plants affect nematodes. *Advances in Ecological Research, 17*, 61–113.

5

J. FRANCO AND G. MAIN

MANAGEMENT OF NEMATODES OF ANDEAN TUBER AND GRAIN CROPS

Fundación PROINPA, Cochabamba, Bolivia

Abstract. The exploitation of different Andean crops, including oca, ulluco, mashwa, quinoa and lupine, is revised, for integrated management of plant parasitic nematodes such as *Globodera* spp., *Nacobbus aberrans* and *Thecavermiculatus andinus*, in potato and other Andean crops. The effects of selected lines and varieties in rotations with potato are discussed, with particular attention to the eggs hatching effects and nematodes reproduction. Knowledge on the relationships (host/non host and trap/antagonist crop) among Andean crops and their most important parasitic nematodes appears very important. Some lines were identified which could play an important role in the future implementation of an integrated management of soil nematodes, with long-term benefits for Andean traditional agricultural systems.

1. INTRODUCTION

Numerous tuber crops have been domesticated in the Andes – one of the major centres of plants domestication in the world – and among them the potato *(Solanum tuberosum* ssp. *andigena)* stands out, although other less known plants were also domesticated, such as oca (*Oxalis tuberosa* Molina), ulluco (*Ullucus tuberosus* Lozano) and mashwa (*Tropaeolum tuberosum* Ruíz & Pavón). Other important crops include the high protein pseudograins, *Chenopodium quinoa* Willd. and *C. pallidicaule* Heller (Chenopodiaceae), and a high protein legume, *Lupinus mutabilis* Sweet (Fabaceae). As a group, these tuber, grain legumes and other crops have been among the primary food sources in the highland Andean region, for centuries.

In Andean agro-ecosystems, such as the highlands (2500–4000 msl) and inter-Andean valleys (1500–2500 msl), several native Andean (e.g. potato, mashwa, olluco, oca, quinoa, amaranth, lupinus) or introduced crops (i.e. broad bean and barley) are either monocropped or grown as a multicrop mosaic, in traditional sequences or rotations. To populate higher areas, cold-tolerant species were adapted, for instance quinoa, which can be grown up to 3900 msl or qañiwa, which thrives at 4000 msl.

Fallow, a practice that is still common in certain regions, and each crop in the sequence, excerce their own effects on the local agro-ecosystem. These include changes in soil properties, such as the physic-chemical (e.g. texture, structure, organic matter content, etc.) and biological (i.e. plant growth promoters, biocontrol agents), consequently affecting the growth and yield of the next crop in the rotation.

A. Ciancio & K. G. Mukerji (eds.), Integrated Management and Biocontrol of Vegetable and Grain Crops Nematodes, 99–117.

Such changes can also affect the diversity, co-existence, incidence and severity of the main soil-borne pests and diseases of tuber crops.

Andean crops are grown within subsistence systems but most of the native crops, which have potential to provide a richer diet, remain under-utilized. This is partly due to a lack of knowledge by farmers about how to manage soil-borne pests and diseases (e.g. bacteria, fungi, insects, nematodes) and to improve soils, together resulting in reduced production, soil sickness and erosion.

Poverty and food insecurity are prevalent in the inter-Andean valleys and altiplano of the Andean region. Potato is the staple food crop, providing 30–50% of total calories consumed by rural highland households. In Bolivia, for example, it is grown by nearly 280,000 small-farm families on 130,000 ha. Most potato farmers are poor and potato is not only their food staple but also their principal cash crop. Although potato is commonly grown in highland production systems, yields are the lowest in the world, due to a complex of biotic factors, such as pests and diseases, poor soils and erosion.

In recent years, as market production has increased, potato cropping has become more intensive and rotations have shortened. These aspects have assisted the spread and establishment of pests as constraints to production. However, although it would be impossible to substitute potato in the Andean agricultural systems, some under-utilized Andean root and tuber and small grain crops, already incorporated in traditional crop rotations in the Andean region, represent real possibilities for partial substitution. Andean tuber crops are important food crops in Colombia, Ecuador, Peru and Bolivia, where nearly all production is for fresh market consumption. Importance for people within and outside the Andean region is the nutritional value of these crops, but also their potential as trade commodities, since there have already been some efforts for their introduction into other regions of the World.

2. ANDEAN TUBER CROPS

The three Andean tubers (oca, mashwa and ullucu) are grown in the same agro-ecological zone and their soil requirements and cultivation practices are very similar to those of the potato. For this reason they are herein dealt together. The traditional form of cultivation is on *melgas*: after cultivation of the potato, the land is divided into three to five plots, each of which is sown with one of the Andean tubers. Nearly all production is for fresh market consumption.

One of the most important factors for people living within and outside the Andean region is the nutritional value of the crops. These three Andean tuber crops have been cultivated in the Andean region for centuries and they continue to be an important food crop in Colombia, Ecuador, Peru and Bolivia today. They are a good source of nutrition and have strong aesthetic appeal due to their wide degree of variation in form and color. A large degree of the diversity of these species has been collected and is available for research and breeding. One of the species, *O. tuberosa,* has spread to Mexico and New Zealand where it is marketed and consumed in numerous dishes. Another, *U. tuberosus,* is now canned in Peru and exported to many major United States cities.

2.1. Oca (Oxalis tuberosa Molina)

This species produces elongate or rounded tubers, slightly roughened from the enlarged scale leaves. Oca is a good source of energy, although the proteins and fat contents are low. The prospects for this crop lie in the possibility of increasing its yield and in its use as an source of flour alternative to wheat. The high yields in dry matter obtained from this crop and the possibility of attaining up to 6 or 7 tonnes of flour per hectare are relevant factors in an agro-industrial research programme. Attacks by pests, such as weevils, may cause the loss of an entire crop. One virus disease has been reported, and postharvest management also need to be improved

2.2. Mashwa or Isaño, Añu (Tropaeolum tuberosum Ruíz & Pavón)

Tubers of this crop are elongate or conical. This crop is grown together with olluco, oca and native potatoes in small plots (30–1000 m^2). Average annual yields vary from 4 to 12 t/ha in Peru but, under experimental conditions, they have reached up to 70 t/ha. Mashwa is important for meeting the food requirements of resource-poor people in marginal rural areas of the high Andes. Due to its unusual flavour, mashwa may have a better chance of more extensive use as animal feed. From an agronomic point of view, mashwa is very hardy because it grows on poor soil, without use of fertilizers and pesticides. Even under these conditions, its yield can be double that of the potato.

2.3. Ullucu or Papalisa, Lisa (Ullucus tuberosus Lozano)

This crop produces smooth spherical tubers 2–10 cm across or curved and elongate to 25 cm long. Of the three Andean tubers, the ullucu is the most popular and has become established on the tables of both the rural and urban population in Ecuador, Peru and Bolivia. Its average protein content is 1.7% in the edible tuber, while the carbohydrate and energy content is slightly less than that of most tubers. Although the ullucu is a hardy plant that is suited to the difficult conditions of the Andes, viral diseases seem to constitute one of its most serious problems.

Another limiting factor is the prolonged cultivation period (seven or eight months to mature). In other words, ullucu plants are exposed longer to drought, frost, pests and diseases and other adverse factors which are frequent in the Andes. The biggest advantage of the ullucu is that it is firmly established among rural and urban people in areas where its supply is almost continuous throughout the year.

2.4. Andean Grain Crops

Important grain crops include the high protein pseudograins, quinoa and cañahua (*Chenopodium quinoa* Willd. and *Ch. pallidicaule* Heller, respectively), amarantus (*Amaranthus caudatus* L.), and a high protein legume, *Lupinus mutabilis* Sweet. As a group, these grain legume crops with the tuber ones have been among the primary food sources in the highland Andean region for centuries.

2.4.1. Quinoa (Chenopodium quinoa)

This small grain is a versatile foodstuff grown in rotations with the tuber crops. At present it is being grown in Colombia, Ecuador, Peru, Bolivia, Chile and Argentina. Its marginalization began with the introduction of cereals such as barley and wheat, which eventually replaced it.

The quinoa parts used as human food include the grain, the young leaves up to where ear formation begins (the protein content of the ear is as much as 3.3 percent in the dry matter) and, less frequently, the young ears. Their nutritional value is considerable: the content and quality of proteins are outstanding, because of their essential amino acid composition (lysine, arginine, histidine and methionine). Its biological value is comparable to casein and it is especially suitable for food mixtures with legumes and cereals. Its use may be extended from the rural to the urban and peri-urban populations.

Among Andean grains, *C. quinoa* is the most versatile from the point of view of culinary preparation. The whole plant is used as green fodder. Harvest residues are also used to feed cattle, sheep, pigs, horses and poultry. Its production potential is good, with adequate crop management and pest control, and yields of more than 3 to 4 tonnes per hectare can be obtained.

In recent years, quinoa was introduced on the international market. The traditional cultivation technique consists of sowing under dry conditions in a crop rotation with potato or on strips in maize crops, with little soil preparation and using only the residual organic fertilizers from the preceding crop. As traditional growers always look for safety in cultivation, they therefore sow several ecotypes at different times and in different locations. There is no pest and disease control.

2.4.2. Lupine or Tarwi (Lupinus mutabilis Sweet)

Cultivation of this crop is greatest in Peru, Bolivia and Ecuador. The crop is generally cultivated in rotation with potato or cereals, without the use of fertilizers or manures. Yields range between 500 and 1000 kg per ha depending on the region and ecotypes used. The Andean lupin is an important source of protein (42% in the dry grain, 20% in the cooked grain and 45% in the flour) and fat. It is used for human consumption and industrially, after the bitter taste has been removed. Its nutritional value and forms of use are not widely known, which is why its consumption is not more widespread among the population.

The alkaloids (sparteine, lupinine, lupanidine, etc.) are used to control ectoparasites and intestinal parasites of animals. In the flowering state, the plant is incorporated into soil as green manure and effectively improves the quantity of organic matter and the structure and moisture retention of soil. Because of its alkaloid content, it is frequently sown as a hedge or to separate plots of different crops, preventing damage which animals might cause. Harvest residues (dry stems) are used as fuel because of their high cellulose content, which provides an appreciable calorific value.

3. MOST IMPORTANT NEMATODES IN TUBER AND GRAIN CROPS

In the Andean region the presence of plant parasitic nematodes such as *Globodera* spp. in potato, *Nacobbus aberrans* and *Thecavermiculatus andinus* in potato and other Andean crops, have been reported. The first two nematodes are commonly known as "potato cyst nematode" and "potato rosary nematode". These two nematodes are widely distributed in most potato growing areas and can cause potato yield losses up to 58 and 63%, respectively (Franco, 1994; Franco et al., 1999a). In Bolivia a prevalence shift by altitude for economic nematodes of potato is observed. *Globodera* predominates on the Alto Plano giving way to *Nacobbus* in much of the valleys and to *T. andinus*. Both species seriously damage potato and can also attack other Andean crops (Céspedez, Franco & Montalvo, 1999; Franco & Mosquera, 1993b) which play an important role in traditional crop rotation systems, thus affecting the production capacity of the Andean farmers. The economic importance of PCN in Europe and elsewhere has ensured a good understanding of the species to which locally important factors can be added. In contrast to PCN, *Nacobbus* is not a well studied nematode. On the other hand, *T. andinus* can easily be misidentified as *Globodera* spp. by the presence of white spherical females bodies attached to the roots of its host plant. However, these females do not change color nor become cysts. The eggs inside female body hatch spontaneously and no dormancy has been established.

3.1. Potato Cyst Nematodes

Globodera spp., commonly known as the "potato cyst nematodes", cause severe yield losses in potato. *Globodera rostochiensis* (Wollenweber, 1923) Behrens, 1975 and *G. pallida* (Stone, 1973) can be easily identified during flowering time by the presence of white or yellow spherical bodies (females) attached to potato roots, respectively. These later will turn brown and become cyst which is the survival stage. Inside the cysts, embrionated eggs will molt into second stage juveniles which remain dormant and protected from adverse conditions until they are stimulated to hatch by root exudates produced by the next potato crop. It has a limited host range and only attacks plants from the *Solanaceae* family.

Globodera rostochiensis and *G. pallida* are two of the most important plant-parasitic nematodes in Bolivia and other countries in the Andean region. These potato cyst nematodes (PCN) are widely distributed in most cultivated areas in Bolivia, causing severe direct yield losses (Franco, 1994). Other indirect losses are due to the rejection of seed potatoes from nematode-infested fields (Dirección Nacional de Semillas, 1996).

Eradication of these nematodes is almost impossible and no management strategies are being utilized or developed to reduce nematode population densities. Thus, soil productivity (sustainability) will increasingly deteriorate. An effective management strategy for PCN in Bolivia is complex, due to mixed field populations consisting of both species and different pathotypes (Franco, Oros, Main, & Ortuño, 1998). Therefore, it is likely that an effective management strategy will require the rational use of several components, in order to preserve the environment and to maintain plant productivity.

Among the tactics to manage PCN, crop rotation plays a very important role in traditional Andean agricultural systems (Herve, 1994). However, the development of rotation systems with antagonistic and non-host crops for nematode control depends not only on yield responses, but on economic, ecological, and other constraints in individual situations. Although it is widely known that crop rotations can aid in nematode management, many producers do not view currently available rotation plans as economically feasible (Rodriguez-Kabana & Canullo, 1992). Therefore, more crops and cultivars should be evaluated against important nematode pests so that the number of more efficient and useful rotation crops will increase. Such crops can either be non-hosts or antagonistic to nematodes. For example, studies in Bolivia have shown that there are lines and varieties within non host crops to *N. aberrans* which can be utilized as trap crops: nematodes hatch and/or root invasion occurs, but nematode multiplication does not proceed further (Céspedez et al., 1999; Franco, Main, Ortuño, & Oros, 1997).

On average, potato is planted once in three years under Central Andean Peruvian conditions, and the most common pattern of rotation is two consecutive years of potato, an Andean tuber-bearing crop, cereals and finally a fallow period. Previously, when human population pressure was slight, fallow periods were often as long as 20 years, but these periods have been reduced in recent decades to an average of 2–3 years and sometimes are left out altogether (Esprella, Herve, & Franco, 1994).

3.2. Potato Rosary Nematode

Nacobbus aberrans induces the formation of root nodules on its host-plants (its original name was: “False Root Knot Nematode”) where the enlarged females are enclosed within a protruding gelatinous matrix containing the eggs, which remain in the soil with small root tissues as survival stages.

This species, also known as potato rosary nematode, is responsible for considerable losses in the potato crop: it is able to multiply in 69 host species belonging to 17 botanical families (Castiblanco, Franco, & Montecinos, 1998), including most Andean tuber and grain crops, on which it causes measurable yield losses. The ability of *N. aberrans* to become established in different environmental conditions complicates the management of this nematode, which is subject to international phytosanitary quarantine regulations in an effort to limit its introduction to other countries (CABI and EPPO, 1997; Manzanilla Lopez et al., 2002).

Nacobbus aberrans is adapted to a wide range of climatic conditions (Alarcón & Jatala, 1977), but many aspects of its ecology are still poorly understood. Morphometrical data for a range of *N. aberrans* populations from different geographical areas have been published by a number of authors (Sher, 1970; Johnson, 1971; Quimí, 1979; Doucet, 1989; Doucet & Di Rienzo, 1991; Manzanilla-López, Harding, & Evans, 1999) and detailed morphological observations of populations of *N. aberrans* from Argentina were reported by Doucet and Di Rienzo (1991), and from Argentina, Bolivia, Mexico and Peru by Manzanilla-López (1997).

In South America, this nematode has been detected mainly in the western countries, including Ecuador, Peru, Bolivia, northern Chile and Argentina. With the exception of Ecuador, where there are no confirmed reports of infections on potato, *N. aberrans* is a major pest of potato and also vegetable and field crops (including sugarbeet) in these South American countries (Franco, 1994). In the temperate highlands of the Andean regions of southern Peru, Bolivia, northern Chile and northern Argentina (Jujuy, Salta and Tucuman provinces), *N. aberrans* is the most common pest of potato and other local tuber crops such as mashua, oca and olluco. In Argentina, it damages vegetable crops in Catamarca, Cordoba, Mendoza and San Juan provinces, and also in the subtropical lowlands in Buenos Aires, Rio Cuarto, and Santa Fe provinces (Doucet, 1989).

For practical purposes and on the basis of the most valuable crops damaged by these various *N. aberrans* populations, three slightly different groups have been established: (1) *the sugarbeet group*, including populations infecting sugarbeet but not potato; (2) *the potato group*, which damages potato and also infects sugarbeet but not chilli pepper; (3) *the bean group*, to include populations that attack beans and chilli pepper but are not able to infect potato or sugarbeet.

However, more precise host range information from different localities is required, although conditions for the host range test and the cultivars of potato, chilli pepper, kochia, sugarbeet, bean or other hosts to be used should be standardized. The use of molecular tools should be encouraged as they have the potential for quick and reliable diagnosis of local races. Besides, much of useful information has been written in Spanish in reports and publications that lack English summaries and are not accessible to the international community.

3.3. Nematodes of Oca

Thecavermiculatus andinus is commonly known as the "nematode of the oca" (Astocaza & Franco, 1983; Golden, Franco, Jatala, & Astocaza, 1983), because of its high multiplication on this crop. It is distributed in farms around the Titicaca lake, also attacking other Andean crops (Franco & Mosquera, 1993a) (Fig. 1).

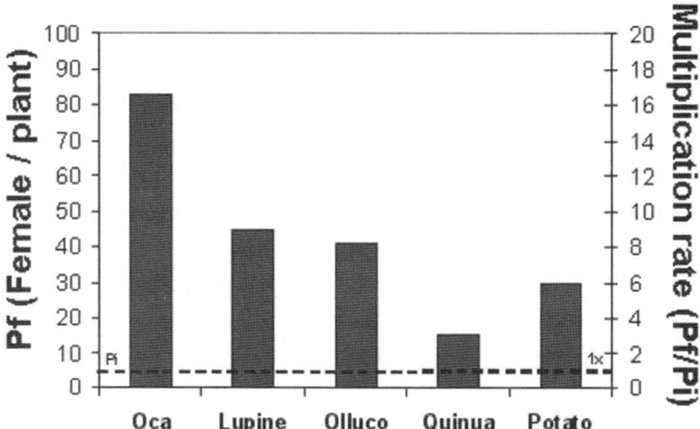

*Figure 1. Number of newly females formed (Pf) and multiplication rate (Pf/Pi) of the "oca nematode" (*Thecavermiculatus andinus*) in roots of different Andean crops.*

Greenhouse studies carried out on the effect of different densities of this nematode on the four Andean crops (lupine, quinua, oca and olluco), showed a negative effect on plant height, stem diameter, leaf fresh and dry weight. Root weight and yield losses were also inversely related to the nematode densities in the soil (individuals per gram of soil). Negative linear regressions of yield losses in relation to nematode density, indicated that this nematode species significantly reduces the yield of quinoa (Fig. 2).

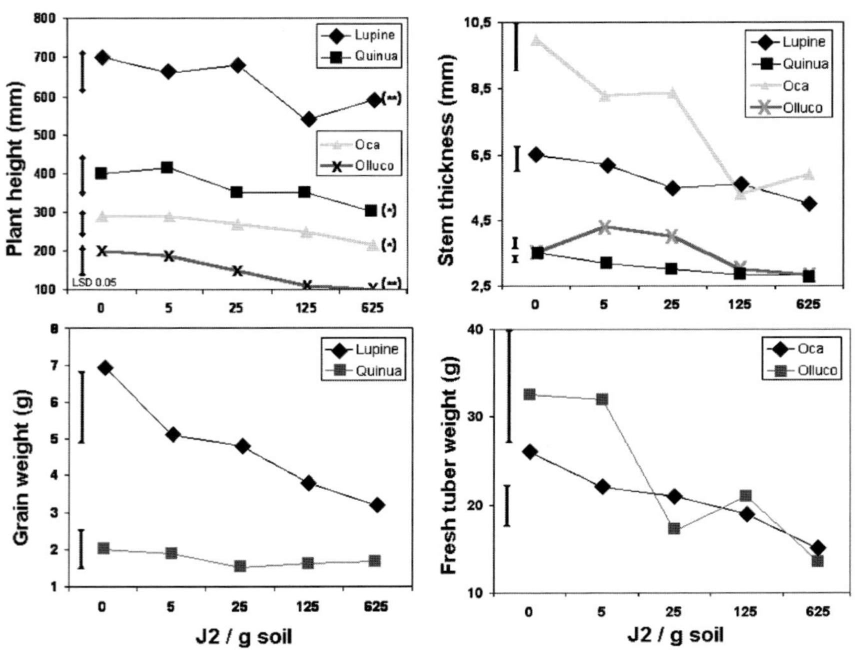

Figure 2. Effect of soil population density of T. andinus *on plant development and production of lupine, quinoa, oca and olluco.*

As indicated above, the nematodes present in Andean countries differ from each other in their morphological and biological characteristics (Table 1). These differences help to identify each species by the morphology of the adult female. Moreover, important differences in host-parasite relationships, development, reproduction on different traditional Andean crops have been found (Table 2), and they will be discussed afterwards.

The first consideration in the host-parasite relationships is to define if a given crop behaves as a host to a parasitic nematode, allowing its feeding, development and reproduction on the plant. Thus the potato crop is a host for all three mentioned nematodes, whereas others such as quinoa (*C. quinoa*), olluco or papalisa (*U. tuberosus*), oca (*O. tuberosa*), lupinos or tarwi (*L. mutabilis*) and mashua (*T. tuberosum*) are host crops only of *N. aberrans* and/or *T. andinus* (Céspedez et al., 1999; Franco & Mosquera, 1993b; Ortuño et al., 1999).

On the other hand, there are different levels of susceptibility and resistance according to nematode multiplication rates – efficient, moderately efficient and not very efficient to each of the three species. Therefore rates of nematode multiplication, combined with the degree of plant tolerance to a given nematode and the growing conditions, will define the degree of crop damage.

Table 1. Morphological and biological characteristic of main plant parasitic nematodes of the Andean region.

Nematodes	*Nacobbus aberrans*	*Thecavermiculatus andinus*	*Globodera* spp.
Common name	"Potato rosary nematode"	"Oca nematode"	"Potato cyst nematode"
Putatative Origin	Andean	Andean	Andean
Distribution	Argentina, Bolivia Chile, Peru, Mexico	Bolivia, Peru	World
Morphology	dimorphism (female enlarged)	dimorphism (female spherical)	dimorphism (spherical cyst)
Host range	17 families and 69 species	11 families and 86 species	Solanaceae
Races	among and within crops	unknown	within crops
Life cycle	2½ generations per season	2 generations	1 generation
Reproduction	crossed	crossed	crossed
Dormancy	facultative	unknown	obligatory
Dissemination	plant tissue	plant tissue	soil
Survival	root residues	root residues	cysts

Among different host-parasite relationships, while quinoa is a non-host of *Globodera* spp., it has been found that certain lines behave as "traps" because their roots are efficient hatching stimulators and roots are invaded, but no nematode development or multiplication occur. A non-host plant will not affect the nematode population density in the soil and a natural decline will occur as response to environmental factors. The finding that certain quinoa lines stimulate hatching of potato cyst nematodes represents a novel contribution (Franco et al., 1999) and adds a key component in the possibility of lowering potato cyst nematode populations in farmer fields.

Table 2. Criteria to define the behavior of plant species in relation to the development and reproduction of N. aberrans *and* Globodera *sp.*

Behavior of plants species		*Hatching*	*Invasion*	*Development*		*Reproduction nodule/cyst*
Among	*Within*		*(J2)*[3]	*(J3–J4)*[3]	*Females*	
Host	Efficient (S)[1]	H [2]	H	H	H	H
	Moderately (R) efficient	H, M	H, M	H, M	M	M
	Not very (R) efficient	H, M	H, M	H, M	No	No
Non Host	Efficient (T)	H	H	H, M, No	No	No
	Moderately efficient	M	M	M, No	No	No
	Not efficient	No	No	No	No	No

[1](S): Susceptible; (R): Resistant; (T): Trap.
[2]H: High; M: Moderate; No: none.
[3]J2–J4: 2nd to 4th nematode development stages.

4. PATHOGENIC RELATIONSHIPS

To establish the most important pathogenic links between Andean crops and their nematodes, the relationship between potato cyst nematodes and tuber and grain crops is discussed. Potato cyst nematodes have a limited host range and only attack plants from the family *Solanaceae*. Nevertheless, various Andean crops were evaluated according the criteria established in Table 2, in the search for non-host crops behaving either as a “trap crop” or as an “antagonist”, in order to significantly reduce the densities of this nematode in soil (Franco et al., 1999).

After evaluating quinoa and lupine lines under greenhouse conditions (1000 cc pots with 80 cysts enclosed in muslin bags and each with a Total Initial Viability, TIV of 100 eggs/cyst), some lines revealed the indicated “trap effect” (Fig. 3). A residual viability value, where the ratio between the final nematode population and the initial nematode population numbers (Pf/Pi) is 1.0, indicates that the crop currently under cultivation exerts no hatching stimulus, as the nematode population at the end of the crop cycle (Pf) is the same as before planting (Pi). Thus, the number of non-hatched eggs remaining inside the cyst as potential inoculum for the next potato crop is similar before and after the crop.

Potato is an efficient host plant and in average stimulates 62% egg hatching, where the residual viability (number of eggs left inside the cyst) averages 38% of the initial population density (Pi). Quinoa and lupine, on the other hand, as non-hosts of *Globodera* spp., usually show a Pf/Pi ratio close to 1.0, because there is no nematode multiplication nor hatching stimulus. Certain lines of quinoa (10), and lupine (57 and 55), however, show a stimulatory hatching effect ("trap effect"). They show a residual viability or Pf/Pi of 0.18, 23 and 27 with 82, 77 and 73% hatchings, which are higher than the potato hatching stimulus, and thus may leave the field with a very low residual eggs viability. Other lines/crops show effects similar to potato but better than fallow.

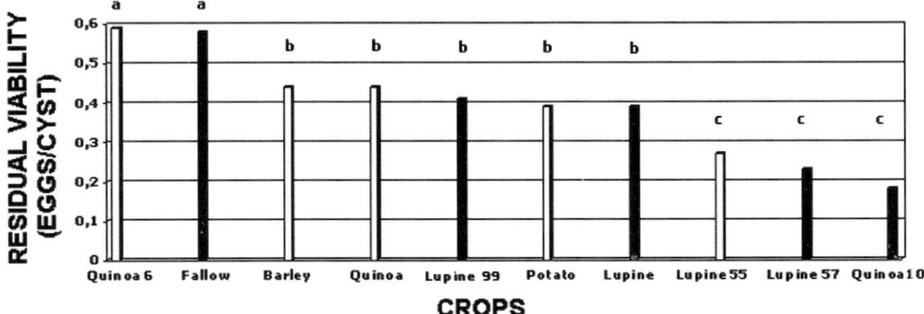

Figure 3. Residual viability (eggs/cyst) of Globodera *sp. as affected by various crops (DMS = 0.1).*

Starting from these results, several studies were set up in order to establish the relationship (stimulatory hatching/inhibitory development effects) between different crops/species and *Globodera* spp. in traditional Andean cropping systems in the valley and Altiplano areas of Cochabamba (Bolivia). The varieties, genotypes, lines or entries identified within Andean crops as trap or antagonistic plants would hence improve the efficiency of alternative crops in existing traditional rotation systems (Table 3). Their effect will be achieved by reducing soil nematode densities through the hatching of viable eggs, without juveniles root penetration and/or nematode development and reproduction, thus improving the yield of the subsequent potato crop.

Table 3. Characteristics of traditional Andean cropping systems in valley and Altiplano regions of Bolivia

Valley (Valles Interandinos)	*Altiplano*
Native and improved potatoes	Native sweet and bitter potatoes
Leguminosae crops (*Vicia*, *Phaseolus*, lupine)	Andean grains (quinoa, amaranthus)
Andean Root and Tuber crops	Andean tuber crops
Fodder crops	Natural Pastures
Ganadería	Ganadería
Short fallow	Long fallow
Irrigated and rainfed lands	Rainfed land

Therefore, under the same described conditions, the effect of several genotypes/lines/cultivars/ecotypes/varieties of different Andean crops (and some other introduced and known as non hosts of *Globodera* spp., see Table 4) was established. The hatching test of a natural mixed cyst population of *G. pallida* and *G. rostochiensis* was performed in small museline bags (Pi = 20 eggs/g of soil). After a six months growing period of different crops, cysts extracted from museline bags were macerated and the eggs counted to establish Final Populations (Pf). By difference between Pi and Pf the number of hatched juveniles as affected by the plant stimulus was established.

According with obtained results, similar interactions have also been identified for each crop according to criteria described in Table 5. Effects were normalized in relation to the % of hatched juveniles in potato plants (100%) and were classified as Efficient (>80%), Moderately Efficient (79–50%) and Poorly Efficient (<49%). In all cases, three different controls were established (potato cv. Waych'a, barley IBTA-80 and fallow), in three replications per treatment.

Table 4. Non host crop species for Globodera *spp. evaluated for their effect on nematode egg hatching.*

Common name	*Scientific name*	*Crop*	*No. lines/genotypes*
Lupine	*Lupinus mutabilis*	Grain	107
Oca	*Oxalis tuberosa*	Tuber	303
Ullucu, Papalisa	*Ullucus tuberosus*	Tuber	85
Quinoa	*Chenopodium quinoa*	Grain	5
Mashwa, Isaño	*Tropaeolum tuberosum*	Tuber	27
Barley	*Hordeum vulgare*	Grain	13

Evaluation began with 107 lines of *L. mutabilis* obtained from a germplasm bank (3 reps/line) and inoculated with *Globodera* cysts as indicated previously (TIV = 100 eggs/cyst). At plant maturity, muslin bags containing cysts were pulled out from the soil and cysts were placed in potato roots exudates (PRE) to stimulate eggs hatching. According with the number of hatched/unhatched eggs in PRE, two parameters were estimated: Infective Viability (IV) = number of free second stage juveniles or hatched eggs, and Residual Viability (RV) = number of unhatched eggs after maceration of cysts. After adding both values (IV + RV), the Total Final Viabilty (TFV) was established. The difference between both values of Total Viabilities (TIV-TFV) gave the effect of each lupine line on the *Globodera* spp. population.

Table 5. Criteria for defining the behavior of plant species in relation to the development and reproduction of Globodera *spp.*

Crops	*Crop*	*Hatching*	*Invasion*	*Development*		*Reproduction*
Between	*Within*		J2	J3-J4	Female	Viable cysts
Host	Highly Efficient (S)**	>80*	>80	>80	>80	>80
	Moderately Efficient (R)	50–79	50–79	50–79	50–79	50–79
	Poorly Efficient (R)	<49	<49	<49	0–49	0–49
Non Host	Highly Efficient (Traps)	>80	>80	>80	0	0
	Moderately Efficient	50–79	50–79	50–79	0	0
	Poorly Efficient (Antagonistic)	<49	<49	<49	0	0

*Percentage of success in each stage of development in relation to an efficient host plant (i.e. potato).
**(S): Susceptible; (R): Resistant.

Results obtained with *Globodera* egg hatching in PRE of the 107 lines of *L. mutabilis* showed three different responses which are summarized in the observed effect of 15 lupine lines (Fig. 4).

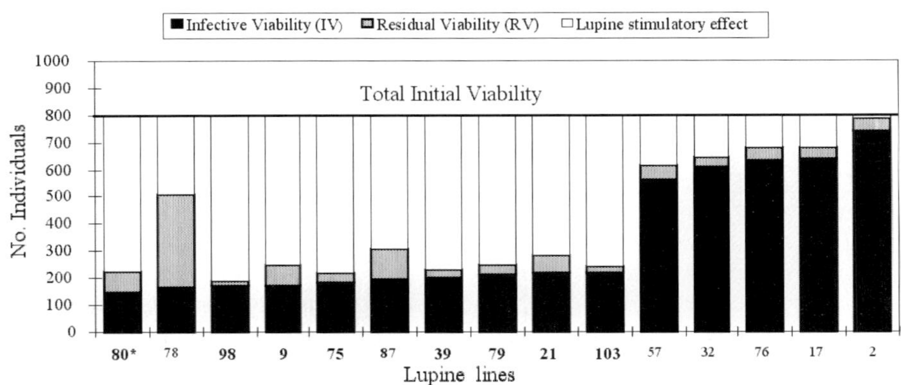

*Figure 4. Effect of nine lupine (*L. mutabilis*) lines selected as "Trap" – No Efficient Host (* = 9, 21, 39, 75, 79, 80, 87, 98 and 103) on hatching, Infective Viability (IV) and Total Initial Viability (TIV) of* Globodera *spp., compared to six "Antagonistic" No Efficient No Host lines of lupine (2, 17, 32, 57, 76 and 78).*

The IV of cysts collected from some lupine lines was high because most of eggs hatched in PRE, indicating no early effect of these lines on hatching of *Globodera* spp. (lines 57, 32, 76, 17 and 2). On the contrary, other few lupine lines showed an early stimulatory effect on hatching, since Residual Viability (RS) was very low as just few juveniles emerged in PRE and also very few remained in the cysts (lines 80, 98, 9, 75, 39, 79, 21 and 103). The third group (lines 76 and 87) showed a very low number of emerged juveniles (IV) in PRE hatching tests but, differing from the second group, a high number of unhatched eggs remained within the cysts (RV). These results show a range of effects on PCN population. The stimulatory hatching effect in the second group of lupine lines ("trap lines") and/or the inhibitory hatching effect of the last group ("antagonistic lines"), represent a response to either roots chemical compounds (direct effect) or possible microorganism activity acting against nematodes. Moreover, the use of lupine foliage as green manure (7–10 t/ha) by some farmers, applied at flowering or after a first legume harvest before a potato crop, appears to reduce PCN densities allowing higher tuber yields (Franco, 1989; Iriarte, 1995).

Results obtained after evaluating 303 lines of oca (*O. tuberosa*) showed eight different groups according with their hatching stimulus on *Globodera* spp. eggs. The first three groups behaved as Efficient including 236, 30 and 14 entries or lines, respectively. The best lines were Bol4028, Bol3992, Bol4046, Bol4095, Bol3898, Bol4114, Bol4110, Bol4190, Bol4113, Bol3919, Bol4012, Bol4565, Bol3873, Bol3991, Bol4024, Bol4038, Bol4042, Bol4058, Bol4151, Bol4162, Bol4185, Bol4336, Bol4363, Bol4416, Bol4422, Bol4505 and Bol4511. In these lines the hatching stimulatory effect was quite similar to that of the potato control (85%), and higher than that of fallow (74%) and barley (56%). The groups 4 to 7 and 8 behaved as Moderately Efficient (Fig. 5).

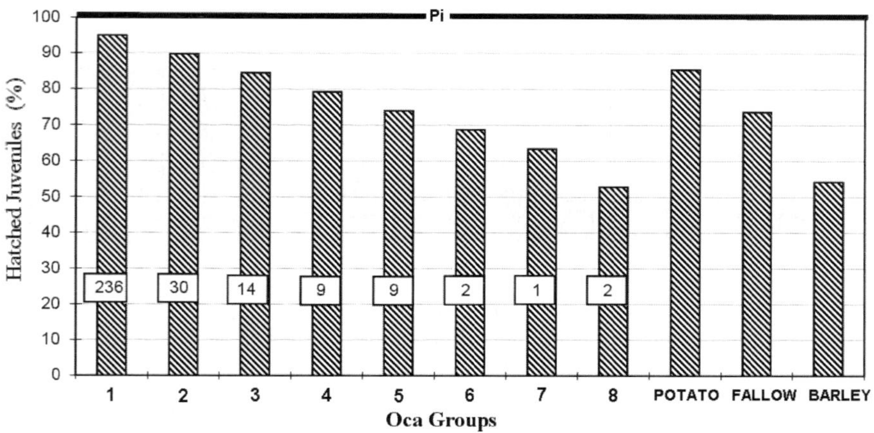

Figure 5. Hatching of Globodera *spp. juveniles in 303 lines of oca (*O. tuberosa*) selected within eight groups (squares), compared to hatching in potato, fallow and barley.*

Within the ollucu (*O. tuberosus*) germplasm, 49 entries/lines in the first group behaved as Efficient because of its high hatching stimulus (89%). The best lines were Bol3975, Bol4213, Bol4388, Bol4395, Bol4389, Bol7003, Bol4322, Bol4479, Bol3963 and Bol4572. Groups 2–4 were Moderately Efficient and groups 5–8 appeared Poorly Efficient (Fig. 6).

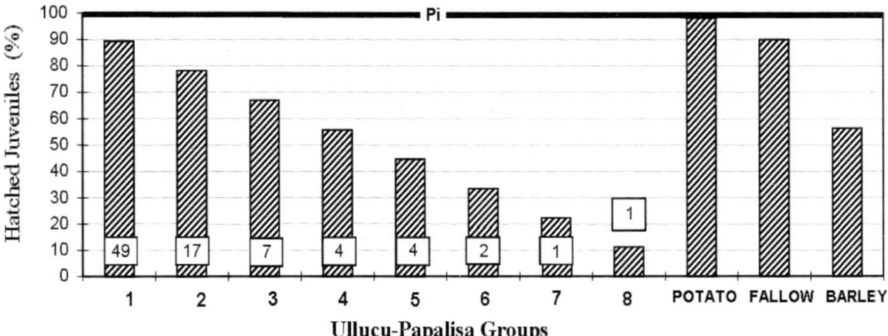

Figure 6. Hatching of Globodera *spp. juveniles in 85 ullucu (*U. tuberosa*) lines selected in eight groups, and compared to potato, fallow and barley.*

In barley varieties (Fig. 7), var. Zapata and line 9-15-92 appeared as Efficient Trap Crops, because hatching was close to that of the potato control, but with other varieties scoring lower and decreasing hatching levels.

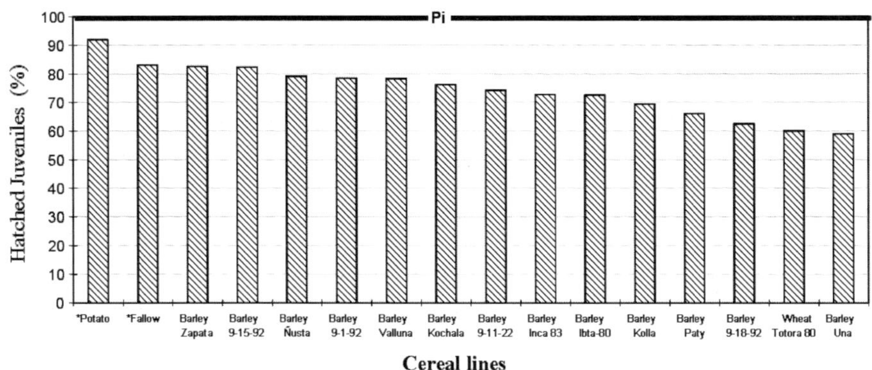

Figure 7. Hatching of Globodera *spp. juveniles in 13 barley* (Hordeum vulgare) *and one wheat* (Triticum sativum) *lines, compared to potato, fallow and barley IBTA 80.*

All evaluated quinoa lines scored an Efficient hatching stimulatory effect when compared to controls (Fig. 8), with highest values for line 1180, widely used as a fodder crop.

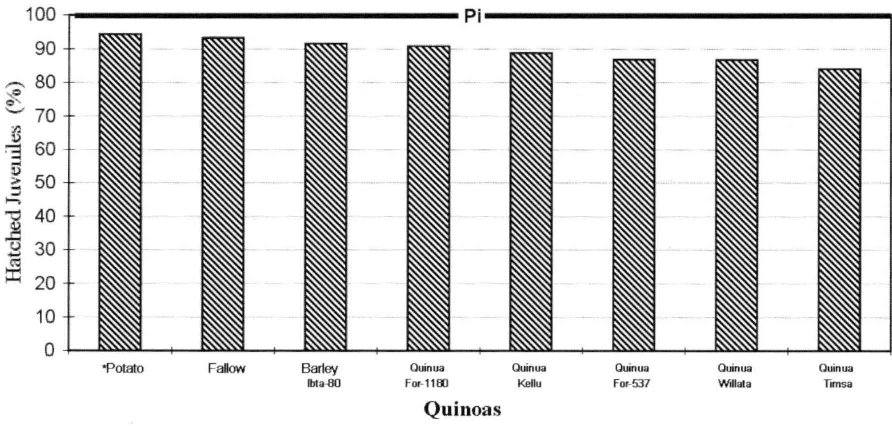

Figure 8. Hatching of Globodera *spp. juveniles in 5 quinoa lines* (C. quinoa) *compared to potato, fallow and barley IBTA 80.*

An efficient hatching stimulatory effect was also observed in the isaño lines Bol4382, Bol4071, Bol4179 and Bol4040. Other 15 lines behaved as Moderately Efficient, with the last eight scoring as Poorly Efficient (Fig. 9).

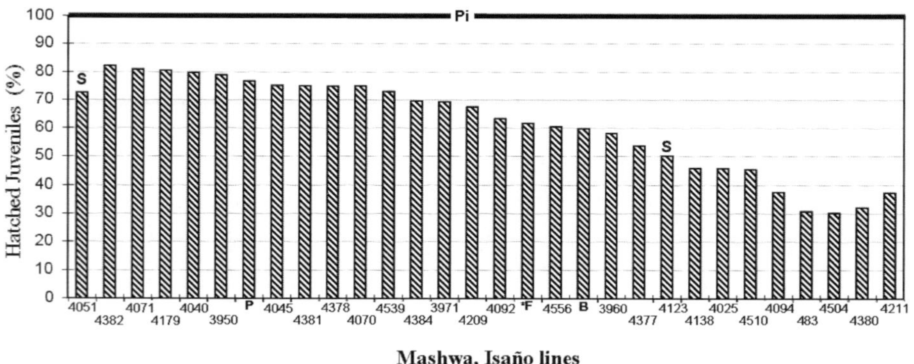

Figure 9. Hatching of Globodera *spp. juveniles in 27 isaño* (T. tuberosum) *lines, compared to potato (P), fallow (F) and barley (B) IBTA 80 (S = Selected line for* N. aberrants*).*

Data obtained on the relationship between Andean crops and *Globodera spp.*, show the variable effect that the cultivars, varieties, genotypes or lines of different Andean crops can exert on the development and multiplication of this nematode. Therefore, within a crop rotation program it is important in first place to identify a No Host Crop and afterwards, an antagonistic or trap genotypes within the crop. Under these conditions, PCN population densities in infested soils will decline more drastically allowing shorter rotation periods between susceptible potato crops. This is particularly important in some cropping systems, since in certain PCN infested areas rotation schemes can take up to 15 years before a new potato crop can be established (Esprella et al., 1994). The use of different genotypes of Andean crops in traditional cropping systems would also favor the recuperation of degraded soil fertility (i.e. lupine crop) as well as the biodiversity conservation and exploitation of Andean minor crops.

5. INTEGRATED MANAGEMENT OF TUBER AND GRAIN NEMATODES

Certain selected lines of Andean crops may be incorporated as important components in the integrated management of nematodes, due to their ability to reduce nematode populations of *Globodera* spp., *N. aberrans* and *T. andinus* (Table 6). While potato, ulluco and quinoa are host-crops of *N. aberrans*, resistant cultivars within each crop can be used to reduce soil nematode population densities (i.e. two quinua resistant *cvs*., Real Kulli and Salustita and the resistant potato *cv* Gendarme have been identified; no information is available on ulluco). With No Host crops to *N. aberrans*, antagonistic or trap lines/cultivars can also be incorporated in strategies for integrated management (IBTA-80 as a trap barley line, but no information is available with oca and lupine lines).

For *Globodera* spp. with the potato crop as unique Host crop, the use of resistant cultivars will help to reduce soil nematode densities (Maria Huanca and Huanquita, Peruvian potato cultivars with partial resistance to certain pathotypes of *G. pallida*). Among No Host crops such quinoa and lupine, trap and/or antagonistic lines can be incorporated in the strategy but within oca, ulluco and cereals, these have to be identified. Finally, no research with Andean crops has been carried out with *T. andinus* and studies should be planned on its distribution and relationships with main crops in traditional systems.

Table 6. Behavior of Andean crops in relation to their main parasitic nematodes.[*]

Crops	*N. aberrans*	*Globodera*	*T. andinus*
Quinoa	Host (S/R)	No Host (T/A)	Host (S/R?)
Potato	Host (S/R)	Host (S/R)	Host (S/R?)
Oca	No Host (T?)	No Host (T?)	Host (S/R?)
Olluco	Host (S/R)	No Host (T?)	Host (S/R?)
Cereals	No Host (T)	No Host (T?)	No Host (T?)
Lupine	No Host (T?)	No Host (T/A)	Host (S/R?)

[*] S: Susceptible; R: Resistant; T: Trap; A: Antagonistic; ?: Not established.

Some options for two different agroecosystems are listed in Table 7. They are: quinoa lines resistant to *N. aberrans* and lines which can be used as trap-crops against *Globodera* spp.; potato *cvs* resistant to *N. aberrans* and *Globodera spp*.; lines of Andean root and tubers, behaving as trap crops to *N. aberrans*, and finally some lines of legume crops, such as lupine, which behave as a trap crop, or as an antagonist plant to *Globodera spp.*, playing also a very important role in improving soil fertility (Franco, 1989). Several attributes of these crops would be applicable in the traditional agricultural systems of the high plateau region of Bolivia.

Table 7. Use of selected Andean crops in crop rotation as a component of the Integrated Management of nematodes, to control nematode population densities in traditional agricultural systems.

Crop behavior	*Agroecosystem*	
	High Andean	*Inter andean valleys*
Resistant	Quínoa, potato, olluco, oca	Quinoa, potato
As a Trap crop	Olluco, oca, quínoa	Quinoa, barley
Antagonistic crop	Quinoa, lupine	Quinoa, lupine
Soil improver	Leguminous: lupine	Bean, pea, vicia

It can be concluded that knowledge on the relationships (host/non host and trap/antagonist crop) of Andean crops with their most important parasitic nematodes is very important. The lines identified could play an important role in the future implementation of an integrated management of soil nematodes, expecially when other traditional tools (i.e. chemicals) may become less efficient, or unsuitable, for potato protection. Appropriate choice of intercropping and rotation schemes with minor crops may hence provide long-term benefits for the Andean population and at a low cost, protecting at the same time the Altiplano environment and local, traditional agricultural systems.

REFERENCES

Alarcón, C., & Jatala, P. (1977). Efecto de la temperatura en la resistencia de *Solanum andigena* a *Nacobbus aberrans*. *Nematropica, 7,* 2–3.

Astocaza, E., & Franco, J. (1983). El "nematodo de la oca" (*Thecavermiculatus andinus* sp.n.) en el altiplano peruano. *Fitopatología, 18,* 39-47.

CABI, & EPPO. (1997). Quarantine pests for Europe (2nd ed., 1425 pp.). Wallingford, UK: CAB Internacional.

Castiblanco, O., Franco, J., & Montecinos, R. (1998). Razas y gama de hospedantes en diferentes poblaciones de *Nacobbus aberrans* (Thorne, 1935), Thorne and Allen, 1944. *Revista de la Asociación Latino Americana de la Papa* (ALAP), 11, 85–96.

Céspedez, L., Franco, J., & Montalvo, R. (1999). Comportamiento de diferentes especies vegetales a la invasión y desarrollo de *Nacobbus aberrans* (Thorne, 1935), Thorne and Allen, 1944. *Nematropica, 28,* 165–171.

Dirección Nacional de Semillas. (1996). Normas específicas de certificación de semilla de papa (13 pp.). Secretaría Nacional de Agricultura y Ganadería, La Paz, Bolivia.

Doucet, M. E. (1989). The genus *Nacobbus* Thorne & Allen, 1944 in Argentina. 1. Study of a population of *N. aberrans* (Thorne, 1935) Thorne and Allen, 1944 on *Chenopodium album* L. from Rio Cuarto, Province of Córdoba. *Revue de Nématologie, 12,* 17–26.

Doucet, M. E., & Di Rienzo, J. A. (1991). El género *Nacobbus* Thorne and Allen, 1944 en Argentina. 3. Caracterización morfológica y morfométrica de poblaciones de *N. aberrans* (Thorne, 1935) Thorne & Allen, 1944. *Nematropica, 21*, 19–35.

Esprella, R., Hervé, D., & Franco, J. (1994). Control del nematodos quiste de la papa (*Globodera* spp.) por el descanso largo controlado comunalmente: Altiplano Central Boliviano. In D. Hervé, D. Genin, & G. Riviere (Eds.), Dinámicas del descanso de la tierra en los Andes (pp. 175–183). La Paz, Bolivia, IBTA, ORSTOM.

Franco, J. (1989). El tarwi o Lupino (*Lupinus mutabilis* Sweet): Su efecto en sistemas de cultivo (54 pp.). Lima, Perú: Centro Internacional de la Papa.

Franco, J. (1994). Problemas de nematodos en la producción de papa en climas templados en la región andina. *Nematropica, 24*, 179–195.

Franco, J., & Mosquera, P. (1993a). Patogenicidad del "Nematodo de la Oca" (*Thecavermiculatus andinus* sp.n.) en cuatro cultivos andinos. *Revista Latinoamericana de la Papa, 5/6,* 30–38.

Franco, J., & Mosquera, P. (1993b). Ampliación de la gama de hospedantes del "Nematodo de la Oca" (*Thecavermiculatus andinus* sp. n., Golden et al., 1983) en los Andes Peruanos. *Revista Latinoamericana de la Papa, 5/6,* 39–45.

Franco, J., Main, G., Ortuño, N., & Oros. R. (1997). Crop rotation: An effective component for the Integrated Management of *Nacobbus aberrans* in potato. *Nematropica, 27*, 110 [abstract].

Franco, J., Oros, R., Main, G., & Ortuño, N. (1998). Potato cyst nematodes (*Globodera* spp.) in South America. In R. J. Marks & B. B. Brodie (Eds.), *Potato cyst nematodes, biology, distribution and control* (pp. 239–269). International Centre for Agriculture and Biosciences, Oxon, U.K, New York, U.S.A: CAB International.

Franco, J., Ramos, J., Oros, R., Main, G., & Ortuño, N. (1999a). Pérdidas económicas causadas por *Nacobbus aberrans* y *Globodera* spp. en el cultivo de la papa en Bolivia. *Revista Latinoamericana de la Papa, 11*, 40–66.

Franco, J., Main, G., & Oros, R. (1999b). Trap crops as a component for the Integrated Management of *Globodera* spp. (Potato cyst nematodes) in Bolivia. *Nematropica, 29,* 51–60.

Golden, A. M., Franco, J., Jatala, P., & Astocaza, E. (1983). Description *of Thecavermiculatus andinus* n. sp. (Meloidoderidae), round cystoid nematode from the Andes mountains of Peru. *Journal of Nematology, 15*, 357–363.

Herve, D. (1994). Desarrollo sostenible en los Andes Altos: Los sistemas de cultivos con descanso largo pastoreado. In Dinámicas del descanso de la tierra en los Andes (pp. 15–36, 356 pp.).

Iriarte, L. (1995). Influencia de la incorporación del haba como abono verde para controlar nematodos y su efecto en la producción de papa cv. Waych'a (S. *tuberosum* ssp. *andigena*). Tesis Ing. Agr. Facultad de Ciencias Agrícolas y Pecuarias "Martín Cárdenas", Universidad Mayor de San Simón. 118 pp.

Johnson, J. D. (1971). The taxonomy and biology of a new species of *Nacobbus* (Hoplolaimidae: Nematoda) found parasitizing spinach (*Spinacia oleracea L.) in Texas*. Ph.D. thesis, Graduate College of Texas A & M University, College Station, Texas, USA, 142 pp.

Manzanilla-López, R. H., Harding, S., & Evans, K. (1999). Morphometric study on twelve populations of *Nacobbus aberrans* (Thorne, 1935) Thorne and Allen, 1944 (Nematoda: Pratylenchidae) from Mexico and South America. *Nematology, 1*, 477–498.

Manzanilla-López, R. H. (1997). Studies on the characterisation and bionomics of *Nacobbus aberrans* (Thorne, 1935) Thorne and Allen, 1944 (Nematoda: Pratylenchidae). Ph.D. Thesis. University of Reading, UK, 395 pp.

Manzanilla-López, R. H., Costilla, M. A., Doucet, M., Franco, J., Inserra, R. N., Lehman, P. S., et al. (2002). The genus *Nacobbus* Thorne and Allen, 1944 (Nematoda: Pratylenchidae): Systematics, distribution, biology and management. *Nematropica, 32*, 149–227.

Ortuño, N., Franco, J., Balderrama, F., Blanco, R., Main, G., & Oros, R. (1999). Identificación de quinuas resistentes a nematodos en Bolivia. In D. L. Danial (Ed.), Tercer Taller de PREDUZA en Resistencia Duradera en Cultivos Altos en la Zona Andina, Cochabamba, Bolivia, pp. 171–175.

Rodriguez-Kabana, R., & Canullo, G. H. (1992). Cropping systems for the management of phytonematodes. *Phytoparasitica, 20*, 211–224.

Quimí, V. H. (1979). Studies on the false root-knot nematode *Nacobbus aberrans*. Ph.D. thesis, University of London, Imperial College, UK, 235 pp.

Sher, S. A. (1970). Revision of the genus *Nacobbus* Thorne and Allen, 1944 (Nematoda: Tylenchoidea). *Journal of Nematology, 2*, 228–235.

6

GREGORY R. NOEL

IPM OF SOYBEAN CYST NEMATODE IN THE USA

U.S. Department of Agriculture, Agricultural Research Service

Abstract. The cropping system in the USA that produces soybean every other year exerts severe pressure on the production system. Unless the system changes dramatically to produce soybean every third or fourth year, severe pressure will be continue and will reduce the likelihood of sustainable soybean production. In the foreseeable future, genetic resistance of soybean coupled with crop rotation will be the foundation of any management system for *Heterodera glycines*. Marker assisted selection will aid in selection and incorporation of specific genes and will increase the efficiency of developing resistant cultivars. However, any source of resistance may not be durable even with sound nematode management practices. Rotation of resistance genes (sources of resistance) has met with some success and may increase durability of sources of resistance. Integrating biological control in the production system with promising organisms such as *Hirsutella rhossiliensis* and *Pasteuria nishizawae* may lead to a truly sustainable system of soybean production in which *H. glycines* is no longer a yield limiting production factor.

1. INTRODUCTION

Although more than 100 species of nematodes have been associated with soybean in North America, only a few species are of economic importance (Donald et al., 1984; Noel, 1999). The soybean cyst nematode (*Heterodera glycines*) and root-knot nematodes (*Meloidogyne arenaria, M. javanica,* and *M. incognita*) are the nematodes responsible for most of the crop loss in soybean (*Glycine max* (L.) Merr.) in the USA., with *M. hapla* rarely causing economic loss. Crop loss caused by *H. glycines* was 2.91 million tonnes (t) in 2003, 3.48 million t in 2004, and 1.94 million t in 2005 (Wrather & Koenning, 2006). These losses were 31%, 28% and 28% respectively of the total crop loss due to diseases and nematodes and exceed any other disease. Crop loss due to individual species of *Meloidogyne* and other nematodes is not readily available. During the years of 2003–2005, the combined crop loss due to *Meloidogyne* spp. and all other species ranged from 106,000 to 139,000 t. Crop loss data is not readily available for the other major soybean producing nations Argentina, Brazil, and China.

In recent years the incidence and distribution of the reniform nematode (*Rotylenchulus reniformis*) have increased in the southern USA. Nearly 100 species of nematodes have been associated with soybean (Schmitt & Noel, 1984; Noel, 1999). The Columbia lance nematode (*Hoplolaimus columbus*) is important in the states of Georgia, North Carolina and South Carolina, but also has been reported from Alabama and Louisiana. *Hoplolaimus galeatus* is found frequently in

A. Ciancio & K. G. Mukerji (eds.), Integrated Management and Biocontrol of Vegetable and Grain Crops Nematodes, 119–126.

association with soybean in the northern USA, but crop loss has not been demonstrated. *Hoplolaimus magnistylus* occurs in Arkansas, Mississippi, and Tennessee, but is considered of minor economic importance. Although several species of lesion nematodes (*Pratylenchus alleni, P. brachyurus, P. coffeae, P. hexincisus, P. neglectus, P. penetrans,* and *P. scribneri*) are associated with soybean, crop loss has rarely been demonstrated with lesion nematodes (Lawn & Noel, 1986). The sting nematodes *Belonolaimus gracilus* and *B. longicaudatus* have been associated with soybean primarily in the southern Atlantic states, but infestations have been reported from Alabama, Arkansas, Kansas, New Jersey, Oklahoma, and Texas. In 2003–2005 crop loss was reported for *Paratrichodorus minor* in Virginia, *Rotylenchulus reniformis* in Alabama, Arkansas and North Carolina, *H. magnistylus* in Louisiana and *H. columbus* in Georgia and North Carolina (Wrather & Koenning, 2006). *Hoplolaimus columbus* also caused crop loss in South Carolina (J. Mueller, pers. comm.). Nematodes other than *H. glycines* are important on a local or regional basis. Since *H. glycines* is of paramount importance in soybean production in the USA and other countries, certain aspects of management and IPM will be discussed herein.

Perhaps the most difficult aspect of nematode control for the soybean farmer is the determination of damage thresholds and action thresholds. In soils typical of much of the midwestern USA, a damage threshold based on experiments in commercial production fields was determined (Noel, 1984). The soil in both fields was a silt loam with 2% organic matter. Damage functions generated under field conditions are illustrated in Fig. 1A and 1B. In order to develop those damage functions, 20 soil cores were taken in each of 24 sites in a grid within fields and the number of cysts and eggs were determined at planting. Additionally, aldicarb was applied in paired comparisons of treated and nontreated soil. Yield and numbers of eggs and cysts were determined at the end of the growing season. The damage functions demonstrate that increasing numbers of eggs at planting resulted in greater crop loss and aldicarb was effective in increasing yield.

Application of aldicarb also illustrated other considerations in management of nematodes in soybean. First, most of the sites in the field where no cysts were recovered at planting had detectable cysts at harvest. Thus, absence of cysts in a soil sample does not mean that *H. glycines* is absent, and even when the nematode is believed to be absent it may in fact cause crop loss, as illustrated by the increase in yield associated with aldicarb when no nematodes or low numbers of nematodes were present at planting. In addition, these fields were infested with *P. scribneri*, but it is not known if the numbers and environmental conditions were conducive to the nematode. Research conducted in microplots established that damage to soybean caused by *H. glycines* is greater in sandy soils as compared to soils with more silt and clay (Koenning & Barker, 1995). Effects of the nematode on yield were ameliorated by irrigation in the sandy clay loam at all population levels, but not in the loamy sand and higher population levels.

Developing a practical and affordable sampling recommendations for *H. glycines* is complicated. Farmers simply cannot afford to pay for sufficient samples to blanket an entire field or spend the number of man hours necessary to develop reliable data as to the distribution and numbers of *H. glycines* (Schmitt, Barker, Noe,

& Koenning, 1990). Also, predictive sampling for *H. glycines* has not been investigated sufficiently. Thus, farmers need to adopt an action threshold of one cyst (i.e. detection of *H. glycines*). The adoption of this action threshold is supported by the damage functions (Fig. 1), which illustrate (as supported by subsequent studies) that yield loss of 15–20% can occur in the absence of visible symptoms (Donald et al., 2006; Wang et al., 2000; Noel & Edwards, 1996). Another practical aspect of predictive sampling is that the vast majority of farmers will take samples after harvest. Samples can be processed during the next one or two months prior to purchase of seed for the next growing season. The number of eggs, especially viable eggs, will be higher in the fall sample.

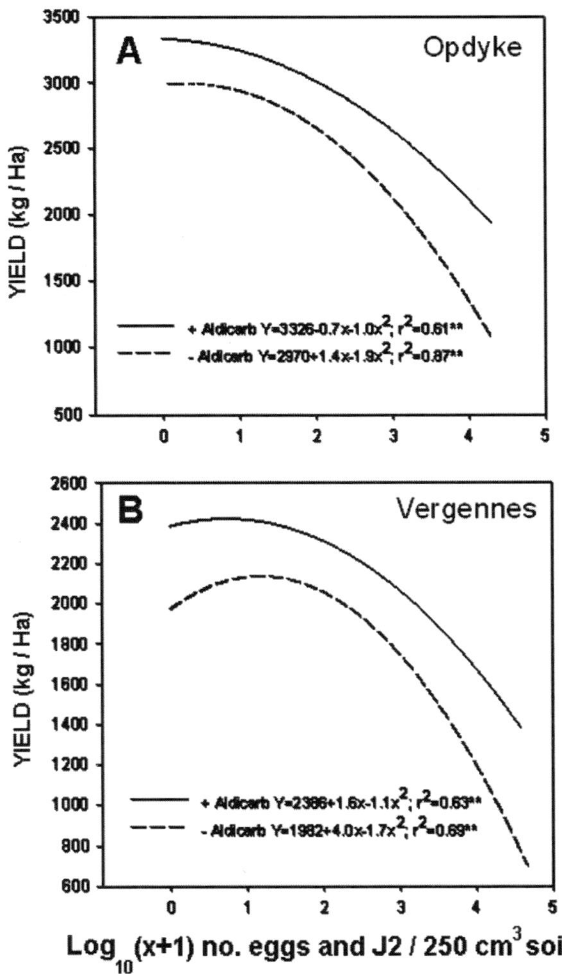

Figure 1. Damage functions for Heterodera glycines *on soybean treated and not treated with aldicarb (11.0 g a.i./100 m of row) at (A) Opdyke and (B) Vergennes, Illinois, USA.*

The recommendation to alternate crop rotation with planting resistant and then susceptible cultivars, to stabilize selection pressure, was never investigated in the field. The recommendation was based on expected reestablishment of the Hardy-Weinberg equilibrium when selection pressure would be removed by planting the susceptible cultivar. Thus, effects on selection for nematode aggressiveness on resistant cultivars would be ameliorated. I know a farmer who has faithfully practiced the maize/resistant soybean/maize/susceptible soybean rotation for 20 years and has averaged a soybean yield of 3,700 kg/ha during that time. Soil samples taken during this time have not indicated any increase in numbers of *H. glycines* on his farm and the yields indicated no increase in nematode virulence has occurred. He has always planted resistant cultivars that carry the PI88.788 source of resistance. In his situation, good nematode management has maintained a high yield and also protected the durability of resistance on his farm. In a long-term study, rotation of cultivars with different sources of resistance was effective in reducing numbers of *H. glycines* below the detection level (Noel & Edwards, 1996).

Management of nematodes in soybean primarily consists of planting resistant cultivars and crop rotation. Application of nematicides is almost nonexistent. Approximately 700 cultivars resistant to *H. glycines* are available (Shier, 2006). Differences in the levels of resistance in these cultivars to *H. glycines* may compromise their effectiveness. Figure 2 illustrates the level of resistance of 300 commercial cultivars stated as being resistant to *H. glycines* Hg Type 0 (race 3). It is obvious that the full level of resistance of PI88.788 has been lost in many cultivars. At present there are about 700 cultivars listed as having resistance to *H. glycines* (Bond, Niblack, & Noel, 2006; Shier, 2006). About 99% of these derive their resistance from PI88.788 and have cv. Fayette in their pedigree (Bernard, Noel, Anand, & Shannon, 1988).

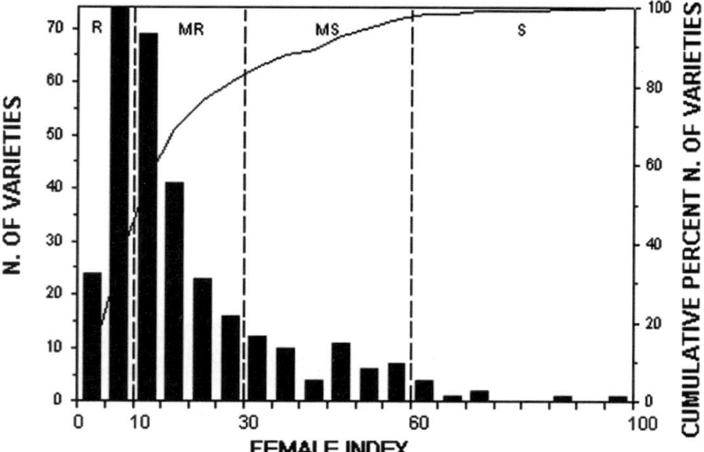

Figure 2. Level of resistance of 300 proprietary varieties to Heterodera glycines *Race 3 (HgType 0), where R is resistant, MR is moderately resistant, MS is moderately susceptible, and S is susceptible (from Noel, 2004).*

Unfortunately, “yield drag”, that is the lower yield potential of resistant cultivars in the absence of *H. glycines* or numbers below the damage threshold, has caused many farmers at some point to abandon the planting of a resistant cultivar. This has had disastrous results for many soybean farmers. In order to ameliorate the lower yield potential of resistant cultivars, some companies have marketed multilines or blends. This management practice is of no value. Yield of the multilines often is less than the resistant cultivar (Table 1). Planting a high yielding resistant variety is an economically more viable option for the farmer.

Application of nematicides is used rarely to control *H. glycines* because planting resistant cultivars is more economically viable (Table 1; Koenning, Coble, Bradley, Barker, & Schmitt, 1998; Noel, 1987). Very few soybean producers know what Hg Type is present in their fields, but tests to determine Hg Types are expensive. Therefore, most farmers plant resistant cultivars without knowledge of the parasitic ability of *H. glycines* in their fields. Rotation of resistance genes by planting cultivars derived from different sources of resistance is promising (Noel & Edwards, 1996).

Table 1. Yield (kg/ha) of soybean cultivars and blends in Heterodera glycines *infested soil either treated (+) or not treated (–) with aldicarb applied in furrow at the rate of 11.0 g a.i./100 m of row, on four farms in Illinois, USA.*

	Bogota		*Opdyke*		*Sidney*	
Cultivar or blend	–	+	–	+	–	+
Union[a] (U)	2,157	2,265	1,183	1,048	2,748	2,816[b]
Fayette[c]	3,152	3,300	1,976	1,908	2,997	3,057
Franklin[d] (FR)	2,661	2,654	1,808	1,552	2,695	2,742
U/FR 3:1	2,661	2,829	1,740	1,693	2,782	2,715
U/FR 1:1	2,695	2,728	1,788	1,680	2,796	2,816
U/FR 1:3	2,480	2,621	1,478	1,572	2,641	2,829
$FLSD_{0.05}$	255		319		160	

[a]Susceptible to *H. glycines*.
[2/3]Century (susceptible) and CN290 (Peking source of resistance) planted at Sidney.
[1/8]PI88.788 source of resistance.
[3/8]Peking sources of resistance.

With the increase in no-till crop production in the USA during the last 10 years, there are unanswered questions regarding soil health, soil suppressiveness, and control of nematodes and plant pathogens. There are conflicting reports concerning the effect of no-till production on *H. glycines*, which may be due to site specific parameters. Chen, Stienstra, Lueschen, and Hoverstad, (2001) reported no effect of tillage, whereas Noel and Wax (2003), showed that numbers of *H. glycines* eggs increased more in no-till, but following rotation with maize there was no long-term effect. In the first few years of adapting no-till production, numbers of *H. glycines* eggs may increase. However, a survey in the midwest USA found that tilled fields with a high clay content supported higher numbers of *H. glycines*. No-till fields with a lower clay content supported larger nematode numbers. In the southern USA several researchers reported lower numbers of *H. glycines* associated with soybean grown no-till (Edwards, Thurlow, & Eason, 1988; Hershman & Bachi, 1995; Koenning, Schmitt, Barker, & Gumpertz, 1995; Tyler, Chambers, & Young, 1987).

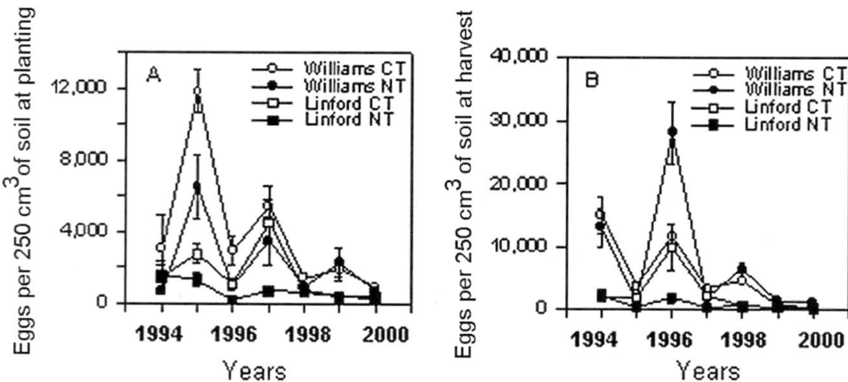

Figure 3. Population dynamics of Heterodera glycines *in a soybean-corn rotation planted either with conventional tillage (CT) or no-tillage (NT). (A) Numbers of eggs at planting, and B) numbers of eggs at harvest. Soybean cvs. Williams 82, susceptible to* H. glycines, *and Fayette or Linford, resistant to* H. glycines, *were planted in 1994 (Fayette) 1996, 1998, 1999, and 2000. Corn was planted in 1995 and 1997. Each datum is the treatment mean, and bars represent the standard error of the mean (from Noel & Wax, 2003).*

Many times, via anecdotal accounts and via personal experience, fields in which soybean yield was severely reduced become profitable. The soils have become suppressive, but capitalizing on this suppressiveness has proven difficult. Chen (2004) published a thorough review of SCN management with biological methods, including soil amendments and various organisms.

Many species of fungi have been isolated from cysts. Various fungal and bacterial species have been evaluated for their potential as biological control agents, but success in the field has been elusive. Chen and Reese (1999) reported the parasitism of *H. glycines* by *Hirsutella rhossiliensis* and population dynamics during a long-term corn/soybean rotation study. Although numbers of *H. glycines* were not reduced below the damage threshold, parasitism by *H. rhossiliensis* could play an integral role in IPM of *H. glycines*. In a microplot study, *Pasteuria nishizawae* was effective in reducing the number of cysts and eggs/cyst of *H. glycines* (Atibalentja, Noel, Liao, & Gertner, 1998; Noel et al., 2005; Fig. 4.). Infested soil from the microplots was used to infest a long-term study on tillage. *Pasteuria nishizawae* was transferred successfully and in years 1, 2 and 3 was associated with management of *H. glycines* and increases in yield of soybean (Noel et al., 2005).

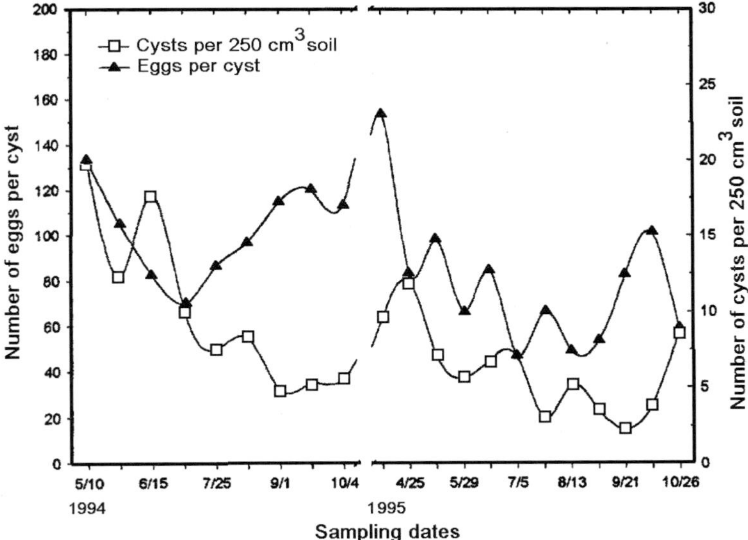

Figure 4. Population dynamics of Heterodera glycines *in a microplot study using soil naturally infested with* Pasteuria nishizawae *(from Atibalentja* et al.*, 1998).*

2. CONCLUSIONS

In the last decades, much progress has been made in the management of *H. glycines* for several aspects, including the application of resistant/tolerant germplasm and deployment of biological and/or agronomic management procedures. However, the intensive cropping system practiced in North and South America presents many challenges, and further research efforts are required for incorporation of IPM in a more sustainable and environment friendly production system.

REFERENCES

Atibalentja, N., Noel, G. R., Liao, T. F., & Gertner, G. Z. (1998). Population changes in *Heterodera glycines* and its bacterial parasite *Pasteuria* sp. in naturally infested soil. *Journal of Nematology, 30*, 81–92.

Bernard, R. L., Noel, G. R., Anand, S. C., & Shannon, J. G. (1988). Registration of 'Fayette' soybean. *Crop Science, 28*, 1028-1029.

Bond, J., Niblack, T. L., & Noel, G. R. (2006). Varietal information program for soybeans. SCN resistance ratings. http://web.aces.uiuc.edu/VIPS/v2CompVar/v2CompVarQ1.cfm?b=y&selPNV=N&selLoc=ALL&selYr=2006&selCompID=All&selMG=All&selType=All&cSO=1&nSO=1&selDT=1&cnt=0&inc=50&pg=1.

Chen, S. Y., Stienstra, W. C., Lueschen, W. E., & Hoverstad, T. R. (2001). Response of *Heterodera glycines* and soybean cultivar to tillage and row spacing. *Plant Disease, 85*, 311–316.

Chen, S. Y. (2004). Management with biological methods. In D. P. Schmitt, J. A. Wrather, & R. D. Riggs (Eds.), *Biology and management of soybean cyst nematode* (2nd ed., pp. 207–242). Marceline, Iowa: Schmitt and Associates of Marceline.

Chen, S. Y., & Reese, C. D. (1999). Parasitism of the nematode *Heterodera glycines* by the fungus *Hirsutella rhossiliensis* as influenced by crop sequence. *Journal of Nematology, 31*, 437–444.

Donald, P., Myers, R. F., Noel, G. R., Noffsinger, E. M., Norton, D. C., Robbins, R. T., et al. (1984). *Distribution of plant-parasitic nematode species in North America.* Hyattsville, MD: The Society of Nematologists.

Donald, P. A., Pierson, P. E., St. Martin, S. K., Sellers, P. R., Noel, G. R., MacGuidwin, A. E., et al. (2006). Assessing *Heterodera glycines*-resistant and susceptible cultivar yield response. *Journal of Nematology, 38*, 76–82.

Edwards, J. H., Thurlow, D. L., & Eason, J. T. (1988). Influence of tillage and crop rotation on yields of corn, soybean, and wheat. *Agronomy Journal, 80*, 76–80.

Hershman, D. E., & Bachi, P. R. (1995). Effect of wheat residue and tillage on *Heterodera glycines* and yield of doublecrop soybean in Kentucky. *Plant Disease, 79*, 631–633.

Koenning, S. R., & Barker, K. R. (1995). Soybean photosynthesis and yield as influenced by *Heterodera glycines*, soil type and irrigation. *Journal of Nematology, 27*, 51–62.

Koenning, S. R., Coble, H. D., Bradley, J. R., Barker, K. R., & Schmitt, D. P. (1998). Effects of a low rate of aldicarb on soybean and associated pest interactions in fields infested with *Heterodera glycines*. *Nematropica, 28*, 205–211.

Koenning, S. E., Schmitt, D. P., Barker K. R., & Gumpertz, M. L. (1995). Impact of crop rotation and tillage system on *Heterodera glycines* population density and soybean yield. *Plant Disease, 79*, 282–286.

Lawn, D. A., & Noel, G. R. (1986). Field interrelationships among *Heterodera glycines*, *Pratylenchus scribneri* and three other nematode species associated with soybean. *Journal of Nematology, 18*, 98–106.

Noel, G. R. (1984). Relating numbers of soybean cyst nematode to crop damage. *Proceedings of the Fifth Cyst Nematode Workshop,* 17–19.

Noel, G. R. (1987). Comparison of 'Fayette' soybean, aldicarb, and experimental nematicides for management of *Heterodera glycines* on soybean. *Journal of Nematology, 19*(4S), 84–88.

Noel, G. R. (1999). Sting nematodes and other nematode diseases. In G. L. Hartman, J. B. Sinclair, & J. C. Rupe (Eds.), *Compendium of soybean diseases.* (4th ed., pp. 50, 51, 57, 93 & 94). St. Paul, MN: APS Press.

Noel, G. R. (2004). Resistance in soybean to soybean cyst nematode, *Heterodera glycines*. In R. C. Cook & D. J. Hunt (Eds.), *Nematology monographs and perspectives (*Vol 2, pp. 253–261). Leiden, NV: Koninklijke Brill.

Noel, G. R., Atibalentja, N., & Bauer, S. (2005). Suppression of soybean cyst nematode populations by *Pasteuria nishizawae*. *Nematropica, 35*, 91 [Abstract].

Noel, G. R., & Edwards, D. I. (1996). Population development of *Heterodera glycines* and soybean yield in soybean-maize rotations following introduction into a noninfested field. *Journal of Nematology, 28*, 335–342.

Noel, G. R., & Wax, L. M. (2003). Population dynamics of *Heterodera glycines* in conventional tillage and no-tillage soybean/corn cropping systems. *Journal of Nematology, 35*, 104–109.

Schmitt, D. P., Barker, K. R., Noe, J. P., & Koenning, S. R. (1990). Repeated sampling to determine precision of estimating nematode population densities. *Journal of Nematology, 22*, 552–559.

Schmitt, D. P., & Noel, G. R. (1984). Nematodes parasites of soybean. In W. R. Nickle (Ed.), *Plant and Insect Nematodes*. New York: Marcel Dekker.

Shier, M. (2006). 2007 SCN resistant varieties. Retrieved February 1, 2007 from http://web.extension.uiuc.edu/livingston/reports/i120/index.html

Tyler, D. D., Chambers, A. Y., & Young, L. D. (1987). No-tillage effects on population dynamics of soybean cyst nematode. *Agronomy Journal, 79*, 799–802.

Wang, J., Donald, P. A., Niblack, T. L., Bird, G. W., Faghihi, J., Ferris, J. M., et al. (2000). Soybean cyst nematode reproduction in the north central United States. *Plant Disease, 84*, 77–82.

Wrather, J. A., & Koenning, S. R. (2006). Estimates of Disease effects on soybean yields in the United States 2003–2005. *Journal of Nematology, 38*, 173–180.

MARCELO E. DOUCET[1], PAOLA LAX[1] AND NORMA CORONEL[2]

THE SOYBEAN CYST NEMATODE *HETERODERA GLYCINES* ICHINOHE, 1952 IN ARGENTINA

[1]*Laboratorio de Nematología. Centro de Zoología Aplicada, Casilla de correo 122, (5000) Córdoba, Argentina*
[2]*Estación Experimental Agroindustrial Obispo Colombres, (4101) Tucumán, Argentina*

Abstract. The damage caused by the soybean cyst nematode *Heterodera glycines* in Argentina is revised, together with possible management strategies. This nematode emerged in the last decade as one of the most important parasites of soybean in the region. The histopathology, population dynamics and dispersal of *H. glycines* and the effects of selected resistant lines and varieties are discussed. Among management tools, observations on the effects of soil fungi on *H. glycines* densities suggest a possible role of natural suppressiveness and biological control. Recommended actions include development of detailed knowledge about the occurrence of the nematode, trainings of experts, development of sound outreach programs and extension activities, evaluation of the nematode incidence on yields and research related to soybean resistance and possible exploitation of natural antagonists.

1. INTRODUCTION

Soybean (*Glycine max* (L.) Merrill) is the crop of greatest production in Argentina. It is the main commodity export, both unprocessed and processed (Weskamp, 2006), generating the greatest foreign currency income in the country (Giorda, 1997). Soybean expansion in the country started in the 1970s, and since then cultivated areas continued to expand up to 15.2 M ha in the 2005/2006 cropping season, with a production estimated in 40.5 M TM and an average yield of 2660 kg/ha. A 127% increase in the cultivated area has been estimated for the last 10 year-period, whereas production increased by 237% (Rossi, 2006). Transgenic cultivars comprise 98% of cultivars planted (Secretaría de Agricultura, Ganadería, Pesca y Alimentos). The Pampas region concentrates 83.68% of the soybean-producing area, the remaining 16.32% being distributed in other provinces (Fig. 1). Among pests affecting soybean yield, several soil nematode species have a particular incidence. To date, 32 genera and 25 phytophagous species (according to the classification criteria of Yeates, Bongers, Goede, Freckman, & Georgieva, 1993) have been detected in soybean-producing areas in Argentina (Table 1), among which *Meloidogyne* spp. and *Heterodera glycines* stand out.

A. Ciancio & K. G. Mukerji (eds.), Integrated Management and Biocontrol of Vegetable and Grain Crops Nematodes, 127–148.

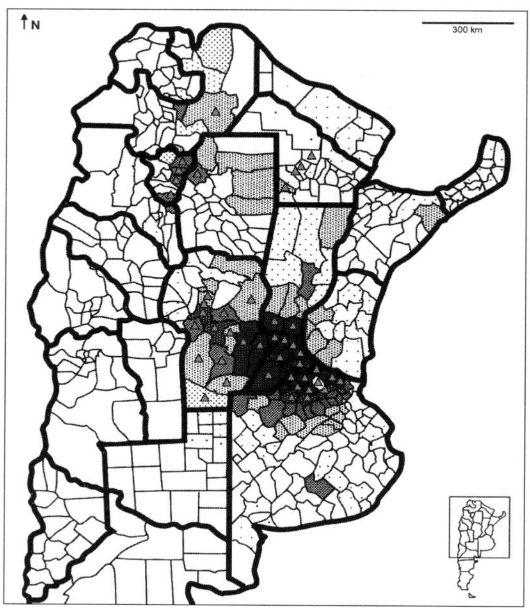

Figure 1. Distribution of the soybean-producing area in Argentina and of the nematode Heterodera glycines. *Each point corresponds to a 500-ha area or fraction larger than 100 ha; triangles indicate the departments (in provinces) where the nematode has been detected. (Adapted from Giorda, 1997).*

2. *HETERODERA GLYCINES*

The soybean cyst nematode (SCN), *Heterodera glycines*, poses the greatest problem to the normal soybean crop development and yield in the principal producing countries worldwide (Ichinohe, 1988; Noel, 1992, 1993; Noel, Mendes, & Machado, 1994; Young, 1996; Kim, Riggs, Robbins, & Rakes, 1997).

This species has a high reproductive potential, an efficient dispersal mechanism, populations with marked variability, and the capacity to survive in the soil for several years in the absence of a suitable host (Riggs & Wrather, 1992).

Up to the end of the 1990s, the greatest problems in soybean crops caused by nematodes in Argentina were attributed only to species of the genus *Meloidogyne* (March, Ornaghi, Beviacqua, Astorga, & Marcellino, 1985; Doucet & Racca, 1986; Doucet, 1993). However, during the 1997/98 cropping season fields strongly attacked (Fig. 2) by representatives of the genus *Heterodera* were detected in different localities of the provinces of Córdoba and Santa Fe (Doucet et al., 1997; Baigorri, Vallone, Giorda, Chaves & Doucet, 1998). Subsequent studies showed that the species attacking soybean was *H. glycines* (Doucet & Lax, 1999).

Table 1. Genera and species of nematodes detected in soils cultivated with soybean in Argentina.

Nematode	References
Aorolaimus sp.	Doucet and Racca, 1986
Aphelenchoides sp.	Doucet and Racca, 1986; Niquén Bardales and Venialgo Chamorro, 2004
Aphelenchus sp.	Niquén Bardales and Venialgo Chamorro, 2004
Boleodorus sp.	Doucet and Racca, 1986
Belonolaimus longicaudatus	Doucet, 1998
Cactodera sp.	Doucet and Racca, 1986; Costilla and Coronel, 1998a, 1999a
C. cacti	Costilla and Coronel, 1998a, 1999a
Coslenchus sp.	Doucet and Racca, 1986
Criconema sp.	Doucet and Racca, 1986
Criconemella ornata	Doucet, 1998
Ditylenchus sp.	Niquén Bardales and Venialgo Chamorro, 2004
Filenchus sp.	Niquén Bardales and Venialgo Chamorro, 2004
Globodera sp.	Niquén Bardales and Venialgo Chamorro, 2004
Helicotylenchus sp.	Doucet and Racca, 1986; Costilla and Coronel, 1999a; Vega and Galmarini, 1970; Coronel, Ploper, Jaldo, and Gálvez, 2004a; Niquén Bardales and Venialgo Chamorro, 2004; Fuentes, Salines, Distéfano, Gilli, and Mazzini, 2006
H. dihystera	Doucet, 1998
H. multicinctus	Costilla and Coronel, 1998a
Hemicycliophora sp.	Niquén Bardales and Venialgo Chamorro, 2004
Heterodera sp.	Baigorri et al., 1998
H. glycines	Costilla and Coronel, 1998a, 1999a, 1999b; Doucet and Lax, 1999
Ibipora sp.	Niquén Bardales and Venialgo Chamorro, 2004
Lelenchus sp.	Doucet and Racca, 1986
Macroposthonia sp.	Coronel et al., 2004a
Meloidogyne sp.	Astorga, Ornaghi, March, Beviacqua, and Marcellino, 1984; Ornaghi, Boito, and López, 1984; Coronel et al., 2004a; Niquén Bardales and Venialgo Chamorro, 2004
M. arenaria	Baigorri et al., 2004
M. incognita	Ornaghi, Beviacqua, March, and Astorga, 1981; March et al., 1985; Doucet and Racca, 1986; Doucet and Pinochet, 1992; González, Cap, and Andreozzi, 1983; Costilla and Coronel, 1998a; 1999a; Fuentes et al., 2006
M. incognita (race 2)	Ornaghi et al., 1984
M. javanica	Doucet and Racca, 1986; Doucet and Pinochet, 1992; Costilla, 1994; Costilla and Coronel, 1998a, 1999a; Fuentes et al., 2006
Neopsilenchus sp.	Niquén Bardales and Venialgo Chamorro, 2004
Nothocriconema sp.	Fuentes et al., 2006
Paratrichodorus sp.	Doucet and Racca, 1986
P. minor	Doucet, 1998
Paratylenchus sp.	Doucet and Racca, 1986; Niquén Bardales and Venialgo Chamorro, 2004
Pratylenchus sp.	Doucet and Racca, 1986; Costilla and Coronel, 1999a; Coronel et al., 2004a; Niquén Bardales and Venialgo Chamorro, 2004

(continued)

Table 1 (continued)

Nematode	References
P. agilis	Doucet and Racca, 1986
P. brachyurus	Doucet and Racca, 1986; Costilla and Coronel, 1998a
P. delattrei	Doucet and Racca, 1986
P. goodeyi	Costilla, 1994
P. hexincisus	Doucet and Racca, 1986; Doucet, 1988
P. neglectus	Doucet and Racca, 1986
P. penetrans	Doucet and Racca, 1986
P. pratensis	Costilla, 1994
P. scribneri	Doucet and Racca, 1986
P. vulnus	Doucet and Racca, 1986
P. zeae	Doucet and Racca, 1986; Costilla, 1994; Costilla and Coronel, 1998a
Psilenchus sp.	Niquén Bardales and Venialgo Chamorro, 2004
Rotylenchus sp.	Doucet and Racca, 1986
Scutellonema sp.	Doucet, 1998; Coronel et al., 2004a; Niquén Bardales and Venialgo Chamorro, 2004
Trichodorus sp.	Coronel et al., 2004a
Tylenchorhynchus sp.	Doucet and Racca, 1986; Coronel et al., 2004a
Tylenchus sp.	Coronel et al., 2004a; Niquén Bardales and Venialgo Chamorro, 2004
Xiphidorus sp.	Doucet and Racca, 1986
X. saladillensis	Luc and Doucet, 1990
Xiphinema sp.	Doucet and Racca, 1986; Coronel et al., 2004a; Niquén Bardales and Venialgo Chamorro, 2004
X. "americanum" (sensu lato)	Luc and Doucet, 1990
X. index	Doucet, 1998
X. krugi	Luc and Doucet, 1990

The knowledge that species had appeared in Brazil in the 1991/92 cropping season should have been a well-founded reason to organize a systematic exploratory search in the main soybean-producing areas in the country, especially in those areas where seeds from Brazil were used. However, because of the lack of preventive policies, it was only in the 1997/98 cropping season that authorities and producers became aware of the occurrence of this nematode in the country. At that time, damage became evident, possibly because of the high population densities.

Since that moment, surveys in different fields devoted to this crop were performed, with the aim of defining infested and pathogen-free areas. At present, the nematode can be considered to have a wide distribution (Fig. 1), occurring in several localities in the provinces of Buenos Aires, Córdoba, Chaco, Salta, Santa Fe, Santiago del Estero and Tucumán (Lax, Doucet, & Lorenzo, 2001). It has been mentioned that between 500,000 ha (Wrather et al., 2001) and 1,500,000 ha would be infested by this pest within the core soybean-producing area of the country (Gamundi, Borrero, & Lago, 2002).

Figure 2. Infested field with Heterodera glycines *in the province of Córdoba, Argentina.*

Several highly infested plots have been detected in some localities, very often exceeding the damage threshold levels accepted by other countries (Table 2).

Table 2. Density of Heterodera glycines *(cysts) in soils, for different provinces of Argentina.*

Province	*Cysts/100 g soil*	*Reference*
Buenos Aires	1–134	Gamundi et al., 1999a
Chaco	2–6	Gamundi et al., 1999a
Córdoba	1–352	Serrano et al., 1999
	17–362	Doucet, Lax, Giayetto, and Di Rienzo, 2001a
	1–403	Gilli et al., 2000
Salta	2	Gamundi et al., 1999a
Santa Fe	0–197	Gamundi, Lago, Bacigalupo, Borrero, and Riart, 1999b
	1–299	Gamundi et al., 1999a
Tucumán	15–300	Costilla and Coronel, 1998a
	1–400	Coronel and Costilla, 1999
	1–498	Coronel, Costilla, Ploper, and Devani, 2001
	3-67	Gamundi et al., 1999a

3. LIFE CYCLE

While the information available indicates that the different stages of the nematode's cycle in Argentina fit with those reported for the species, there are no accurate data on the duration of each stage, for the different soybean-producing areas. The areas are distributed in phytogeograhical regions with very different climatic, pedological and plant cover characteristics. It is possible therefore that, depending on the sites and the maturity groups for the soybean cultivars that can be cultivated, the species' life cycle length may vary considerably. It has been mentioned that the cycle length mainly depends on soil temperature (Lawn & Noel, 1990; Noel, 1993), and, in some places, three to seven generations may occur in a single cropping season (Burrows & Stone, 1985).

Research conducted in an infested plot in the province of Córdoba revealed that by the end of the cropping season, a great number of seeds fell to the ground during harvest. Driven by favourable climate conditions, new soybean plants grew some time later and continued to develop until early winter. Although the development of the aerial part of these plants is limited, analyses performed in the roots demonstrated the presence of white females with egg masses and cysts, indicating that the nematode development continues, even at very low soil temperatures in July (6.2–10ºC). Thus, although environmental conditions are not optimal, a population in these conditions would complete at least one new generation between autumn and early winter (Lax, 2003). It would therefore be very important to gather accurate information, in the country, about the situation in other soybean cultivated areas where the SCN is present.

4. POPULATIONS AND RACES

As in other phytophagous nematode populations, several aspects of *H. glycines* show a considerable variability. The evaluation and comparison of certain morphometrical characters corresponding to second-stage juveniles, males, females, and cysts showed significant differences among populations corresponding to different races (Lax & Doucet, 2001a, 2001b, 2001c, 2002; Lax, 2003).

Some level of intra and inter-population variability were recorded in morphological characters at these stages, not being possible to detect characters that clearly differentiated the populations examined (Lax & Doucet, 2001a, 2001b, 2001c, 2002). Although SCN was detected in numerous localities of different provinces, only a few populations have been properly characterized with respect to both the parameters used to identify the species, and different aspects of its biology.

The species has several races, each one showing a particular preference for certain soybean cultivars. To date, races 1, 3, 5 (HG type 2, 5, 7), 6, 9, and 14 (Table 3) have been recognized by the differential host test (Riggs & Schmitt, 1988; Niblack et al., 2002) in the country, race 3 appearing as the most frequent.

Moreover, the levels of variability were analysed in two populations of the nematode from different provinces, using Random Amplified Polymorphic DNA markers, with the aim of evaluating the genetic population structure of this species. This study revealed an important degree of genetic differentiation between both populations, probably as a consequence of limited gene flow between them or because each population was under different management practices at its site of origin (Lax, Rondan Dueñas, Gardenal, & Doucet, 2004).

Table 3. Races of Heterodera glycines *in different localities of Argentine provinces where soybean is cultivated.*

Province	Locality	Race	References
Córdoba	Córdoba	3	Gilli et al., 2000
	Corral de Bustos	3	Gilli et al., 2000
	Laguna Larga	1, 3	Silva et al., 1999
	Marcos Juárez	3	Dias, Silva, and Baigorri, 1998; Silva et al., 1999
Santa Fe	Armstrong	3	Silva et al., 1999
	Las Parejas	3	Dias et al., 1998; Rossi, Nari, and Cap, 1999
	Maciel	3	Zelarrayán, 1999
	Santa Fe	3	Gilli et al., 2000
	Tortugas	3, 14	Dias et al., 1998
	Totoras	1	Silva et al., 1999
	Totoras	3	Rossi et al., 1999; Zelarrayán, 1999
Tucumán	Burruyacú	3	Coronel and Costilla, 1999
	Cruz Alta	5 (HG type 2, 5, 7)	Coronel and Costilla, 1999; Coronel, Ploper, Devani, and Galvez, 2005
	Garmendia	3	Coronel et al., 2001
	La Virginia	3	Zelarrayán, 1999
	Los Hardoy	3	Zelarrayán, 1999
	Los Pereyra	5	Coronel et al., 2001
	San Agustín	5	Coronel et al., 2001
	San Luis de las Casas Viejas	6	Coronel et al., 2001
	Taruca Pampa	6	Coronel et al., 2001
Unknown	Unknown	9	Gamundi et al.,2002

5. HOST-NEMATODE RELATIONSHIPS

5.1. Histological Alterations

Histological changes induced by this nematode species in roots of cultivars usually used in the soybean-producing areas in Argentina were analyzed. Depending on the cultivar, such alterations may be of major or minor intensity. In susceptible plants, both in the primary and lateral roots, cell necrosis in the cortex can be observed, which is produced by the lesion generated by juveniles penetrating and migrating and by females subsequently establishing inside tissues. Females are located near the central cylinder, which is the most affected region because of the parasite feeding site (syncytium) formed here. Syncytia can occasionally project to the cortex or form in that region. When two or more syncytia develop in neighbouring areas, they can occupy almost the entire central cylinder, which can be reduced to a small group of cells. Functional syncytia (Fig. 3), associated with females with egg masses, are composed of cells of variable shape and show different degrees of hypertrophy. Cell walls are slightly thickened and show interruptions in some sectors, allowing cytoplasm movement in neighbouring cells.

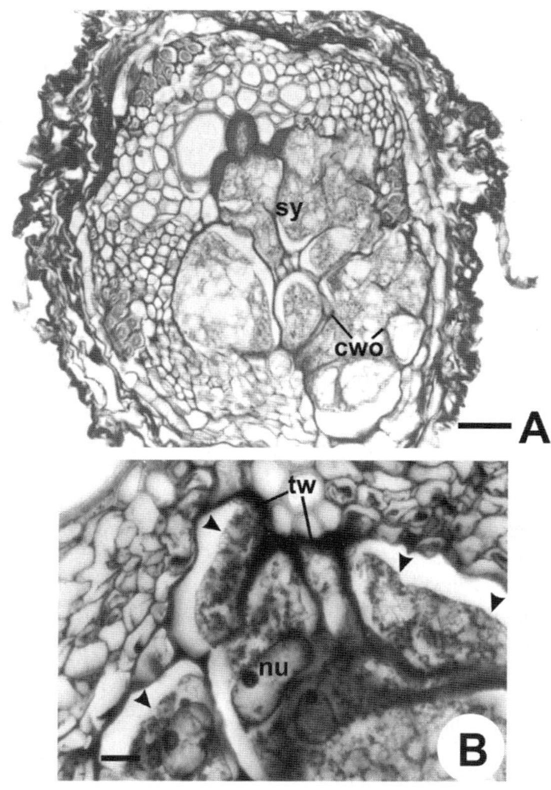

Figure 3. Histological section of root of soybean parasited by a Heterodera glycines *population, race 1. (A) Cultivar Pioneer 9501, functional syncytium in the central cylinder. (B) Cultivar Asgrow 5401, details of the syncytium showing cell plasmolysis (arrows). Abbreviations: cwo: cell wall opening; nu: nucleus; tw: thickened wall; sy: syncytium. Scale bars: A = 25 μm; B = 10 μm. (Adapted from Tordable, 2004).*

The syncitia cytoplasm is dense, of granular aspect, with different degrees of vacuolization and with large-sized nuclei of spherical, ovoid, or lobulated shape and prominent nucleoli. Walls adjacent to xylem vessels usually show a good development of irregular wall thickenings and of rugose texture. Non-functional syncytia can also be distinguished (Fig. 4), frequently associated with cysts. Dead cells show different levels of disorganization and are non-functional (Tordable, 2004).

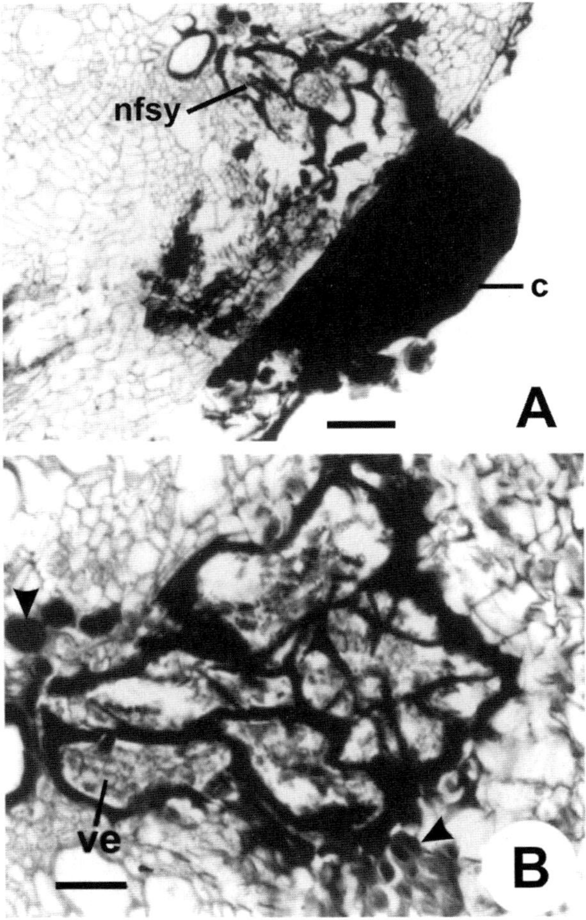

Figure 4. Histological section of root of the soyben cultivar Pioneer 9501, parasitised by a Heterodera glycines *population, race 1. (A) Non-functional syncytium related to a nematode cyst. (B) Detail of non-functional syncytium showing occluded cells of xylem and modified fibres (arrows). Abbreviations: c; cyst; nfsy: non-funcional syncytium; ve: vesicle. Scale bars: A = 50 μm; B = 25 μm. (Adapted from Tordable, 2004).*

Nodules of *Bradyrhizobium japonicum* with functional syncytia were occasionally detected at an early differentiation stage. The nematode-induced syncytium was located in the cortical area of the nodule, close to the conductive tissue (Fig. 5). It was composed of slightly hypertrophied cells of thin, partially fragmented walls.

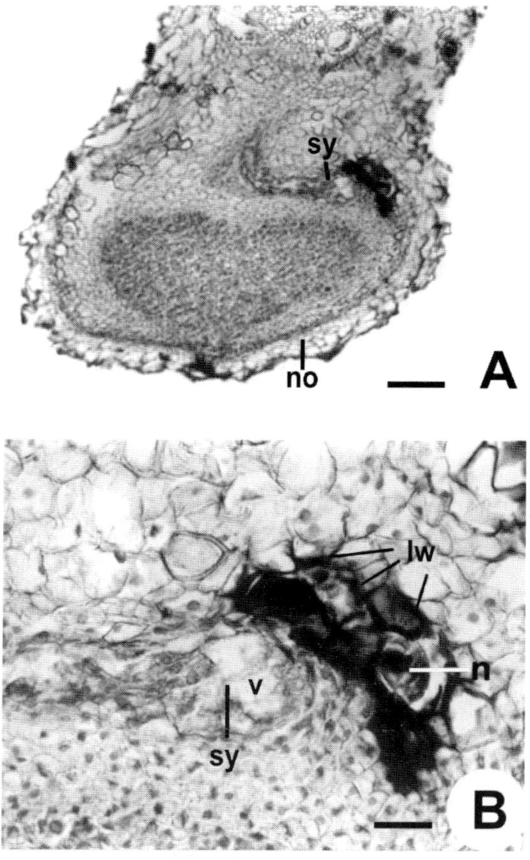

Figure 5. Histological section of a nodule of Bradyrhizobium japonicum *in the soybean cultivar Pioneer 9501, associated to a* Heterodera glycines *population, race 1. (A) Functional syncytium in the nodule cortical zone. (B) Detail of syncytium. Abbreviations: lw: lignified wall; n; nematode; no: nodule; sy: syncytium; v: vacuole. Scale bars: A = 50 μm; B = 25 μm. (Adapted from Tordable, 2004).*

The cytoplasm was granulose, slightly dense, and with a large vacuole in some cells, and nuclei were spherical, somewhat hypertrophied, with prominent nucleoli (Tordable, Lorenzo, & Doucet, 2003). Infested nodules degenerate rapidly and represent one of the reasons for plants losing their efficiency in atmospheric nitrogen fixation (Khan, 1993).

Specific studies showed that, under local crop conditions, both transgenic cultivars – Asgrow 5435 RG (Tordable et al., 2003), Pioneer 94B01, Asgrow 5901 (Tordable, 2004) – and non-transgenic cultivars – Pioneer 9501 (Doucet, Tordable, & Lorenzo,

2003), Asgrow 5401 (Lorenzo, Doucet, & Tordable, 1999), Asgrow 5402, Asgrow 5409, NK 641, Torcacita (Tordable, 2004) – were susceptible to a *H. glycines* population of race 1, in the province of Córdoba. It should be noted that Asgrow 5435 RG, indicated as resistant to SCN in soybean seed catalogues (Nidera Semillas, 1998/1999), behaved as a susceptible cultivar (Tordable et al., 2003). Only one of the cultivars evaluated, Asgrow 5153 (non-transgenic) showed to be resistant to this population.

The histopathological analysis conducted in this cultivar did not reveal the presence of syncytia in the roots. Some modifications were observed (cell necrosis, wall thickenings, hypertrophied nuclei being among the main ones), not directly related, however, to the presence of juveniles or another stage of *H. glycines* (Tordable, 2004).

5.2. Response of Cultivars to the Attack of SCN

In Argentina, works on genetic improvement of soybean are carried out with the aim of gathering more informations on yield, tolerance to herbicides, behaviour in the presence of different pathogens, among the main aspects (Baigorri et al., 2004). Since 1980, the Instituto Nacional de Tecnología Agropecuaria (INTA) has been leading the National Network for the Evaluation of Soybean Cultivars (Red Nacional de Evaluación de Cultivares de Soja).

Annual studies are conducted to evaluate yield, agronomic characteristics, and response of new cultivars of different maturity groups to certain pests, as well as new cultivars' development in different soybean-producing regions in the country (north, northern pampas, and southern pampas) (Baigorri et al., 2000, 2004; Fuentes et al., 2005). To date, the degree of susceptibility to SCN has been evaluated only for some cultivars in nursery, and the response of some of them has been evaluated in the field only exceptionally. The races to which the cultivars considered are susceptible/resistant were not considered either.

The response of a group of soybean cultivars usually cultivated in the northwestern region of the country to the action of SCN populations belonging to races 5 and 6 was evaluated in greenhouse conditions. Most of these cultivars behaved as susceptible and moderately susceptible to these races, whereas few of them showed to be moderately resistant (Coronel et al., 2001; Coronel, Ploper, Devani, Galvez, & Jaldo, 2003a; Coronel & Devani, 2006). Table 4 summarizes the cultivars that were found to be moderately resistant (MR) and resistant (R) to populations of different races of SCN in Argentina.

Table 4. Soybean cultivars indicated as moderately resistant (MR) and resistant (R) to Heterodera glycines *populations in Argentina. Nematode race is indicated in parenthesis.* [*]*Names of cultivars correspond to those given by authors in the references.*

Cultivar[*]	Response to SCN	References
A 4602 RG	R (3, 14)	Vallone, 2002
A 5428 RG	R (3)	Vallone, 2002
A 6401 RG	MR (6)	Coronel et al., 2003a

(continued)

Table 4. (continued)

Cultivar*	Response to SCN	References
A 6040 RG	MR (6); R (3)	Vallone, 2002; Coronel et al., 2003a
A 6411 RG	MR (5)	Coronel and Devani, 2006
AW 4902 RR	R (3, 14)	Vallone, 2002
AW 5581 RR	R (3, 14)	Vallone, 2002
Asgrow 4004	R (3)	Gamundi, Bodrero, Mendez, Lago, and Lorenzatti, 1998
Asgrow 4501 RG	MR (3)	Gamundi et al., 1998
Asgrow 5153	R (3)	Gamundi et al., 1998
Asgrow 5435 RG	R (3)	Gamundi et al., 1998
Asgrow 5634 RG	R (3)	Gamundi et al., 1998
Asgrow 6001	MR (3)	Lago, Riart, Gamundi, Bodrero, and Midula, 1999
Asgrow 6444 RG	R (3)	Gamundi et al., 1998
Asgrow 6445 RG	MR (3, 14)	Gamundi et al., 1998
Anta 82 RR	MR (6)	Coronel et al., 2003a
Campeona 6.4	MR (5, 6); R (3, 14)	Coronel et al., 2001, 2003a; Vallone, 2002
NK Coker 6738 SC	MR (6); R (1, 3)	Gamundi et al., 1998
NK Coker 8.1	MR (5, 6); R (1, 3)	Gamundi et al., 1998; Coronel et al., 2001, 2003a
Forrest	MR (5)	Coronel et al., 2001
GR 80	MR (6)	Coronel et al., 2003a
Hartwig	R (5)	Coronel et al., 2001; Coronel, Ploper, Jaldo, and Gálvez, 2004b
Leo 56 RR	MR (3)	Ferrarotti and Roldán, 1999
Leo 81 RR	MR (3)	Ferrarotti and Roldán, 1999
Leo 240 RR	R (3)	Ferrarotti and Roldán, 1999
Leo 558 RR	R (3)	Ferrarotti and Roldán, 1999
Leo 10074	MR (3)	Ferrarotti and Roldán, 1999
Leo 10364	R (3)	Ferrarotti and Roldán, 1999
Leo 10448	R (3)	Ferrarotti and Roldán, 1999
Leo 10560	MR (3)	Ferrarotti and Roldán, 1999
Leo 10888	MR (3)	Ferrarotti and Roldán, 1999
Mágica 7.3 RR	MR (6); R (3)	Vallone, 2002; Coronel et al., 2003a
Maleva 42 RR	R (3; 14)	Vallone, 2002
Maravilla 45 RR	MR (6)	Coronel et al., 2003a
Nativa 46 RR	R (3)	Vallone, 2002
Nueva Maria 55 RR	MR (5)	Coronel and Devani, 2006
PI 437654	R (5)	Coronel et al., 2001, 2004b
Pioneer 94B01	R (3, 14)	Gamundi et al., 1998
Pioneer 94B41	R (3, 14)	Gamundi et al., 1998
Pioneer 9492	R (3, 14)	Gamundi et al., 1998
Qaylla RR	MR (5, 6)	Coronel et al., 2003a; Coronel and Devani, 2006
TJ 2046	R (1)	Vallone, 2002
TJ 2055 RR	MR (5)	Coronel and Devani, 2006
TJ 2070 RR	MR (5)	Coronel and Devani, 2006

5.3. Relationship of SCN with the Environment

It is well known that population density of *H. glycines* fluctuates throughout the development of the soybean crop in response to different environmental factors, temperature and humidity having the greatest influence (Schmitt, 1992). A single three-year study has been conducted on the basis on this information, which evaluated the possible correlations between the factors mentioned, a given sequence of susceptible cultivars, and the density of the nematode population in an infested field in Córdoba province (Lax & Doucet, 2004).

Only a small proportion of the variation detected in the number of individuals (second-stage juveniles and cysts) in soil was explained by temperature and humidity. With respect to the relationship with the cultivars used, density of individuals gradually decreased over time, although cultivars were susceptible plants. Observations in cysts revealed that many of them had eggs attacked by fungi (Fig. 6) which behaved as biological control agents of this SCN population. Eggs of the nematode attacked by pathogenic fungi were also observed in fields of the province of Tucumán (Costilla & Coronel, 1998a, 1998b). Another situation of a decrease in a *H. glycines* population was also observed during 1998/2004 in some plots of the province of Santa Fe, but the apparent causes were not mentioned (Lago, Borrero, & Gamundi, 2006).

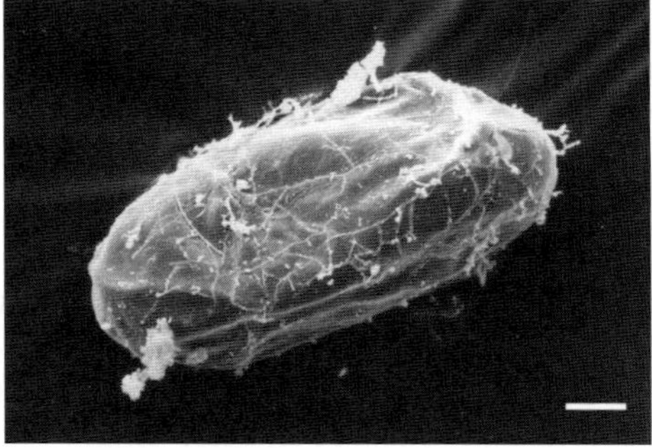

Figure 6. Scanning electron microscopy. Egg of Heterodera glycines *attacked by fungi. Scale bar: 10 μm.*

6. LOSSES

Losses produced in Argentina by *H. glycines* have been estimated only in some areas. In the localities of Marcos Juárez (province of Córdoba) and Totoras (province of Santa Fe) losses of 1400 and 2300 kg/ha, respectively, were recorded for the 1997/1998 cropping season (Gamundi et al., 2002). Losses of 30% were estimaded in plots located in central Córdoba (Doucet & Lax, 1999). For localities of southern Córdoba, yield reductions up to 64% were recorded (Gamundi, 1999). For the same cropping season and in the same provinces, a yield reduction of more than 58% was recorded, and production losses of about 55,000 tons were estimated in the country (Wrather et al., 2001). However, the values indicated were taken from publications that did not provide the methods used for losses evaluation.

The yield of different soybean commercial cultivars was evaluated in microplots infested with SCN in the northwestern region of the country (Coronel et al., 2003b). In all cases yields were significantly higher in non-infested than in infested plots. Losses ranged from 24.4 to 36.0%, and tended to be greater in the highest yielding cultivars (Table 5).

Table 5. Evaluation of yields of commercial soybean cultivars in non-infested and infested plots of the province of Tucumán, Argentina (From Coronel, Ploper, Jaldo, Galvez, & Devani, 2003b).

Cultivar	*Yield (kg/ha)*			
	Non-infested plots	*Infested plots*	*Yield losses (kg/ha)*	*% Yield losses*
A 8000 RG	2958	1991	967**	32.1
A 6040 RG	2806	1809	997*	35.5
Munasqa RR	2767	1771	996**	36.0
A 5409 RG	2722	1827	895*	32.9
Anta 82 RR	2208	1556	652*	29.5
Qaylla RR	2133	1533	600**	28.1
Mágica 7.3 RR	1974	1493	481*	24.4

*, ** Means significantly different between non-infested and infested plots at $P \leq 0.05$ and $P \leq 0.01$, respectively.

7. MANAGEMENT

Heterodera glycines has some characteristics that make its management complex: a short life cycle, high reproductive potential, populations with remarkable physiological variability, resistance stage (cyst) and very efficient dispersal mechanisms (wind, water, animals, contaminated seed bags). In Argentina, management of the species should be based on an integrated control approach. This involves having a thorough knowledge of all those aspects related to the biology and ecology of both the parasite and the host. It is surprising to note that in some works addressing integrated control of soybean pests, nematodes are not included (Satorre et al., 2003; Aragón & Flores, 2006).

7.1. Early Nematode Detection

Knowing if SCN is present or absent in a soil prior sowing is the most important step in the management of possible problems. If the species is absent, it will be possible to cultivate any crop suitable for the region. If it is present, depending on the race and nematode population density, it will be necessary to take measures in order to protect the plants. Thus, early detection of the pathogen is crucial. Given the particular life cycle of this species, infective juveniles, cysts, and possibly males may be found in the soil. While the first two stages appear more frequently, cysts will always be detected. The soil analysis prior to seeding will determine what steps must be followed.

Different soil extraction techniques are used for nematodes, depending on the stages to be detected. Vermiform specimens (second-stage juveniles and males) are obtained by centrifugal-flotation technique (Jenkins, 1964), whereas cysts are extracted by the traditional flotation technique (Fenwick, 1940). While detecting cysts is of greatest concern, studies on population dynamics require observation and counting of the three stages mentioned. This requires having soil samples to extract the different stages separately, which in turn involves investing a considerable amount of time and effort. Combining both methods mentioned offers the possibility of extracting the three stages from the same soil sample, with the advantage of ensuring a more efficient cyst recovery (Doucet, Lax, Di Rienzo, & Suarez, 2001b).

7.2. Identification of Races

Once the presence of *H. glycines* in a field is confirmed, the race must be defined with the aim of selecting resistant/tolerant cultivars that may be cultivated in that field. Although 6 races have been identified so far, given the extensive area devoted to soybean in Argentina the information available is probably very limited. Since studies on this particular topic were carried out at specific sites and at a given moment, they do not necessarily show the real situation in the area. Besides, it is important to evaluate the race identity regularly at a given place, since the continuous use of the same resistant cultivar may, through the insurgence of a selection pressure, bring about the appearance of a different race that may cause severe damage to such plant.

7.3. Chemical Control

Given the vast areas devoted to soybean crops in Argentina, the use of nematicides is economically unfeasible (besides producing environmental pollution and other problems). However, an assay was carried out in the province of Santa Fe to evaluate the action of carbofuran and aldicarb, in a plot attacked by the nematode and cultivated with susceptible cultivars. The results showed that applying the product did not provide increased yields (Gamundi et al., 1998).

7.4. Crop Rotation

When the presence of one or several nematode races is detected, measures must be taken so that crop production is not reduced significantly. This means that soybean will coexist with the SCN. Hence the SCN population density will have to be reduced to levels that do not produce severe damage to the crop, which is accomplised by using different medium/long term crop rotation schemes in the cultivated plots (non-host plants, resistant or susceptible cultivars).

Among the crops most commonly grown in the Pampas, the following have been mentioned as non-hosts: maize, sorghum, sunflower, peanut, cotton, sugarcane, alfalfa and safflower (Gamundi et al., 1998). Observations in three infested plots showed that density of viable cysts decreased by 75% with maize, 90% with sorghum, and 80% with sunflower followed by pasture seeding in autumn (Gamundi et al., 1998).

The incidence of different crop sequences on the cyst population was evaluated in different fields of Santa Fe (Lago et al., 1999; Méndez, Bodrero, Gamundi, & Lago, 1999; Gamundi et al., 2002, 2004). The sequences of susceptible soybean-sunflower, maize-susceptible soybean, susceptible soybean-moderately resistant soybean reduced the amount of cysts in soil (Lago et al., 1999). When SCN density exceeds 10 viable cysts/100 gr of soil it is considered necessary to plant a non-host crop (corn or resistant soybean) for a period longer than a cropping season, in order to reduce the SCN population to levels below the damage treshold. When populations are lower, alternating a non-host crop or resistant soybean with susceptible soybean or with the wheat-susceptible soybean sequence contributes to a decrease of cysts in soil, and therefore yields are not so much affected. Wheat/susceptible soybean double crop was the most effective way to place the susceptible materials in the rotations to avoid the appearance of new races and the growth of the population. The lowest cyst infestation levels were achieved with maize as a monoculture (Gamundi et al., 2002; Gamundi, Bodrero, Lago, Mendez, & Capurro, 2004). However, it is very important to remember that once the nematode was established in a field, its eradication has not been possible, so far.

7.5. Preventive Measures Against Cyst Dispersal

Preventing this species' spread represents, as in many other cases, one of the most efficient strategies to tackle the problems it causes. Among the measures that are advocated, no-till is usually considered. Through this system much less soil is removed than through conventional tillage, reducing wind-borne cyst dispersal (Andrade & Asmus, 1997).

Using nematode-free seeds is another preventive measure of great importance. This system contributes to the preservation of crop development and to prevent the pathogen from establishing in the plots to be cultivated. If the SCN is already present, its population may be increased with the presence of new cysts (of the same or another race) together with seeds contaminated with soil particles.

In Argentina, soybean is frequently cultivated on roadsides along roads heavily used even by soybean seed transporting trucks. This practice is strongly discouraged, since it may contribute to the appearance of new infestation sources as seeds and nematode-contaminated soil fall on the ground. To avoid transport of infested soil, agricultural machinery used in a given plot should be carefully washed before being used in another plot. Using hired machinery without taking into account this recommendation is one of the many causes of nematode dispersal and colonization of new soils.

Preventing *H. glycines* dispersal requires not only practices like the ones mentioned above, but also implies the implementation of efficient phytosanitary control programs by responsible institutions, both at the national and provincial levels, as well as strong awareness of the problem by technicians and producers.

8. CONCLUSIONS

H. glycines is widely dispersed in Argentina. However, little attention has been given to basic aspects of the species biology, mode of recognition, races, behaviour, natural antagonists, ecology, interactions with other nematode species, such as *Meloidogyne* spp., among other aspects that are essential for selecting more efficient management strategies. Enhancing the knowledge of these aspects is of great importance, since this crop has been expanded enormously in the country in the last years. Different soybean cultivars show a notable adaptation capacity to diverse climate and pedological conditions; therefore, colonization of new soils by the crop is very likely to be coupled with dispersal of this nematode species.

Therefore, the following actions are highly recommended:

(1) *Knowledge of the areas free of and infested by SCN.* Thus, cultivating new cultivars of high yield in non-infested soils and defining necessary rotation schemes in infested soils will be feasible. This implies developing specific prospective programs in all those localities where the crop is cultivated in the country. Furthermore, this previous knowledge will prevent soils attacked by the nematode from being cultivated with cultivars that contribute to increase the pathogen population.

(2) *Training of experts in the subject.* It is very important that experts are very well trained in this issue. The occurrence of morphologically similar species like *H. trifolii* in the country highlights the need for training people who are provided with all the necessary information in the subject, to perform sound and reliable diagnosis.

(3) *Development of sound outreach programs.* Despite the time elapsed since this nematode was first detected and the damage it produces to the soybean crop, some technical experts and producers still do not have basic robust information on the topic. The institutions responsible for this problem must ensure efficient outreach mechanisms to reverse this situation.

(4) *Evaluating the magnitude of the impact of SCN on the crop production.* Determining the nematode incidence on yields, both at the local and national levels, is very important. Up to the present, results published on this issue would correspond to specific trials lacking a rigorous experimental design; hence, the information obtained would have a relative significance. The estimation of losses caused by the nematode in numerous regions of the country requires adapting different methods according to the places, the most widely used cultivars, the race, and the pathogen population density.

(5) *Conducting research related to possible natural antagonists.* As it has been already indicated, some natural antagonists naturally occurring in the soil (e.g., fungi) would contribute to the reduction of the nematode population density. The diversity that characterizes the different environments where the crop is cultivated suggests that there may be other natural antagonists that could be employed with the same purpose.

Contrarily to what was assumed, population densities of the SCN seem to have decreased in different soybean-producing areas since it was detected. Up to the present, the action of antagonists is considered to be the reason of such reduction. However, whether this trend will continue or not is uncertain. For this reason, the nematode occurrence in the principal soybean-producing areas in the country is a serious threat, which implies the urgent need for further basic and applied research aimed at gathering useful information to manage the problem.

REFERENCES

Andrade, P. J. M., & Asmus, G. L. (1997). Disseminação do nematóide de cisto da soja (*Heterodera glycines*) pelo vento durante o preparo do solo. *Nematologia Brasileira*, *21*, 98–99.

Aragón, J. R., & Flores, F. (2006). Control integrado de plagas en soja en el sudeste de Córdoba. In *Soja: Actualización 2006*. Informe de Actualización Técnica, N° 30, 19–23.

Astorga, E. M., Ornaghi, J. A., March, G. J., Beviacqua, J. E., & Marcellino, J. (1984). Estudios de difusión e incidencia de nematodos causantes de agalla, *Meloidogyne* spp. en cultivos de soja. *Oleico*, *25*, 45.

Baigorri, H., Vallone, S. D., Giorda, L. M., Chaves, E., & Doucet, M. (1998). Estrategias para el control de una superplaga. *Supercampo*, *41*, 78–81.

Baigorri, H., Croatto, D., Piatti, F., Fossati, J., Bodrero, M., Macor, L., et al. (2000). Resultados de la red nacional de evaluación de cultivares de soja en la región pampeana norte Campaña 1997/2000. *Soja: resultados de ensayos de la campaña 1999/2000*. INTA: Estación Experimental Marcos Juárez, N° 63, 9–19.

Baigorri, H., Robinet, H., Iriarte, L., Galván, M., Lizondo, M., Erazu, L., et al. (2004). Resultados de la red nacional de evaluación de cultivares de soja en la región pampeana norte y pampeana sur. Campañas 2001/02 a 2003/04. In: H. Baigorri & L. Segura (Eds.), *Soja: Actualización 2004*. Información para Extensión, N° 89, 1–19.

Burrows, P. R., & Stone, A. R. (1985). *Heterodera glycines*. *C.I.H. Descriptions of plant parasitic nematodes*, Commonwealth Agricultural Bureaux, Set. 8 Nº 118, 4 pp.

Coronel, N. B., & Costilla, M. A. (1999). *Heterodera glycines* Ichinohe, en soja en el noroeste argentino. Resúmen de trabajos y conferencias presentadas. Mercosoja 99. 21–25 Junio, Rosario, Argentina. 35–36.

Coronel, N. B., Costilla, M. A., Ploper, L. D., & Devani, M. (2001). El nematodo del quiste de la soja: distribución, caracterización de poblaciones y reacción de cultivares en el NOA. *Avance Agroindustrial*, *22*, 33–37.

Coronel, N. B., Ploper, L. D., Devani, M. R., Galvez, M. R., & Jaldo, H. (2003a). Reaction of soybean cultivars to *Heterodera glycines* race 6 in Tucumán, Argentina, 2002. Biological and cultural test for control of plant disease (online.) Report 18:P005. DOI: 10.1094/BC 18. St. Paul, MN, USA: The American Phythopatological Society [Abstract].

Coronel, N. B., Ploper, L. D., Jaldo, H., Galvez, M. R., & Devani, M. R. (2003b). Yields of soybean cultivars in microplots infested with soybean cyst nematode, 2001/2002. Biological and cultural test for control of plant disease (online.) Report 18:P004. DOI: 10.1094/BC 18. St. Paul, MN, USA: The American Phythopatological Society [Abstract].

Coronel, N. B., Ploper, L. D., Jaldo, H. E., & Gálvez, M. R. (2004a). Nematodes associated with soybean in Tucumán, Argentina. Abstracts VII World Soybean Research Conference, IV International Soybean Processing and Utilization Conference, III Congresso Brasileiro de Soja. February 29–March 5, Foz do Iguassu, PR, Brazil. 166 [Abstract].

Coronel, N. B., Ploper, L. D., Jaldo, H. E., & Gálvez, M. R. (2004b). Reaction of soybean cultivars to *Heterodera glycines* race 5. Abstracts VII World Soybean Research Conference, IV International Soybean Processing and Utilization Conference, III Congresso Brasileiro de Soja. February 29–March 5, Foz do Iguassu, PR, Brazil. 93 [Abstract].

Coronel, N. B., Ploper, L. D., Devani, M. R., & Galvez, M. R. (2005).Caracterización de una población de *Heterodera glycines* mediante el Test Tipo HG. Resúmenes XIII Congreso Latinoamericano de Fitopatología. Córdoba, Argentina. 492 [Abstract].

Coronel, N. B., & Devani, M. (2006). Reacción de cultivares de soja al nematodo del quiste de la soja (*Heterodera glycines* Ichinohe, 1952) raza 5. Resúmenes XII Jornadas Fitosanitarias Argentinas 28–30 Junio, Catamarca, Argentina. 153–154.

Costilla, M. A. (1994). Nematodes endoparásitos asociados al cultivo de soya, maíz y poroto en el noroeste argentino. *Nematropica, 24*, 77 [Abstract].

Costilla, M. A., & Coronel, N. B. (1998a). Presencia del nematode del quiste *Heterodera glycines* Ichinohe, 1952, y su importancia en el cultivo de soja en el Noroeste Argentino. *Avance Agroindustrial, 74*, 24–28.

Costilla, M. A., & Coronel, N. B. (1998b). Presencia de nematodos Heteroderidae en cultivo de soja en el noroeste argentino. *Nematropica, 28*, 124 [Abstract].

Costilla, M. A., & Coronel, N. B. (1999a). Ocurrence of nematodes Heteroderidae in soils from northwestern Argentina, with special reference to soybean cyst nematode, *Heterodera glycines* Ichinohe, 1952. Proceedings World Soybean Research Conference VI. Chicago, Illinois, USA. 628–629.

Costilla, M. A., & Coronel, N. B. (1999b). Presencia del nematode del quiste *Heterodera glycines*, Ichinohe 1952, y su importancia en el cultivo de la soja en la provincia de Tucumán. Resúmenes X Jornadas Fitosanitarias Argentinas. 7–9 Abril, Jujuy, Argentina. 103.

Dias, W. P., Silva, J. F. V., & Baigorri, H. (1998). Survey of *Heterodera glycines* races in Argentina. *Nematropica, 28*, 125 [Abstract].

Doucet, M. E., & Racca, R. R. (1986). Estudio preliminar de los nematodos fitófagos asociados al cultivo de soja (*Glycine max* (L.) Merrill) en la provincia de Córdoba, República Argentina. *IDIA*, 50–56.

Doucet, M., E. & Pinochet, J. (1992). Ocurrence of *Meloidogyne* spp. in Argentina. *Journal of Nematology*, 24, 765–770.

Doucet, M. E. (1993). Consideraciones acerca del género *Meloidogyne* Goeldi, 1887 (Nemata: Tylenchida) y su situación en Argentina. Asociaciones y distribución. *Agriscientia,* VX, 63–80.

Doucet, M. E., Baigorri, H. E. J., Giorda, L. M., Ornaghi, J., Chaves, E., & Vallone, S. D. de. (1997). Nematodos. In: L. M. Giorda & H. E. Baigorri (Eds.), *El cultivo de la soja en Argentina.* INTA, Centro Regional Córdoba, (EEA Manfredi-EEA Marcos Juárez). Coordinación Subprograma Soja. Editorial Editar, San Juan, Argentina, 291–308.

Doucet, M. E. (1998). Soil nematodes associated with soybean in the Argentina Republic. *Nematropica, 28*, 126 [Abstract].

Doucet, M. E., & Lax, P. (1999). Presence of the nematode *Heterodera glycines* (Nematoda: Tylenchida) associated with soybean in Argentina. *Nematology, 1*, 213–216.

Doucet, M. E., Lax, P., Giayetto, A., & Di Rienzo, J. (2001a). Diferente apariencia en cultivares de soja atacados por *Heterodera glycines* en Córdoba, Argentina. *Nematologia Mediterranea, 29*, 59–62.

Doucet, M. E., Lax, P., Di Rienzo, J. A., & Suarez, R. (2001b). Higher efficiency in processing soil samples for the assessment of *Heterodera glycines* populations. *Nematology, 3*, 377–378.

Doucet, M. E., Tordable M. del C., & Lorenzo, E. (2003). Response of the soybean cultivar Pioneer 9501 to *Heterodera glycines*. *Nematologia Mediterranea, 31*, 15–20.

Fenwick, D. W. (1940). Methods for the recovery and counting of cysts of *Heterodera schachtii* from soil. *Journal of Helminthology, 18*, 155–172.

Ferrarotti, J. S. R., & Roldán, D. (1999). Interacción de líneas avanzadas de soja con *Heterodera glycines* raza 3. Resúmen de trabajos y conferencias presentadas. Mercosoja 99. 21–25 Junio, Rosario, Argentina. 9–10.

Fuentes, F. H., Robinet, H., Iriarte, L., Galván, M., Lizondo, M., Erazzú, L., et al. (2005). Resultados de la red nacional de evaluación de cultivares de soja en la región norte, pampeana norte y pampeana sur. Campañas 2002/2003 a 2004/05. In *Soja: Actualización 2005*. Información para Extensión, N° 97, 1–29.

Fuentes, F., Salines, L., Distéfano, S., Gilli, J., & Mazzini, P. (2006). Evaluación de cultivares de soja frente al nematodo de la agalla. In *Soja: Actualización 2006*. Informe de Actualización Técnica, N° 3, 11–14.

Gamundi, J. C., Bodrero, M., Mendez, J. M., Lago, M., & Lorenzatti, S. (1998). Algunos aspectos biológicos y de manejo del "Nematodo del quiste de la soja" *Heterodera glycines*. Soja para mejorar la producción. Campaña 1997/1998. N° 8. Estación Experimental Agropecuaria Oliveros INTA.

Gamundi, J. C. (1999). Resultados de las investigaciones sobre *Heterodera glycines* en Argentina. Resúmen de trabajos y conferencias presentadas. Mercosoja 99. 21–25 Junio, Rosario, Argentina. 31–34.

Gamundi, J. C., Lago, M., Bodrero, M., Capalbo, J., Serrano, R., Piatti, F., et al. (1999a). Relevamiento del "nematodo del quiste de la soja" *Heterodera glycines* en las principales áreas sojeras de la Argentina. Resúmen de trabajos y conferencias presentadas. Mercosoja 99. 21–25 Junio, Rosario, Argentina. 47–48.

Gamundi, J. C., Lago, M., Bacigalupo, S., Bodrero, M., & Riart, S. (1999b). Infestación del "nematodo del quiste de la soja" en el centro-sur de Santa Fe, determinado por el servicio de identificación de la E. E. A. Oliveros-INTA. Resúmen de trabajos y conferencias presentadas. Mercosoja 99. 21–25 Junio, Rosario, Argentina. 39–40.

Gamundi, J. C., Bodrero, M., & Lago, M. E. (2002). Nematodo del quiste de la soja. *IDIA* XXI, 3, 83–87.

Gamundi, C., Bodrero, M., Lago, M., Mendez, M., & Capurro, J. (2004). Influence of crop sequences on the population of "soybean cyst nematode", *Heterodera glycines*, in the South of Santa Fe, Argentina. Abstracts VII World Soybean Research Conference, IV International Soybean Processing and Utilization Conference, III Congresso Brasileiro de Soja. February 29–March 5, Foz do Iguassu, PR, Brasil. 165 [Abstract].

Gilli, J., Gadbán, L., Baigorri, H., Croatto, D., Piatti, F., & Guerra, G. (2000). Nematodos parásitos del cultivo de soja en Argentina. Estación Experimental Agropecuaria Marcos Juárez. *Soja: Resultados de Ensayos de la campaña 1999/2000*. Información para Extensión, N° 63, 47–53.

Giorda, L. M. (1997). La soja en Argentina. In L. M. Giorda & H. E. Baigorri (Eds.), *El cultivo de la soja en Argentina*. INTA, Centro Regional Córdoba, (EEA Manfredi-EEA Marcos Juárez). Coordinación Subprograma Soja. Editorial Editar, San Juan, Argentina. 11–26.

González, S., Cap, G., & Andreozzi, R. (1983). Interacción entre micorrizas arbúsculo-vesiculares y *Meloidogyne incognita* en soja. V Jornadas Fitosanitarias Argentinas. 7–9 Setiembre, Rosario, Argentina [Abstract].

Ichinohe, M. (1988). Current research on the major nematode problems in Japan. *Journal of Nematology, 20*, 184–190.

Jenkins, W. R. (1964). A rapid centrifugal-flotation technique for separating nematodes from soil. *Plant Disease Reporter*, *48*, 692.

Kim, D. G., Riggs, R. D., Robbins, R. T., & Rakes, L. (1997). Distribution of races of *Heterodera glycines* in the Central United States. *Journal of Nematology*, *29*, 173–179.

Khan, M. W. (1993). *Nematode interactions* (377 pp.). London, New York: Chapman and Hall.

Lago, M., Riart, S., Gamundi, J. C., Bodrero, M., & Midula, G. (1999). Efecto de cultivos no hospedantes y cultivares de soja, sobre el nematodo del quiste de la soja: análisis de casos en lotes de productores. Resúmen de trabajos y conferencias presentadas. Mercosoja 99. 21–25 Junio, Rosario, Argentina. 33–34.

Lago, M., Bodrero, M. L., & Gamundi, J. C. (2006). Relevamiento del "nematodo del quiste de la soja" *Heterodera glycines* en el sur de la provincia de Santa Fe. III Congreso de Soja del Mercosur. 27–30 Junio, Rosario, Argentina. 449–450.

Lawn, D. A., & Noel, G. R. (1990). Effects of temperature on competition between *Heterodera glycines* and *Pratylenchus scribneri* on soybean. *Nematropica, 20*, 57–69.

Lax, P., & Doucet, M. E. (2001a). *Heterodera glycines* Ichinohe, 1952 (Nematoda: Tylenchida) from Argentina: 1. Morphological and morphometrical characterisation of second-stage juveniles. *Nematology*, *3*, 165–170.

Lax, P., & Doucet, M. E. (2001b). *Heterodera glycines* Ichinohe, 1952 (Nematoda: Tylenchida) from Argentina: 2. Morphological and morphometrical characterisation of males. *Nematology, 3*, 543–549.

Lax, P., & Doucet, M. E. (2001c). *Heterodera glycines* Ichinohe, 1952 (Nematoda: Tylenchida) from Argentina: 3. Morphological and morphometrical characterisation of females. *Nematology, 3*, 699–704.

Lax, P., Doucet, M. E., & Lorenzo, E. (2001). El nematodo *Heterodera glycines* Ichinohe, 1952 y su importancia para el cultivo de soja en Argentina. *Boletín de la Academia Nacional de Ciencias*, 66, 87–98.

Lax, P., & Doucet, M. E. (2002). *Heterodera glycines* Ichinohe, 1952 (Nematoda: Tylenchida) from Argentina: 4. Morphological and morphometrical characterisation of cysts. *Nematology*, *4*, 783–789.

Lax, P. (2003). Estudio de poblaciones del nematodo *Heterodera glycines*, 1952 (Nematoda: Tylenchida) asociado al cultivo de soja. Tesis Doctoral. Facultad de Ciencias Exactas, Físicas y Naturales, Universidad Nacional Córdoba, Córdoba, Argentina. 131 pp.

Lax, P., Rondan Dueñas, J. C., Gardenal, C. N., & Doucet, M. E. (2004). Genetic variability estimated with RAPD-PCR markers in two populations of *Heterodera glycines* Ichinohe, 1952 (Nematoda: Heteroderidae) from Argentina. *Nematology*, *6*, 13–21.

Lax, P., & Doucet, M. E. (2004). Fluctuación de *Heterodera glycines* (Nematoda) en un lote cultivado con soja. II Reunión Binacional de Ecología (XXI Reunión Argentina de Ecología, XI Reunión de la Sociedad de Ecología de Chile). 31 Octubre-5 Noviembre, Mendoza, Argentina. 401 [Abstract].

Lorenzo, E., Doucet M. E., & Tordable, M. del C. (1999). Reacción de un cultivar de soja al ataque de *Heterodera glycines* (Nematoda: Tylenchida). *Kurtziana*, *27*, 285–291.

Luc, M., & Doucet, M. E. (1990). La Familia Longidoridae Thorne, 1935 (Nemata) en Argentina. *Revista de Ciencias Agropecuarias*, *7*, 19–25.

March, G. J., Ornaghi, J. A., Beviacqua, J. E., Astorga, E. M., & Marcellino, J. (1985). Comportamiento de cultivares de soja frente al nematodo causante de agallas *Meloidogyne javanica* (Treub) Chiitwood. *IDIA*, 441–444, 70–77.

Méndez, J. M., Bodrero, M. L., Gamundi, J. C., & Lago, M. (1999). Influencia de la secuencia de cultivos sobre la población del "nematodo del quiste de la soja", *Heterodera glycines*. Resúmen de trabajos y conferencias presentadas. Mercosoja 99. 21–25 Junio, Rosario, Argentina. 41–42.

Niblack, T. L., Arelli, P. R., Noel, G. R., Opperman, C. H., Orf, J. H., Schmitt, D. P., et al. (2002). A revised clasiffication scheme for genetically diverse populations of *Heterodera glycines*. *Journal of Nematology*, *4*, 279–288.

Nidera Semillas. (1998/1999). *Genética de avanzada*. Campaña 1998/1999. 20 pp.

Niquén Bardales, E. C., & Venialgo Chamorro, C. A. (2004). Presencia de nemátodos, en cultivos de soja en suelos con labranza cero, del Dpto. 9 de Julio, Chaco. Universidad Nacional del Nordeste, Comunicaciones Científicas y Tecnológicas. Available at http://www.unne.edu.ar/Web/cyt/com2004/5-Agrarias/A-019.pdf.

Noel, G. R. (1992). History, distribution, and economics. In R. D. Riggs & J. A.Wrather (Eds.), *Biology and management of the soybean cyst nematode*. APS PRESS, USA, 1–13.

Noel, G. R. (1993). *Heterodera glycines* in soybean. *Nematologia Brasileira*, *17*, 103–121.

Noel, G. R., Mendes, M. L., & Machado, C. C. (1994). Distribution of *Heterodera glycines* races in Brazil. *Nematropica, 24*, 63–68.

Ornaghi, J. A., Beviacqua, J. E., March, J., & Astorga, E. M. (1981). *Meloidogyne incognita* (Kofoid & White) Chitwood en soja: su identificación, evaluación del grado de infestación, rango natural de hospedantes y asociación a *Sclerotium rolfsii* Sacc. IV Jornadas Fitosanitarias Argentinas. 19–21 Agosto, Córdoba, Argentina. 52.

Ornaghi, J. A., Boito, G. T., & López, A. B. (1984). Identificación de especies y razas de diferentes poblaciones de *Meloidogyne* en cultivos de soja en el Dto. Río Cuarto. *Oleico*, *25*, 45 [Abstract].

Riggs R. D., & Schmitt, D. P. (1988). Complete characterization of the race scheme for *Heterodera glycines*. *Journal of Nematology*, *20*, 392–395.

Riggs, R. D., & Wrather, J. A. (1992). *Biology and management of the soybean cyst nematode* (186 pp.). USA: APS Press.

Rossi, R. L., Nari, C., & Cap, G. (1999). Identificación de razas del nematodo del quiste en dos localidades de la provincia de Santa Fe. Resúmen de trabajos y conferencias presentadas. Mercosoja 99. 21–25 Junio, Rosario, Argentina. 37–38.

Rossi, R. L. (2006). Impactos recientes de la soja en Argentina. III Congreso de Soja del Mercosur. 27–30 Junio, Rosario, Argentina. 115–118.

Satorre, E. H. (2003). *El libro de la soja*. Servicios y Marketing Agropecuario, Buenos Aires, Argentina. 264 pp.

Schmitt, D. P. (1992). Populations dynamics. In R. D. Riggs & J. A. Wrather (Eds.), *Biology and management of the soybean cyst nematode*. APS PRESS, USA, 51–59.

Serrano, R., Piatti, F., Guerra, G., Baigorri, H., Croatto, D., Gilli, J., et al. (1999). Situación del nematodo del quiste de la soja en la provincia de Córdoba. Estación Experimental Agropecuaria Marcos Juárez. *Soja: Resultados de Ensayos de la campaña 1998/1999*. Información para Extensión, N° 59, 42–47.

Silva, J. F. V., Asmus, G. L., Dias, W. P., Baigorri, H., Serrano, R., & Ferraz, L. C. B. (1999). Levantamento de raças do nematóide de cisto da soja na Argentina. Anais Congresso Brasileiro de Soja, 17–20 Maio, Londrina, Brazil. 448 [Abstract].

Tordable, M. del C., Lorenzo, E., & Doucet M. E. (2003). Histopathology of Asgrow 5435 RG soybean roots induced by *Heterodera glycines* race 1, in Córdoba, Argentina. *Nematologia Brasileira, 27*, 55–60.

Tordable, M. del C. (2004). Evaluación del comportamiento del cultivares de soja (*Glycine max* (L.) Merrill) ante la acción del nematodo fitoparásito *Heterodera glycines* Ichinohe, 1952. Tesis Doctoral. Facultad de Ciencias Exactas, Físico-Químicas y Naturales, Universidad Nacional de Río Cuarto, Córdoba, Argentina. 183 pp.

Vallone, S. D. de. (2002). Enfermedades de la soja. *IDIA* XXI, N° 3, 68–74.

Vega, E., & Galmarini, H. R. (1970). Reconocimiento de los nematodes que parasitan los cultivos hortícolas de los departamentos de San Carlos y Tunuyán, Mendoza (Argentina). *IDIA, 272*, 17–41.

Weskamp, A. (2006). La importancia de ser parte de los mercados para la efectiva comercialización de los granos: "todos los granos a los mercados". III Congreso de Soja del Mercosur. 27–30 Junio, Rosario, Argentina. 49–53.

Wrather, J. A., Anderson, T. R., Arsyad, D. M., Tan, Y., Ploper, L. D., Porta-Puglia, A., et al. (2001). Soybean disease loss estimates for the top ten soybean producing-countries in 1998. *Canadian Journal of Plant Pathology*, *23*, 115–121.

Yeates, G. W., Bongers, T., Goede, R. G. M. De, Freckman, D. W., & Georgieva, S. S. (1993). Feeding habits in soil nematode Families and Genera: An outline for soil ecologists. *Journal of Nematology*, *25*, 315–331.

Young, L. D. (1996). Yield loss in soybean caused by *Heterodera glycines*. *Journal of Nematology*, *28*, 604–607.

Zelarrayán, E. L. (1999). Determinación de la raza presente en cuatro poblaciones del nematodo del quiste de la soja *Heterodera glycines*. Resúmen de trabajos y conferencias presentadas. Mercosoja 99. 21–25 Junio, Rosario, Argentina. 6–8.

8

A. F. ROBINSON

NEMATODE MANAGEMENT IN COTTON

USDA – ARS, 2765 F & B Road, College Station, TX 77845, USA

Abstract. The five most important cotton-producing countries are China, United States, India, Pakistan, and Brazil. There are many other important cotton producing regions in Asia, Australia, Africa and the Americas. Cotton is grown entirely in tropical, subtropical, and warm-temperature climates, and the major nematodes of cotton are well adapted to warm environments. Globally, the most damaging nematodes of cotton are *Meloidogyne incognita* races 3 and 4 and *Rotylenchulus reniformis*. These nematodes are of concern in the United States, India, Pakistan, Egypt and Brazil. Additional nematodes of major importance in relatively restricted areas include *Hoplolaimus columbus* and *Belonolaimus longicaudatus* in the southeastern United States and *Pratylenchus brachyurus* in Brazil. *Meloidogyne incognita* frequently is involved in a cotton disease complex with Fusarium wilt that has far more impact on the crop than the nematode or the fungus alone. Until very recently, the primary strategies used for nematode management in cotton have been the application of fumigants and cholinesterase inhibitors, rotation with *Zea mays*, *Arachis hypogaea* or *Glycine max* and incorporation of soil amendments. The primary concern over *P. brachyurus* in Brazil is its potential to damage *Z. mays* or *G. max* grown in rotation with cotton. Promising seed treatments containing avermectin or harpin proteins have recently become available. Several cultivars resistant to *Melodogyne incognita* races 3 and 4 have been released. Currently there is intense research toward the introgression of resistance to *R. reniformis* into upland cotton, *Gossypium hirsutum* from other *Gossypium* species. During the last two years DNA markers for major genes for resistance to *Meloidogyne incognita* and *Rotylenchulus reniformis* have been discovered in upland cotton and offer great potential in the development of resistant cultivars suitable for the wide range of growing conditions where cotton is produced.

1. INTRODUCTION

The most economically important nematode pathogens of cotton are *Meloidogyne incognita* (host races 3 and 4) and *Rotylenchulus reniformis*. Other species known to damage cotton include *Belonolaimus longicaudatus*, *Hoplolaimus columbus*, *Pratylenchus brachyurus* and *Meloidogyne acronea*. Additional nematodes associated with cotton include *Hoplolaimus aegypti*, *H. galeatus*, *H. indicus*, *H. seinhorsti*, *Longidorus* sp., *Paratrichodorus* sp., *Rotylenchulus parvus* (Louw, 1982), *Scutellonema* sp., and *Xiphinema* sp. Previous reviews of cotton nematodes include Blasingame (1994); Bridge (1992); Da Ponte, Jilho, Lordello, and Lordello (1998); Garber, DeVay, Goodel, and Roberts (1996); Heald and Orr (1984); Koenning et al. (2004); Lawrence and McLean (2001); Mueller and Lewis (2001); Overstreet and McGawley (2001); Robinson et al. (2001); Sasser (1972); Starr (1998); Starr and Page (1990); Thomas and Kirkpatrick (2001) and Veech (1984).

A. Ciancio & K. G. Mukerji (eds.), Integrated Management and Biocontrol of Vegetable and Grain Crops Nematodes, 149–182.

Starr's (1998) review provides an excellent, detailed comparison of the biology of the four major nematodes of cotton.

For those unfamiliar with nematodes, they comprise the animal phylum Nematoda and are commonly known as roundworms. They are unsegmented, multicellular animals with several hundred neurons and several simple organ systems (Maggenti, 1981). Most are microscopic. More than 10 000 species of nematodes occupy a diverse variety of terrestrial, marine, and parasitic niches. They are the most ubiquitous of all multicellular terrestrial animals. In cultivated fields, virtually every liter of soil will contain many nematodes, and usually several species. Most nematodes are vermiform (worm-like) throughout life, but parasitic stages of some species are swollen or even globose. Plant-parasitic nematodes have a stylet with which they perforate plant cells and ingest nutrients. Nematode stylets are minute, and most have a bore small enough to serve as a bacterial filter. Most plant parasitic nematodes are obligate plant parasites and can only feed on roots or foliage of vascular plants.

2. GEOGRAPHICAL DISTRIBUTION AND ECONOMIC IMPACT

Meloidogyne incognita (the southern root-knot nematode) and *Rotylenchulus reniformis* (the common reniform nematode) occur in tropical, subtropical, and warm temperate soils throughout most of the world, generally within 35° of the equator (Robinson et al., 2001; Taylor & Sasser, 1978). One or both species are present in most cotton-producing regions and are considered to be serious problems in cotton production wherever they occur.

In the United States, *M. incognita* is found on cotton in all cotton-producing states, and *R. reniformis* occurs only in states east of New Mexico (Heald & Robinson, 1990; Koenning et al., 2004; Lawrence & McLean, 1996; Robinson, 2007). There is current concern in the United States regarding recent increase in incidence and severity of *R. reniformis* infestations in the central cotton belt of the United States (Blasingame & Patel, 1987; Gazaway & McLean, 2003; Overstreet & McGawley, 2000; Robinson, 2007). The two remaining economically important cotton nematodes in the United States, *H. columbus* (the Columbia lance nematode) and *B. longicaudatus* (the sting nematode), both occur primarily in sandy soils in the Coastal Plain regions extending across North Carolina, South Carolina, and Georgia.

Meloidogyne incognita has been reported on cotton in numerous areas in Brazil, Africa, the Middle East, India, and China. A related root-knot nematode, *M. acronea*, is known to damage cotton in the Shire valley of Malawi (Africa) and in Cape Providence, South Africa (Starr & Page, 1990). A very high incidence (94%) of *Pratylenchus brachyurus* in cotton is of great concern in Mato Grosso do Sul, Brazil, due to its potential impact on corn and soybean grown in rotation with cotton (Da Silva et al., 2004). Of 184 samples collected from 15 'municípios' (roughly comparable to counties) in Mato Grosso do Sul State in Brazil, 28% and 17% were positive for *M. incognita* and *R. reniformis*, with 45% and 32% of those samples, respectively, above the damage threshold (Asmus, 2004).

Worldwide cotton yield losses due to nematodes were estimated to be 10.7% by Sasser and Freckman (1987), which was equivalent to 1.9 million metric tons of cotton lint worth $US 4 billion at 1987 prices. United States losses were estimated by the National Cotton Council of America (Blasingame, 2006) to be 1 178 000 bales (4.7%), valued at approximately $US 550 million.

Figure 1. Meloidogyne incognita *damage to cotton, Georgia, U.S.A. (courtesy of R. F. Davis).*

3. SYMPTOMATOLOGY

3.1. Meloidogyne spp.

Distributions of *M. incognita* within fields usually are uneven and scattered (Blasingame, 1994; Thomas & Kirkpatrick, 2001). Infested areas (Fig. 1) are oblong in the direction of cultivation and are often 7–13 m long and 3–10 m wide. Infested areas within a field typically suffer 75–100% damage while other areas in the same field will show no symptoms. Earliest and greatest damage occurs on plants under water stress in the sandiest parts of a field. Severely infected plants often are half the height of normal plants, tend to appear nitrogen-deficient and wilt under drought stress, several days before symptoms appear in uninfected plants. Co-infection with *M. incognita* and the Fusarium wilt fungus, *Fusarium oxysporum* f. sp. *vasinfectum*, often kills plants; plants infected with only *M. incognita* rarely die. Symptoms of *M. incognita* are usually expressed first in sandier areas of the field.

The most obvious symptoms are galls on secondary roots (Blasingame, 1994; Shepherd & Huck, 1989; Thomas, 2001). Galls on cotton are typically smaller than on tomato and okra. Taproots and secondary roots often branch prematurely or abort, forming terminal galls (Fig. 1). Root systems are deficient in fibrous feeder roots and typically grow less than half as deep as root systems of healthy cotton plants. The galls mimic natural physiological sinks and compete with the rest of the plant for photosynthetic assimilates (Abrão & Mazzafera, 2001; Esau, 1977; Jones & Northcote, 1972).

Meloidogyne acronea, which has been reported only from Africa, has been studied less than *M. incognita*, but field symptoms and damage are generally similar to that observed for *M. incognita*. A notable difference is that the sedentary adult females of *M. acronea* protrude from the root surface, unlike the adult females of *M. incognita*, which generally are embedded in the root tissue (Page, 1985).

3.2. Rotylenchulus reniformis

Distributions and stunting symptoms within fields tend to be irregular in new infestations but uniform in old ones, so that the existence of a problem is less obvious (Blasingame, 1994; Heald & Heilman, 1971; Lawrence & McLean, 2001). In the United States, yield losses in infested fields commonly are less than 10% but may exceed 50% if the crop has been water-stressed. Under ideal growing conditions, it is possible for infected plants to exhibit no foliar symptoms at all. More often, however, plants stop rapid growth at the three or four leaf stage (Fig. 2), as if stunted hormonally, leaves take on a light or off green color, typical of potassium deficiency, and flowering and fruit set are delayed two nodes up the main stem.

Figure 2. Cotton fields infested with Rotylenchulus reniformis *in Alabama (A) and Louisiana (B) (courtesy W. S. Gazaway and C. Overstreet).*

Nematodes along roots can be detected with the unaided eye only by observing clumps of sand grains adhering to the gelatinous egg masses surrounding sedentary gravid females protruding from the root surface (Heald & Orr, 1984). Dirt particles remain after gently rinsing roots, making roots look dirty. Otherwise, nematode-infected root systems can appear more or less normal on casual inspection, perhaps with some loss of secondary roots, but without any galls, severely stunted taproots, or forked secondary roots characteristic of root-knot nematode infection on cotton. The primary symptoms, which are visible only with a microscope following special tissue preparation and staining, are extensive hypertrophy and dysfunction of the endodermal and pericyclic cell layers enclosing the vascular cylinder, and consequent blockage of water and nutrient uptake (Cohn, 1973).

3.3. *Hoplolaimus columbus*

This nematode occurs in sandy soils in the southeastern United States. Distributions within fields, like those of *M. incognita*, are uneven and scattered. Infested areas are usually oblong in the direction of cultivation, 7–17 m long and 3–10 m wide (Blasingame, 1994; Mueller & Lewis, 2001). In the United States, yield losses within infested fields are typically 10–25% but may exceed 50% in sandy fields under water stress. Severely infected plants may be stunted 50% or more and wilt under drought stress several days prior to uninfected plants. Leaves often exhibit nutrient deficiency symptoms – in particular, slight to moderate chlorosis characteristic of nitrogen deficiency. *Hoplolaimus columbus* typically feeds on the root tip of the radicle immediately following seed germination and on secondary roots as they develop, resulting in severely stunted root systems that penetrate only 7–10 cm deep, compared to 30 cm in healthy plants.

Other species of *Hoplolaimus* are not generally considered to be pathogenic to cotton but comparative studies are lacking. *Hoplolaimus galeatus* and *H. magnistylus* appear to be commonly encountered in cotton in the United States. Careful studies in Arkansas showed that *H. magnistylus* was not a serious pest of cotton (Robbins, McNeely, & Lorenz, 1998).

3.4. *Belonolaimus longicaudatus*

This nematode is limited primarily to soils containing more than 85% sand (Esser, 1976; Graham & Holdeman, 1953; Robbins & Barker, 1974). Heavy infestations in cotton fields cause stunted, chlorotic growth followed by premature wilting and senescence. Root systems are poorly developed and have dark, sunken lesions along the root axis that can spread laterally to girdle the root and cause it to break off. Although *B. longicaudatus* has a much more restricted geographical distribution than *M. incognita*, *R. reniformis* and *H. columbus*, it is a devastating parasite of cotton where it occurs in the United States, often killing all or most of the plants in large areas of infested fields (Sasser, 1972). It is particularly damaging in fields where *F. oxysporum* f. sp. *vasinfectum* is present.

4. BIOLOGY AND EPIDEMIOLOGY

4.1. *Meloidogyne incognita and M. acronea*

Important differences between root-knot nematodes of cotton are noted, as appropriate.

4.1.1. *Life Cycle*

Meloidogyne incognita and *M. acronea*, like other plant-parasitic nematodes, have four juvenile stages between the egg and the adult (Taylor & Sasser, 1978). Molting in *M. incognita* occurs between stages as the nematode increases in size. One molt occurs within the egg and three subsequently. The second-stage juvenile (J2) that emerges from the egg is the motile, infective stage. It is vermiform, 0.30–0.4 mm. long, developmentally arrested, non-feeding and contains a large reserve of lipid stores within the cells of the intestine, which sustains it while in the soil.

After hatching from eggs, J2 move through soil and invade root tissue, usually near the root tip (Taylor & Sasser, 1978) or invade in the zone of elongation and migrate through the cortex toward the root tip (McClure & Robertson, 1973), where they stop and feed permanently on several continguous protoxylem cells. In susceptible plants, feeding results in the transformation of these cells into greatly hypertrophied and globose nurse cells, referred to as giant cells. These cells are characterized by intense nuclear, ribosomal and mitochondrial proliferation indicative of accelerated metabolic activity, and they have been likened to the transfer cells involved in phloem loading and unloading in fruiting structures and other metabolic sinks in numerous plants (Bird, 1996; Esau, 1977; Jones & Northcote, 1972; Pate & Gunning, 1972). Each giant cell is multinucleate as a result of nuclear without cytoplasmic division.

During the next several weeks, the surrounding tissue differentiates and undergoes extensive hypertrophy and hyperplasia, forming a gall. The nematode molts three times, greatly enlarging during successive molts into a sausage-like shape and then a spheroid shape, about the size of a pin head. When this stage is reached, the female body of *M. incognita* remains almost completely embedded in root tissue, with only the posterior tip exposed, whereas much of the body of *M. acronea* protrudes from the root. Between 500 and 3 000 eggs (0.08 × 0.04 mm) are laid into a gelatinous matrix secreted at the root surface. The final molt in *M. incognita* occasionally produces a male, which is vermiform and many times larger than the J2 from which it grew. Adult males of *M. incognita* do not feed and are not required for reproduction. The eggs produced by females develop into embryos following mitotic parthenogenesis. At favorable temperatures (~30°C) the life cycle is complete in 21–30 days (Taylor & Sasser, 1978).

The life cycle of *M. acronea* is generally similar to that of *M. incognita*; however, *M. acronea* reproduces sexually. Males of *M. acronea* are common and protrusion of adult females from roots probably facilitates insemination (Jepson, 1987; Page, 1985).

4.1.2. Interactions

M. incognita aggravates fungal seedling diseases by providing portals of entry and delaying taproot growth (Heald & Orr, 1984; Walker et al., 1998, 1999). More importantly, *M. incognita* greatly increases the severity and incidence of Fusarium wilt by the fungus *F. oxysporum* f. sp. *vasinfectum* (Devay et al., 1997; Martin et al., 1956; Starr, Wheeler, & Walker, 2001). How predisposition to the fungus occurs is uncertain. However, galling causes extensive longitudinal cracking of the epidermis and cortex of cotton roots, which may facilitate fungal invasion (Shepherd & Huck, 1989). The *Fusarium* wilt/root-knot nematode disease complex frequently results in death of many plants in a field. In fields where only the nematode is present, plants are stunted but seldom killed (Blasingame, 1994). *Meloidogyne incognita* also increases susceptibility to Verticillium wilt, but the effect is less pronounced than with Fusarium (Katsantonis, Hillocks, & Gowen, 2003).

4.1.3. Genetic Variability

The reproductive potentials of isolates of *M. incognita* vary differentially on different plant genotypes. Isolates differing in host specificity can come from widely separated localities or from the same field. Crop rotation practices in Atlantic coast states of the United States revealed isolates that could be assigned to one of four host races, depending on their ability to reproduce on cotton and the resistant tobacco (*Nicotiana tabacum*) cultivar NC-95 (Taylor & Sasser, 1978). Populations reproducing only on cotton were considered race 3 and those reproducing on both cotton and NC-95 were considered race 4. Most populations of *M. incognita* on cotton in the United States are race 3, whereas race 4 is commonly reported from South Africa and India (Jaskaran et al., 2000). There is evidence of variability among populations in California in ability to reproduce on resistant NemX (Ogallo et al., 1997, 1999). The reproductive rates of populations from Texas on cotton also differ (Zhou, Wheeler, & Starr, 2000).

4.2. Rotylenchulus reniformis

A second species of *Rotylenchulus*, *R. parvus*, has been reported on cotton from Africa. However, this discussion is restricted primarily to *R. reniformis*.

4.2.1. Life Cycle

Preparasitic stages of the reniform nematode are similar in size to the J2 of *M. incognita*, and the reproductively mature females also are sedentary parasites of the stele (Gaur & Perry, 1991; Robinson, 2002, 2007; Robinson et al., 1997). However, there are important differences in their life cycles.

After hatching from the egg, the J2 remains vermiform and undergoes three additional molts before it can invade plant tissue and feed. Each molt yields a slightly smaller worm (Bird, 1983), and the final molt produces a vermiform, sexually differentiated adult. Populations of *R. reniformis* encountered on cotton are obligately amphimictic, and equal numbers of males and females are produced. Only females feed and they invade the cortex of roots that have already undergone primary differentiation, most commonly in the zone of elongation, although nematodes are found all along roots (Birchfield, 1962; Cohn, 1973; Heald, 1975; Rebois, Madden, & Eldridge, 1975; Robinson & Orr, 1980). The vermiform female does not migrate through cortical tissue along the length of the root like the J2 of *M. incognita*, but rather enters the cortex perpendicular to the root axis and comes to rest with the stoma pressed to the outer tangential wall of a single, usually endodermal, cell on which it feeds. This cell and a curved sheet of contiguous cells of the pericycle, undergo cell wall dissolution and slight hypertrophy without hyperplasia, producing a simple syncytium that nurses the developing female. In contrast to the localized, globose giant cells induced by *M. incognita*, the *R. reniformis* syncytium often extends several root diameters along the root axis and a gall is not formed. Syncytial cells have enlarged nuclei and nucleoli, safraninophilic cytoplasm, and extensive proliferation of rough endoplasmic reticulum indicative of accelerated metabolism (Rebois et al., 1975).

Many cells, perhaps more than 100, can be involved in a single syncytium, so that when multiplied by the hundreds or thousands of females feeding on a single plant, the cumulative effect can be extensive. Within 6–14 days of root penetration, depending on temperature, the female becomes reproductively mature and, if inseminated, begins to deposit eggs into a gelatinous egg matrix (Rodríguez-Fuentes & Añorga-Morles, 1977; Sivakumar & Seshadri, 1971) similar to that of *M. incognita*. However, the neck elongates sufficiently that the swollen, kidney-shaped posterior two thirds of the body remains completely outside the root, exposed to the soil. The fully grown adult female is less than half the size of an *M. incognita* female, and the total number of eggs produced (60–200) is correspondingly smaller (Sivakumar & Seshadri, 1971). Paradoxically, *R. reniformis* usually occur in soil at population densities several times higher than *M. incognita*, which likely results from *R. reniformis* having a greater effective biotic potential due to a faster life cycle and greater number of potential feeding courts. There are more potential feeding sites for *R. reniformis* than for *M. incognita* on a cotton root system because *R. reniformis* interferes less with the development of fibrous roots, by initiating feeding sites within root zones that have already undergone primary differentiation.

Rotylenchulus reniformis is notorious for its ability to survive desiccation (Sehgal & Gaur, 1988, 1989; Womersley & Ching, 1989). The life cycle of *R. parvus* is similar to that of *R. reniformis* except that *R. parvus* reproduces parthenogenetically, and males are rare (Louw, 1982).

4.2.2. Interactions

Rotylenchulus reniformis can increase the incidence and severity of seedling diseases (Palmatee, Lawrence, VanSanten, & Morgan-Jones, 2004; Sanaralingham & McGawley, 1994) and may increase the incidence and severity of Fusarium wilt, although not to the extent of *M. incognita* (Brodie & Cooper, 1964; Khadr et al., 1972; Neal, 1954). It also has been reported to increase the incidence of Verticillium wilt, caused by the fungus *Verticillium dahliae* (Prasad & Padeganur, 1980). *Hoplolaimus columbus* appears to suppress *M. incognita* but not *R. reniformis* in sandy soils of the southeastern United States (Blasingame, 1994; Mueller & Lewis, 2001). *Rotylenchulus reniformis* may occur in sandy soils as well as finely textured soils but tends to occur at high populations in Texas only in soils with less than 40% sand (Robinson, Heald, Flanagan, Thames, & Amador, 1987; Starr, Heald, Robinson, Smith, & Krause, 1993). Pot studies confirm competition between *M. incognita* and *R. reniformis* (Diez, Lawrence, & Lawrence, 2003; Koenning, Walters, & Barker, 1996).

4.2.3. Genetic Variability

Less research has been done examining genetic variability in *R. reniformis* than in *M. incognita.* Some populations in India can and others cannot reproduce on both castor (*Ricinus communis*) and upland cotton, and these two groups of populations have been designated as races (Dasgupta & Seshadri, 1971a, 1971b). One population from India reproduces on sugarcane (Mehta & Sundara, 1989), a crop species immune to *R. reniformis* populations in Hawaii (Linford & Yap, 1940), Louisiana (Birchfield & Brister, 1962) and Puerto Rico (Ayala, 1962; Roman, 1964). Differences in reproduction and damage caused by 17 populations of *R. reniformis* from the United States on certain cultivars of cotton and soybean also have been observed (McGawley & Overstreet, 1995).

4.3. Hoplolaimus columbus

Hoplolaimus columbus is one of the largest nematodes in the genus *Hoplolaimus* (Mueller, 1993). It reproduces parthenogenetically, like *M. incognita*, and males are rare. However, it remains vermiform throughout life and feeds ectoparasitically as well as endoparasitically while migrating through roots. Root damage results from mechanical destruction of tissue, induction of necrosis and production of portals of entry for fungi and bacteria (Mueller, 1993; Mueller & Lewis, 2001).

4.4. Belonolaimus longicaudatus

Belonolaimus longicaudatus feeds ectoparasitically on root tips and cortical tissues and remains vermiform throughout life (Mueller & Sullivan, 1988). Although it does not invade tissue, adults are very large (1.6–2.6 mm) and have a large stylet that is more than half as long as the entire body of the J2 of *M. incognita* (Overstreet & McGawley, 2001; Robbins & Barker, 1974; Thorne, 1961). *Belonolaimus longicaudatus* is amphimictic.

5. MANAGEMENT

5.1. Sampling and Economic Thresholds

5.1.1. Meloidogyne incognita

Diagnostic soil and plant samples taken during midseason and fall are generally considered the best option for making management decisions (Starr, 1998). The presence of galling after midseason correlates highly with the distribution of *M. incognita* in a field, permitting areas that may require chemical treatment to be identified (Blasingame, 1994). Determining whether populations are sufficiently high to warrant treatment is more difficult. Both eggs and J2s contribute to over winter survival (Starr & Jeger, 1985). Eggs of *M. incognita* in the soil usually reach maximum numbers at harvest (Starr & Jeger, 1985) whereas J2 populations in soil continue to climb as eggs hatch during the fall, becoming more numerous than eggs, then decline faster than eggs during the winter so that eggs and J2 densities in the soil are similarly low and often undetectable by spring. As a consequence, there are more eggs than J2 between May and October, and more J2 than eggs between November and April.

As a practical matter, a decision to apply nematicide or plant a resistant cultivar needs to be made well before planting time, and if the population in a field is not evaluated until near the end of the winter, both eggs and J2 will be at such low densities that a reliable measure of the density in the field may not be obtained. Field studies in California (Roberts & Matthews, 1984) and micro-plot studies in Texas (Starr, Jeger, Martyn, & Schilling., 1989) estimated spring time damage thresholds of 0.05–0.1 J2/cm^3 soil. This is such a low density that some consultants consider detection of a single J2 in the spring sufficient basis for treatment. Because populations are so low in the spring, J2 population densities in the fall are often used as a predictor of damage resulting from the *M. incognita* eggs and J2 that will survive the winter. The recommended fall damage threshold for applying nematicides to control *M. incognita* in most cotton producing regions of the southeastern United States is 0.5–1 J2/cm^3 soil. Populations this high at planting are quite damaging. Pot studies in soil naturally infested with *M. incognita* at 0.96 and 1.08 J2/g soil in India, for example, showed yield responses to carbofuran nematicide (2.0 kg a.i./ha) of 10% and 18%, respectively.

The University of California offers farmers and consultants a mathematical crop damage function relating J2 population density – at or soon after harvest – to percentage yield loss in the following spring. Losses predicted by this model for 0.15, 0.3 and 0.6 J2/g soil, respectively, are 5%, 11% and 22% (Garber et al., 1996; Goodell, McClure, Roberts, & Thomas, 1996). Alternatively, California farmers are offered a weighted gall rating technique whereby plants with 0%, 1–25%, 26–50%, 51–75% and 76–100% of their root systems galled are assigned ratings, respectively, of 0, 1, 2, 5, and 7. Numeric ratings for a collection of randomly selected root systems are averaged to obtain an index value that is used for recommending treatment (Garber et al., 1996; Goodell, et al., 1996).

Variable rate application of nematicides for *M. incognita* control in cotton is intriguing because damage is typically patchy within a field, but obtaining requisite nematode population data cost-effectively may be impossible (Wheeler, Baugh, Kaufman, Schuster, & Siders, 2000; Wrather, Stevens, Kirkpatrick, & Kitchen, 2002). Measurement of soil electrical conductivity, however, shows promise as a tool for rapidly and cheaply characterizing the distribution of projected damage from *M. incognita* across large fields (Wolcott et al., 2005). Recent technological innovations have led to increased use in cotton of yield mapping during harvesting, by means of harvester-mounted lint sensors integrated to a computer and global positioning sensors on board the farm tractor. In fields where nematode damage appears to occur in the same spots of the field every year, it is possible to fumigate test strips across a large field perpendicular to row direction before bedding and planting. Yield mapping of the crop produced can then be used to provide a database for site-specific nematicide application to the next year's crop.

5.1.2. Rotylenchulus reniformis

Several field and pot studies show that damage from *R. reniformis* to cotton can be expected when soil populations during seedling growth are between 1 and 10 nematodes/cm^3 soil, equivalent to 0.8 and 8 nematodes per gram soil at 1.25 specific gravity (Elgawad, Ismail, & El-Metwally, 1997; Gilman, Jones, Williams, & Birchfield, 1978; Palanisamy & Balasubramanian, 1983; Patel, Patel, & Thakar, 2004; Sud, Varaprasad, Seshadri, & Kher, 1984; Thames & Heald, 1974). Because survival over winter is high, end-of-season samples are reliably used as the basis for nematode management decision in the next year's cotton crop. Treatment thresholds in use by consultants and farmers in the United States vary with growing conditions from about 8 to 16 nematodes/g soil collected at the end of the previous season (Koenning, 2002; Komar, Wiley, Kermerait, & Shurley, 2003; Overstreet, 2001; Sciumbato, Blessitt, & Blasingame, 2004). When spring samples are used, the treatment threshold employed is 20% that in the fall, i.e. between 1.6 and 3.2 nematodes/g, and thus very similar to the values observed in quantitative studies. Diagnostic labs often are overwhelmed with samples in the fall; however, studies in Alabama (Lawrence et al., 2005a) have shown that a sample can be stored at 4°C for up to 180 days, and the original nematode density at the time when placed in storage can be calculated.

5.1.3. Hoplolaimus columbus

Soil populations of *H. columbus* in the restricted areas of the eastern United States where it occurs, typically do not decline until late winter (Blasingame, 1994; Mueller & Lewis, 2001). The economic threshold for undisturbed soil in the fall and early winter is one nematode/cm^3 soil, and in the spring 0.3 nematode/cm^3 soil. If the field has been disked or plowed, the threshold is 0.1 nematode/cm^3 soil. *H. columbus* occurs in high numbers in roots but during much of the year sufficient numbers are present in soil to use extracted nematodes as an acceptable indicator of the total number present (Davis & Noe, 2000).

5.1.4. Belonolaimus longicaudatus

This nematode is very large and strong, and is devastating to a cotton crop at very low population densities. Thresholds in Florida field experiments ranged from 0.015–0.039 nematodes/cm^3 soil, and a soil population of 0.8 nematodes/cm^3 was sufficient to expect 100% yield loss (Crow, Weingartner, McSorley, & Dickson, 2000b). In controlled environmental chambers, 0.08 and 0.40 nematodes/cm^3 soil reduced cotton fine roots by 39% and 70%, respectively (Crow, Dickson, Weingartner, McSorley, & Miller, 2000a).

5.2. Control

5.2.1. Natural Physical Factors

The temperature, texture, compaction and moisture of soil can profoundly influence the survival and population dynamics of nematodes parasitizing cotton. *M. incognita* and *R. reniformis* generally occur at latitudes within 35° of the equator and have optimum temperatures for movement (Robinson, 1989, 1994; Robinson & Heald, 1989, 1991, 1993) and reproduction (Rebois, 1973; Taylor & Sasser, 1978) between 27°C and 32°C. However, survival in fallow soil is greatly prolonged at 10°C, while temperatures exceeding 45°C are lethal to hydrated eggs and juveniles (Heald & Robinson, 1987). In some regions soil texture is differentially correlated with the distributions of these two species. In Texas, high population densities of *M. incognita* were commonly found in soils with a wide range of sand content (10–90%) but were infrequently found in soils with more than 60% clay (Robinson et al., 1987; Starr et al., 1993) and high population densities tended to occur only in sandy soils. High population densities of *R. reniformis*, by comparison, tended to occur in soils with less than 50% sand.

In the United States, *B. longicaudatus* appears to be limited to the Coastal Plain of the Atlantic seaboard and occurs at damaging population densities almost exclusively in soils with greater than 85% sand (Esser, 1976). The vermiform stages of *R. reniformis* but not the J2 of *M. incognita* can survive for several years in soils dried below the permanent wilting point for plants (Birchfield & Martin, 1967; Heald & Inserra, 1988; Rodríguez-Fuentes, 1980; Tsai & Apt, 1979; Womersley & Ching, 1989). The ensheathed juveniles of *R. reniformis* may be better adapted for desiccation survival than the exsheathed vermiform females or the newly hatched vermiform J2.

5.2.2. Nematicides

The statement made by Sasser (1972) that "Chemical control of plant pathogenic nematodes in cotton is by far the most expedient and widely used method" has yet to be disproven, even though most of the means to achieve it are gone. In the United States, the primary nematicides remaining available for use in cotton include the fumigants 1,3-dichloropropene and metam sodium and the cholinesterase inhibiters aldicarb and oxamyl (Gazaway et al., 2001; Koenning et al., 2004; Lawrence & McLean, 2000; Lawrence, McLean, Batson, Miller, & Borbon, 1990, Lawrence et al., 2005). Where it occurs naturally, cotton is a perennial with a fast-growing, deep taproot. Today as in 1972 (Sasser), the nematicide strategy for all nematodes in cotton is to save nematicide costs by focusing on protection of the young plant, and target the soil zone that the taproot will grow through during the first few weeks. This means fumigant placement 25–45 cm deep under the center of the bed, or granular nematicide either in the seed furrow or else band-incorporated over the top of the planting bed, with the option to also side-dress later.

5.2.2.1. Conventional Nematicides

Many nematicide efficacy tests have been conducted in cotton and the population suppression and yield responses obtainable with labeled rates have been well characterized, as have the economics (Gazaway et al., 2001; Kinloch & Rich, 2001; Lawrence et al., 1990; Overstreet & Erwin, 2003; Palanisamy & Balasubramanian, 1983; Thames and Heald, 1974; Zimet, Rich, LaColla, & Kinloch, 1999; Zimet, Smith, Kinloch, & Rich, 2002). The nematicide applications usually recommended for managing nematodes in cotton in the United States are similar for *R. reniformis*, *M. incognita* and *H. columbus*. Nematicide application is recommended only when losses are expected to exceed 5%. Recommended rates of Nemacur (fenamifos) 15G and Temik 15G brand aldicarb are the same, 5.6–7.8 kg a.i./ha applied at-plant, often in-furrow. The rate usually recommended for 1,3-dichloropropene (Telone II) is 28 liters/ha applied 10–14 days pre-plant, preferably in the temperature range 16–25°C and usually followed by a low rate of Temik 15G at plant. Soil temperature and moisture optimal for planting are ideal for fumigation but a delay is required to avoid phytotoxicity (Heald & Orr, 1984). In California, a second application of 5.6–7.8 kg a.i./ha Temik 15G is sometimes side-dressed at first square (Garber et al., 1996; Goodell, et al., 1996). In Brazil, terbufos 150G (2.55 kg a.i./ha) has been found to be highly effective against *R. reniformis* and *P. brachyurus*, suppressing populations 93% and 97%, and increasing yields 38–49% (Gonçalves de Oliveira, Kubo, Siloto, & Raga, 1999). Foliar applications of oxamyl have also been used in cotton with good success in some regions (Lawrence & McLean, 2000).

Figure 3. Cotton yields in R. reniformis *infested fumigated soils in Texas (A, C) and Louisiana (B, D) (plants on left in each photo were fumigated). B, D: courtesy C. Overstreet.*

In the United States, granular in-furrow application of a sub-nematicidal rate of aldicarb is widely used prophylactically for early season insect control, and the cost of stepping up the rate (to 5.6–7.8 kg a.i./ha) for nematode control is low. Unfortunately the benefits typically are inferior to fumigation (Gazaway et al., 2001), consistent with the rule of thumb in cotton, that fumigants (Fig. 3) are more effective than granular nematicides (Orr & Brashears, 1977). Nonetheless, at appropriate rates both aldicarb and 1,3-dichloropropene can be profitable, though economically risky, for management of *M. incognita* and *R. reniformis* (Zimet et al., 1999, 2002). A serious recent concern is the development of aldicarb-degrading microflora following long-term use as a prophylactic, demonstrated recently to be occurring in Alabama (McLean & Lawrence, 2003). It is unlikely that suppression of seedling disease-causing microflora by 1,3-dichloropropene and aldicarb is an important component of yield responses to fumigation observed in fields infested with nematodes, because careful studies have shown that the use of 1,3-dichloropropene and aldicarb in cotton fields does not significantly impact plant pathogenic fungi or saprophytic fungal populations (Baird, Carling, Watson, Scruggs, & Hightower, 2004). Site specific application of aldicarb for *M. incognita* management in cotton has been explored, but was found to be less cost-effective than uniform application (Wheeler et al., 1999; Wrather et al., 2002).

5.2.2.2. Novel Nematicides

Several recent tests explored the potential of strategic placement of anhydrous ammonia, a widely used nitrogen fertilizer formulation, for *R. reniformis* management in cotton. Significant yield improvements over the alternative nitrogen control were measured, but consistent suppression of nematode populations was not obtained (McLean, Lawrence, Overstreet, & Young, 2003). During the last 2 years, seed coat formulations of the anthelmithic avermectin-B1 have been extensively tested and are now commercially offered (Cochran, Long, Beckett, Payan, & Belles, 2006; Faske & Starr, 2006; Kemerait et al., 2006; Schwarz, Graham, & Kleyla, 2006). Commercially available formulations of resistance-inducing harpin proteins as seed and foliar treatments also have recently been evaluated, and may find a place in nematode management in cotton (French et al., 2006).

5.2.2.3. Yield Potential Recoverable with Nematicides

Yield increases in recent years in fields infested with *R. reniformis* have often been only 5 or 10%, contrasted to the 40–60% yield suppressions measured in early studies examining the impact of *R. reniformis* on cotton (Birchfield & Jones, 1961; Jones, Newsom, & Finley, 1959), which were done with obsolete but highly effective fumigants. Studies in cotton fields infested with *R. reniformis* show that when conventional fumigation is used, most nematodes are killed 5 cm below and directly above the point of placement up to the soil surface, but populations always quickly rebound during the first half of the crop season and at harvest are often comparable to those in untreated plots (Gazaway et al., 2001; Kinloch & Rich, 2001; Lawrence & McLean, 1996). This has been attributed to the high biotic potential of *R. reniformis*, and recolonization of the upper soil layer by nematodes deeper in the soil. In some fields, more than half of the *R. reniformis* inoculum in the field is deeper than 45 cm (Newman & Stebbins, 2002; Robinson, Cook, Westphal, & Bradford, 2005a; Robinson, Gutierrez, LaFoe, McCarty, & Jenkins, 2005b).

Occurrence of *R. reniformis* deep in the soil raises the question of nematicide efficacy relative to other possible nematode management options, such as crop rotation, biological control and host-plant resistance. If roots are deep and deep roots are important, then less than 100% of the total yield potential should be recovered by nematicide treatment, because nematicides treat only the upper portion of the soil profile. Analyses by Zimet et al. (1999, 2002) indicated that a substantial fraction of the yield potential could be tapped by fumigating shallow sandy soils in the Florida panhandle, which are relatively easy to penetrate due to the large pore space, low diffusive resistance and shallow roots. In comparison, tests in deep soils in Texas where cotton root growth may exceed 2 meters indicated deep placement of fumigant was needed to obtain maximum yield (Cook et al., 2003; Robinson et al., 2005; Westphal, Robinson, Scott, & Santini, 2004). Placing fumigant 81–100 cm below the surface in these fields, suppressed populations 90 cm deep in the soil throughout the season, strongly promoted deep root growth and increased yield by 100%, contrasted to 57% increase obtained by only fumigating 43 cm deep. The additional yield boost obtained by fumigating deeply is very important because it is part of the yield potential that might be tapped by planting a *R. reniformis*-resistant cultivar, should one become available.

5.2.3. Biological Control

Biological control has not yet been adopted as a standard practice for managing nematode problems in cotton production systems. However, *M. incognita* has been the subject of many studies examining the possible use of natural enemies of nematode in crops other than cotton.

Organisms investigated include the parasitic fungus *Hirsutella rhossiliensis*, nematodes trapping fungi (*Monacrosporium cionopagum* and *M. ellipsosporum*) (Robinson & Jaffee, 1996), mycorrhizal fungi (Sikora, 1979), egg parasitizing fungi, such as *Paecilomyces lilacinus*, the obligately parasitic bacterium *Pasteuria penetrans*, strains of *Gluconacetobacter diazotrophicus* (Bansal, Dahaya, Narula, & Jain, 2005) and predaceous nematodes (Stirling, 1991; Robinson & Jaffee, 1996).

In addition, several unidentified and known organisms have shown good potential against *R. reniformis* in controlled experiments. An unidentified fungus isolated from soybean cyst nematode, *Heterodera glycines,* was found to consistently suppress Arkansas populations of *R. reniformis* in pots by up to 98% (Wang, Riggs, & Crippen, 2004). Three of 117 isolates of *Pochonia chlamydosporia* that were tested, parasitized and suppressed an Arkansas population of *R. reniformis* in pots by up to 77% (Wang, Riggs, & Crippen, 2005). Isolates of *Paecilomyces lilacinus* also have been tested against *R. reniformis* in pots (Jayakuma, Ramakrishnan, & Rajendran, 2002).

Among G- (Gram-negative) bacteria, *Pseudomonas fluorescens* suppressed Indian populations of *R. reniformis* by up to 70% (Jayakumar, Ramakrishnan, & Rajendran, 2003). Unidentified agents in three soils from cotton fields in the Texas Lower Rio Grande Valley, whose effects were removable by autoclaving, suppressed populations of *R. reniformis* in sand by 80 and 95% when field soil was added to sand at ratios of 1:20 and 1:10, respectively (A. Westphal and A. F. Robinson, unpublished data). Exploitation of knowledge regarding biological control of nematodes is an opportunity for the future.

5.2.4. Cultural Control

Any practice that tends to reduce water and nutrient stress tends to reduce yield losses due to nematodes. A very old deep tillage practice that is called in-row subsoiling in the southeastern United States and precision tillage in California (Garber et al., 1996), is often used to allow cotton roots to penetrate deeper than normal and thereby partly offset deleterious effects of *M. incognita* on water and nutrient uptake. This practice involves pulling a ripping shank through the soil where the future cotton beds will be located, generally at least 45 cm deep, and is made more precise by the advent of laser guided tractors and global positioning systems. Plowing up old roots in the fall also tends to reduce populations.

Organic soil amendments can be considered for nematode management in cotton if materials are abundant and labor is cheap. Toxic amendments have been explored for management of *R. reniformis* and *M. incognita* on cotton in pots in India, and several are highly effective, including presmud, fresh Azolla, farm yard manure and neem cake (Patel, Patel, & Thakar, 2003). Identification of active components in highly effective amendments could lead to new nematicide chemistry with application to regions where the raw products for making those amendments are not available. In the United States, municipal solid waste application consistently improved tilth, suppressed *H. columbus* populations and increased cotton yield for 3 years in a row in South Carolina (Khalilian et al., 2002). The populations of *H. columbus* in cotton also were suppressed by poultry litter in North Carolina (Koenning & Barker, 2004) and Georgia (Riegel & Noe, 2000). Incorporation of shellfish waste and crop residues that contain chitin or generate biofumigants, highly effective against plant nematodes, has shown economic potential in cotton in Alabama (Hallmann, Rodríguez-Kábana, & Kloeper, 1999). One goal of chitin addition is the attack of chitin in nematode egg shells by augmented chitinolytic microflora, but chitin amendments also alter the C:N ratio and other components of microbiotic interactions in soil. Because *R. reniformis* symptoms in cotton mimic potassium deficiency, potassium supplementation has been explored as an amelioration strategy, but without significant effects on yields (Pettigrew, Meredith, & Young, 2005).

5.2.5. Crop Rotation

In the southeastern United States, the primary crop rotations recommended for managing nematodes in cotton are peanut or groundnut (*Arachis hypogaea*), American corn (*Zea mays*) and soybean (*Glycine max*) (Brathwaite, 1974; Gazaway, Akridge, & Rodriguez-Kabana, 1998; Gazaway, Akridge, & McLean, 2000; Davis, Koenning, Kemerait, Cummings, & Shurley, 2003; Thames & Heald, 1974). Peanut is excellent because *M. incognita*, *R. reniformis* and *H. columbus* all reproduce poorly on peanut (Blasingame, 1994), and the peanut root-knot nematode,

M. arenaria, reproduces insignificantly on cotton (Starr, 1998). Where they are an economic option, sorghum (*Sorghum bicolor*) and small grains can be used to suppress *R. reniformis*.

Great care must be taken in using corn and soybean rotations for nematode management in cotton. Most corn hybrids are highly resistant to *R. reniformis* (Windham & Lawrence, 1992) but support good reproduction by *M. incognita* (Davis & Timper, 2000). Soybean cultivar selection is complex because cultivars differ with respect to resistance to *M. incognita*, *R. reniformis* and soybean cyst nematode, *Heterodera glycines* (Gilman et al., 1978, 1979; Hartwig & Epps, 1977; Harville, 1985; Robbins et al., 2001; Westphal & Scott, 2005). In addition, soybean cultivars fall into maturity groups adapted to specific latitudes, necessitating availability of a specific nematode resistance gene in a specific maturity group for use in nematode management in a specific growing region. In Brazil, sorghum and velvet bean (*Stizolobium deeringianum*) can be used to manage *R. reniformis* (Farias, Barbosa, Vieira, Sánchez-Vila, & Ferraz, 2002). In North American fields infested with *M. incognita*, clover (*Trifolium* spp.) and vetch (*Vicia* sp.) winter cover crops can increase nematode populations, but rye (*Secale cereale*) (McBride, Mikkelsen, & Barker, 1999; Timper, Davis, & Tillman, 2006) and the vetch cultivar Cahaba White (Timper et al., 2006) do not, and have been shown to be acceptable winter cover crops in infested fields.

In California, peanut and nematode-resistant cultivars of soybean and corn are generally not considered suitable rotational crops. However, most varieties of alfalfa (*Medicago sativa*) grown in California and Arizona are resistant to *M. incognita* and can be used in rotation with cotton (Garber et al., 1996; Goodell, et al., 1996). Grain sorghum is recommended for reducing *M. incognita* populations in Arkansas but not on the High Plains of Texas (Blasingame, 1994; Thomas & Kirkpatrick, 2001). Rotational crops recommended for control of *M. acronea* in Africa include pearl millet (*Pennisetum typhoides*), finger millet (*Eleusine coracana*), corn (*Zea mays*), peanut, guar bean (*Cyanopsis tetragonoloba*), and leucaena (*Leucaena glauca*) (Starr & Page, 1990).

5.2.6. Sanitation/Weed Management

Meloidogyne incognita has a wide host range and reproduces on more than 1 000 plant species. In the United States, *M. incognita* reproduces on many of the weeds commonly encountered in cotton fields. *R. reniformis* has a similarly wide host range, reproducing on 87% of more than 350 plant species tested as potential hosts (Birchfield & Brister, 1962; Linford & Yap, 1940; Robinson et al., 1997). Many common weeds in all cotton production regions of the world support prolific reproduction by *R. reniformis* (Carter, McGawley, & Russin, 1995; Inserra, Dunn, McSorley, Langdon, & Richmer, 1989; Lal, Yadav, & Nandwana, 1976; Quénéhervé, Drob, & Topart, 1995). Thus, effective weed management in the crop, during winter fallow, and during crop rotations is critical for managing both of these nematodes in cotton.

In practice, recognizing and removing weeds that interfere with nematode management has been shown to be easier than might be predicted by the host ranges of *M. incognita* and *R. reniformis*. Recent studies in Alabama and Georgia have shown that only a small proportion of weeds found in cotton fields are problematic to nematode management in cotton production because, although most dicotyledonous weeds in cotton in the southeastern United States are hosts, only a few are better hosts than cotton (Davis & Webster, 2005; Dismukes, Lawrence, Price, Lawrence, & Akridge, 2006). In some cases, plants that support high populations of one nematode do not support the other nematode, so it is important to know which weed is a good host for which nematode. Some are sufficiently good hosts to sustain populations in fields planted to non-host corn (Dismukes et al., 2006).

The best hosts among 28 weeds tested in Alabama were three *Ipomoea* spp. (morning glory). Mixed morning glory species sustained the second highest *R. reniformis* populations during a corn rotation in microplots and sicklepod (*Senna obtusafolia*) sustained the highest. Other potentially problematic weeds in corn rotation included coffee senna (*Cassia occidentalis*), common ragweed (*Ambrosia artemisiifolia*) and velvetleaf (*Abutilon theophrasti*). Among 11 weeds examined in Georgia (Davis & Webster, 2005), only purple netsedge (*Cyperus rotundus*), sicklepod, Florida beggar weed (*Desmodium tortuosum*) and smallflower morninglory were comparable to cotton, with populations (expressed as a percentage of that on cotton) of 454%, 81%, 73% and 33%, respectively.

Also *M. incognita* reproduced well (35% of cotton) on purple nutsedge, but not as well as *R. reniformis*. Smallfower morning glory and ivyleaf morning glory were both good hosts for *M. incognita* (70% and 211%) of cotton, but only prickly sida (*Sida spinosa*), which had 407% of *M. incognita* and only 10% of *R. reniformis*, differed strikingly in its suitability as a host for the two nematodes. Weed species also differ through the year. Thus, weed management for nematode control in cotton requires knowing which weeds support which nematodes as well as when they are present during the cropping cycle. Weed hosts, however, do not appear to explain the gradual increase that has been documented in the incidence of *R. reniformis* in the central part of the United States cotton belt (Robinson, 2007).

One other important point regarding sanitation should be made. The use of cotton seed hulls or husks produced by cotton gins and oil mills as a soil amendment and or as cattle feed is practiced in many parts of the world and has been documented to spread infestations of Fusarium (Hillocks & Kibani, 2002).

5.3. Genetic Resistance to Nematodes in Cotton

Host plant resistance is the most efficient way to manage nematodes (Starr, Bridge, & Cook, 2002). There are very few nematode-resistant cotton cultivars available, and those are not widely adapted. Currently, host plant resistance is an area of intense investigation in cotton nematology research.

5.3.1. Terminology

In principle, nematode resistance and tolerance should be clearly distinguished as genetic traits (Cook & Evans, 1987). Resistance refers to the ability of a plant to limit a nematode's reproduction; tolerance has no necessary relationship to resistance and refers to the ability of the plant to grow and yield in soil where the nematode is present (Cook & Evans, 1987). In practice, tolerance in cotton often appears to be accompanied by partial resistance (Davis & May, 2003, 2005), and resistance makes plants more tolerant (Zhou & Starr, 2003).

5.3.2. Resistance and Tolerance Mechanisms

In resistant plants, the *M. incognita* J2 invades roots, migrates through tissue, and attempts to feed on the same cells as in susceptible plants. However, a hypersensitive response by the plant results in the collapse and death of cells probed by the nematode (Carter, 1981; Creech, Jenkins, Tang, Lawrence, & McCarty, 1995; Jenkins et al., 1995; Shepherd & Huck, 1989). Normal root penetration by *R. reniformis* with failure to induce a syncytium also characterizes the resistant response of *Gossypium longicalyx* and *G. hirsutum* hybrids carrying *R. reniformis* resistance from *G. longicalyx* (Agudelo, Robbins, Stewart, Bell, & Robinson, 2005a).

In cotton roots infected by *M. incognita*, toxic terpenoid aldhydes accumulate around the nematode head more rapidly in resistant than in susceptible plants (Veech & McClure, 1977; Veech, 1978, 1979). Roots of the susceptible cotton germplasm line M-8 develop more extensive cracking of the epidermis and cortex as they grow than roots of the resistant line Auburn 623 RNR (Shepherd & Huck, 1989). The resultant leachates and increased physical access to the root interior were hypothesized to explain the greater susceptibility of M-8 to the root-knot nematode and Fusarium-wilt disease complex (Shepherd & Huck, 1989).

An *M. incognita* resistance-specific protein (*MIC-3*) produced in response to nematode infection, has been sequenced and determined to belong to a novel gene family with six members (Zhang et al., 2002). The existence of numerous host differentials between *M. incognita* and *R. reniformis* suggests different resistance mechanisms.

Bacillus thuringiensis delta-endotoxin transgenes encoding for the *Cry 1 Ac* protein do not confer nematode resistance in cotton, and resistance may be lost when transgenes are transferred into resistant genotypes, if resistance inheritance is not monitored during backcrossing (Colyer, Kirkpatrick, Caldwell, & Vernon, 2000). The *Cry 2 Ab* toxins await testing.

5.3.3. Resistant and Tolerant Cultivars and Resistance Sources

5.3.3.1. M. incognita

Early germplasm evaluations (Jones et al., 1958) focused on the *M. incognita* and Fusarium wilt disease complex. Finding resistance to the complex was confusing since resistance to the nematode and resistance to the fungus were inherited independently. Thus, cultivars with resistance to the disease complex in the field often supported high levels of nematode reproduction (Starr & Martyn, 1991; Starr & Veech, 1986) and genotypes that were wilt-resistant, when stem-inoculated with the fungus in the greenhouse, often showed little resistance to the fungus under field conditions if the nematode was present (Shepherd, 1986a; Shepherd & Kappelman, 1986). Wilt resistance has been incorporated into many cultivars, and *M. incognita* resistance remains an important component of resistance to the disease complex in contemporary resistant cultivars (Koenning et al., 2004).

Today, there are at least five independently developed sources of resistance to *M. incognita* available in agronomic types of *G. hirsutum*. The first one came from a cross made by R.L. Shepherd in Alabama in the 1960s (Shepherd, 1974a, 1974b) between the Fusarium wilt resistant cv. Clevewilt 6, and a root-knot nematode-resistant primitive *G. hirsutum* accession from Mexico, registered in the USDA National Cotton Collection as Wild Mexican Jack Jones. The F_{10} selection Auburn 623 RNR, which was more resistant than either parent, was backcrossed to wilt-resistant Auburn 56. Nematode resistance was recovered in the selection Auburn 634, which in turn was backcrossed to obsolete cvs. Deltapine 61, Coker 201, Coker 310 and Stoneville 213 to produce the highly resistant Auburn M lines (M. Robinson et al., 1997; Sheperd 1982, 1986b, 1989; Shepherd & Huck, 1989). This material was used as the source of resistance for Arkot 9111, recently released by the Arkansas Agricultural Experiment Station (Bourland & Jones, 2005), as well as GA161, released by the Georgia Agricultural Experiment Station (May, Davis, & Baker, 2001), and several breeding lines to be released by Mississippi State University in cooperation with USDA. Resistance in the Auburn M lines appears to be inherited as a two-gene system, one dominant and one partial. Significant progress has been made toward discovery of DNA markers suitable for marker-assisted selection and mapping of the resistance genes (Hinchliffe et al., 2005; Shen et al., 2006).

A second source of resistance developed in Brazil has Auburn 56 as a key parent. Auburn 56 is also in the background of the Auburn M lines, and is generally thought be a source of Fusarium but not root-knot nematode resistance in those lines. Genotypes within the resistant Brazil group include the highly resistant IAC/414 and the moderately resistant IAC98/708 and IAC98/732 (Carneiro et al., 2005). The cultivar CD405, also developed for Brazil but apparently unrelated to IAC/414, is reported nematode tolerant (Bélot et al., 2005).

A third source of root-knot nematode resistance came from a cross made by J. E. Jones (Jones, Wright, & Newsom, 1958; Jones, Beasley, Dickson, & Caldwell, 1988; Jones et al., 1991) in Louisiana between Clevewilt 6 and Deltapine 15, ultimately leading to the moderately root-knot nematode-resistant Bayou 7769 (Jones & Birchfield, 1967). Bayou 7769 was in turn crossed with Deltapine 16 and nematode-resistant progeny selected from this cross led to LA 434-RKR, crossed in turn with DES 11–9 to produce the once widely planted, moderately root-knot nematode-resistant cultivar Stoneville LA 887 and the related Paymaster (formerly Hartz) 1560 (Jones et al., 1991). These cultivars combined root-knot nematode resistance with high yield, high fiber quality, medium maturity, high lint percentage and wide adaptation. They also showed excellent field resistance to the *M. incognita* and Fusarium wilt disease complex. A closely related nematode-resistant transgenic cultivar, Stoneville 5599 BR, is currently commercially available and planted in the central United States cotton belt. LA 434-RKR was also used to develop four breeding lines (Jones et al., 1988) adapted to Louisiana that combine root-knot nematode resistance with reniform nematode tolerance. Three of those lines were utilized in turn, to develop seven additional, high yielding breeding lines adapted to South Texas growing conditions, which also are root-knot nematode resistant and reniform nematode tolerant (Cook, Namken, & Robinson, 1997; Cook, Robinson & Namken, 1997; Cook & Robinson, 2005; Koenning, Barker, & Bowman, 2000). Stoneville LA887 and the Auburn M line 240 RNR were both used as parents in the development of GA96-211, recently released by the Georgia Agricultural Experiment Station (May, Davis, & Baker, 2004).

A fourth independently derived source of root-knot nematode resistance in cotton was developed by Angus Hyer with the USDA in California, leading to the development of a resistant breeding line C-225, which was released after his death as the cultivar NemX by California Planting Cotton Seed Distributors (CPCSD) and the University of California (Ogallo et al., 1997). The details of this rather complex lineage are given by Robinson et al. (2001). NemX meets fiber quality as well as yield standards of the Acala cotton types grown in California (Garber et al., 1996) and is resistant to the *M. incognita* and Fusarium wilt disease complex. However, it is not widely adapted and cannot be grown for profit in most other production regions of the United States, but might be adaptable to regions of the world with conditions similar to California. NemX has been the subject of intensive recent investigations in California that have identified DNA markers for the resistance and made substantial progress toward fine-mapping of the recessive resistance gene (Wang, Matthews, & Roberts, 2006). Markers will enable marker assisted selection, facilitating the use of NemX as a source of resistance for cultivars adapted to conditions outside California, and mapping could eventually lead to resistance gene cloning.

A fifth source of resistance, effective against *M. incognita* race 4, is the South African cultivar Gamka, developed from N9311, thought to have come out of Gus Hyer's breeding program in California, but of uncertain relation to NemX.

There is already some evidence of the development of populations of *M. incognita* able to reproduce on NemX in California. NemX was compared with LA887 and an Auburn line in North Carolina (Koenning, Barker, & Bowman, 2001) and suppressed the field population tested well. At least one population on the Texas High Plains, however, induces galls and reproduces on NemX (T. A. Wheeler, personal communication). Thus, it seems likely that resistance-breaking populations will develop, and additional resistance sources may prove invaluable. Most primitive accessions of *G. hirsutum* are good hosts for *M. incognita* but resistance to the nematode is not uncommon in primitive *G. hirsutum*. Additional sources of resistance to *M. incognita* have been identified among accessions of *G. hirsutum* from Mexico, Central America and the Caribbean Basin. Eighteen of 471 accessions examined by the USDA at Mississippi State University in 1983 had a level of resistance intermediate to the moderately resistant Clevewilt 6 and the highly resistant Auburn 623 RNR (Shepherd, 1983). Twelve of those accessions were used to develop day-neutral isolines by crossing with cv. Deltapine 16, and selecting for day neutrality across recurrent backcrosses onto primitive accessions (Shepherd, 1988). Nine more resistant accessions were discovered in 1996 (Robinson & Percival, 1997). It is not known yet if the resistance genes in these sources differ.

5.3.3.2. R. reniformis

Resistance to *R. reniformis* has been hard to find. Of 2 000 *G. hirsutum* genotypes evaluated in the search for resistance (Robinson, Bridges, & Percival, 2004; Robinson & Cook, 2001; Robinson, Cook, & Percival, 1999; Robinson & Percival, 1997; Yik & Birchfield, 1984), only 19 were scored as potentially resistant in the first examination, of which nine (Yik & Birchfield, 1984) were reclassified as susceptible in a subsequent screen (Robinson & Percival, 1997), and four (TX-110, TX-502, TX-1347, TX-1348) were reclassified as *G. barbadense*, leaving only six moderately resistant *G. hirsutum* accessions (TX-25, TX-748, TX-1586, TX-1828, TX 1860, TX-2469). Several additional weakly to moderately resistant primitive accessions of *G. hirsutum* have been found recently (D. Weaver, personal communication).

Stronger levels of resistance than in *G. hirsutum* occur in one or more accessions of *G. barbadense*, *G. arboreum*, *G. herbaceum*, *G. longicalyx*, *G. somalense* and *G. stocksii* (Carter, 1981; Robinson & Percival, 1997; Stewart & Robbins, 1995, 1996; Yik & Birchfield, 1984), which in some cases suppress *R. reniformis* populations in pots 90–100%, compared to susceptible upland cotton, and also suppress populations in the field (Robinson et al., 2006). Currently, projects are underway at the University of Arkansas, Auburn University, Mississippi State University, Texas A&M University and three laboratories of the Agricultural Research Service of the USDA, to introgress resistance from primitive *G. hirsutum*, *G. barbadense*, *G. longicalyx* and *G. arboreum* into agronomic *G. hirsutum* (Avila, Stewart, & Robbins, 2006; Bell & Robinson, 2004; Dighe et al., 2007; Moresco, Morgan, Ripple, Smith, & Starr, 2004; Robinson et al., 2005; Silvey, Ripple, Smith, & Starr, 2003; J. N. Jenkins, E. Sacks, D. Weaver, L. D. Young, personal communication). These are challenging, long term projects, as the requisite transfers of DNA within the genus *Gossypium* are complex involving differences in ploidy and different genomes and sub-genomes, with in many cases low or no intercompatibility, due to chromosome inversions, deletions, etc. (Percival, Wendel, & Stewart, 1999).

Probably the most advanced of the projects, which is being carried out by the USDA at College Station, Texas, in collaboration with Texas A&M University, has employed two tri-species hybrids of *G. hirsutum* (Bell & Robinson, 2004) with *G. longicalyx*, and either *G. armourianum* or *G. herbaceum* as bridges, to introgress virtual immunity to *R. reniformis* from diploid *G. longicalyx* into allotetraploid *G. hirsutum* (Robinson, Jenkins & McCarty, 2007). Introgression was accomplished by recurrent backcrosses to *G. hirsutum* with cytogenetic analysis of early backcross generations to assess progress toward the euploid state ($2n = 52$) for *G. hirsutum*, selection for nematode resistance at each generation, and examination of self progeny at the first, third, sixth and seventh backcross, to identify and eliminate lineages with undesired recessive traits.

By making literally thousands of attempted crosses, 689 first-backcross generation progeny were generated from the two male-sterile hybrids. A small number of these were both resistant and fertile. Introgression was then pursued from 28 resistant backcross-one plants, each of which was backcrossed again three to six times to *G. hirsutum* to derive agronomically suitable types, selecting for nematode resistance by bioassay at each backcross, as well as within segregating progeny from selfed plants at selected generations. This was an arduous process, involving the evaluation of nematode resistance in ca. 3000 plants. The resistance trait was consistently inherited in ratios (resistant:susceptible) of 1:1 in backcross progeny and 3:1 in self progeny, in repeated generations with no loss of resistance across generations and full recovery of resistance in plants where the resistance trait was fixed. This inheritance pattern indicates a single dominant gene, or a block of non-recombinant alien DNA that behaves like a single gene, providing plant breeders with an easy genetic system for development of resistant cultivars. Hundreds of backcross plants were indistinguishable from agronomic cotton, as were 12 progeny sets in the field in 2006, which were descended from selfed mother plants with the resistance trait in the homozygous, fixed condition. Thus, the trait has been fixed genetically in the absence of any known unwanted characteristics. Fiber quality data are excellent.

More than 500 segregating phenotyped plants in the USDA-Texas A&M University *G. longicalyx* project were utilized to discover six SSR markers co-segregating with the trait. One of the markers is co-dominant, allowing it to be used to distinguish homozygous from heterozygous resistant plants, and resides within ca. 1 centiMorgan of the resistance gene (Dighe et al., 2007). Seed of two genetic stocks, Lonren-1 and Lonren-2 were released by USDA, Texas A&M University, and Cotton Incorporated in May of 2007.

In other introgression projects, resistance to *R. reniformis* is being transferred into cotton from *G. barbadense* TX-110 (Moresco et al., 2004), *G. barbadense* GB-713 (Robinson et al., 2005), *G. arboreum* (Avila et al., 2006; E. Sacks, personal communication), *G. barbadense* TX-1348 (L. D. Young, personal communication), or being approached via transgressive segregation within *G. hirsutum* (D. Weaver, personal communication; A. F. Robinson, unpublished data). Bringing resistance into cultivated cotton from different sources is important because the likelihood of ultimately confronting linkage drag between resistance genes and agronomically unacceptable traits in each case is high. Moreover, multiple resistance sources may prove an invaluable resource if and when resistance-breaking nematode populations or races are encountered or develop. There is already ample evidence of much variability within *R. reniformis* (Agudelo, Robbins, Stewart, & Szalanski, 2005b; Dasgupta & Seshadri, 1971b; McGawley & Overstreet, 1995; Nakasono, 1983).

5.3.3.3. H. columbus and B. longicaudatus

Agronomically useful levels of resistance to the sting and Columbia lance nematodes have not been found in *G. hirsutum*. However, one or more cultivars with good tolerance to the Columbia lance nematode have been identified in North Carolina (Koenning, Edmisten, Barker, & Morrison, 2003). Tolerance is found in both late and early maturing cultivars. However, among late cultivars the highest yielding were the most tolerant, whereas among early cultivars, high yielding cultivars were least tolerant (Koenning & Bowman, 2005).

REFERENCES

Abrão, M. M., & Mazzafera, P. (2001). Efeitos do nível de inóculo de *Meloidogyne incognita* em algodoneiro. *Bragantia Campinas, 60,* 19–6.

Agudelo, P., Robbins, R. T., Stewart, J. McD., Bell, A., & Robinson, A. F. (2005a). Histological observations of *Rotylenchulus reniformis* on *Gossypium longicalyx* and interspecific cotton hybrids. *Journal of Nematology, 37,* 444–447.

Agudelo P., Robbins R. T., Stewart, J. McD., & Szalanski, A. L. (2005b). Intraspecific variability of *Rotylenchulus reniformis* from cotton-growing regions in the United States. *Journal of Nematology, 37,* 105–114.

Asmus, G. L. (2004). Ocorrência de nemátoides fitoparasitos em algodoeiro no estado de Mato Grosso do Sul. *Nematologia Brasileira, 28,* 77–86.

Avila, C. A., Stewart, J. McD., & Robbins, R. T. (2006). Introgression of reniform nematode resistance into upland cotton. *Proceedings of the Beltwide Cotton Conferences,* Memphis, TN: National Cotton Council of America, 154.

Ayala, A. (1962). Pathogenicity of the reniform nematode on various hosts. *Journal of Agriculture of the University of Puerto Rico, 46,* 73–82.

Baird, R. E., Carling, D. E., Watson, C. E., Scruggs, M. L., & Hightower, P. (2004). Effects of nematicides on cotton root mycobiota. *Mycopathologia, 157,* 191–199.

Bansal, R. K., Dahaya, R. S., Narula, N., & Jain, R. K. (2005). Management of *Meloidogyne incognita* in cotton using strains of the bacterium *Gluconacetobacter diazotrophicus. Nematologia Mediterranea, 33,* 101–105.

Bell, A., & Robinson, A. F. (2004). Development and characteristics of triple species hybrids used to transfer reniform nematode resistance from *Gossypium longicalyx* into *Gossypium hirsutum. Proceedings of the Beltwide Cotton Conferences,* Memphis, TN: National Cotton Council of America, 422–426.

Bélot, J.-L., Carraro, I. M., Vilela, P. C. A., Papin, O., Martin, J., Silvie, P., et al. (2005). De nouvelles variétés de cotonnier obtenues au Brésil 15 ans de collaboration entre la coopérative centrale de recherche agricole (Coodetec) et la Cirad. *Cahiers Agricultures, 14,* 249–254.

Birchfield, W. (1962). Host-parasitic relations of *Rotylenchulus reniformis* on *Gossypium hirsutum. Phytophathology, 52,* 862–865.

Birchfield, W., & Brister, L. R. (1962). New hosts and nonhosts of the reniform nematode. *Plant Disease Reporter, 46,* 683–685.

Birchfield, W., & Jones, J. E. (1961). Distribution of the reniform nematode in relation to crop failure of cotton in Louisiana. *Plant Disease Reporter, 45,* 671–673.

Birchfield, W., & Martin, W. J. (1967). Reniform nematode survival in air-dried soil. *Phytopathology, 57,* 804 [Abstract].

Bird, A. F. (1983). Growth and moulting in nematodes: Changes in the dimensions and morphology of *Rotylenchulus reniformis. International Journal for Parasitology, 13,* 201–206.

Bird, D. McK. (1996). Manipulation of host gene expression by root-knot nematodes. *Journal of Parasitology, 82,* 881–888.

Blasingame, D. (1994). *Know your cotton nematodes: Cotton foundation publication no. 11255.* Memphis, TN: National Cotton Council of America.

Blasingame, D. (2006). 2005 cotton disease loss report. *Proceedings of the Beltwide Cotton Conference,* Memphis, TN: National Cotton Council of America, 155–157.

Blasingame, D., & Patel, M. V. (1987). A population and distribution survey of cotton nematodes in Mississippi. *Plant Disease Dispatch.* Mississippi State, MS: Mississippi Cooperative Extension Service.

Bourland, F. M., & Jones, D. C. (2005). Registration of Arkot 9111 germplasm of cotton. *Crop Science, 45,* 2127–2128.

Brodie, B. B., & Cooper, W. E. (1964). Relation of plant parasitic nematodes to postemergence damping-off of cotton. *Phytopathology, 54,* 1023–1027.

Bridge, J. (1992). Nematodes. In R. J. Hillocks (Ed.), *Cotton diseases* (pp. 331–353). Wallingford, UK: CAB International.

Carneiro, R. M. D. G., Das Neves, D. I., Facão, R., Paes, N. S., Cia, E., & Grosse-de-Sá, M. F. (2005). Resistencia de genótipos de algodeiro a *Meloidogyne incognita* raça 3: Reprodução e histopatologia. *Nematologia Brasileira, 29,* 1–10.

Carter, C. H., McGawley, E. C., & Russin, J. S. (1995). Reproduction of *Rotylenchulus reniformis* on weed species common to Louisiana soybean fields. *Journal of Nematology, 27*, 494–495.

Carter, W. W. (1981). Resistance and resistant reaction of *Gossypium arboreum* to the reniform nematode, *Rotylenchulus reniformis*. *Journal of Nematology, 13*, 368–374.

Cochran, A., Long, D., Beckett, T H., Payan, L., & Belles, D. (2006). Efficacy of Avicta Complete Pak against nematodes and seedling diseases in western cotton. *Proceedings of the Beltwide Cotton Conferences,* Memphis, TN: National Cotton Council of America, 131.

Cohn, E. (1973). Histology of the feeding site of *Rotylenchulus reniformis*. *Nematologica, 19,* 455–458.

Colyer, P. D., Kirkpatrick, T. L., Caldwell, W. D., & Vernon, P. R. (2000). Root-knot nematode reproduction and root galling severity on related conventional and transgenic cultivars. *Journal of Cotton Science, 4*, 232–236.

Cook, R., & Evans, K. (1987). Resistance and tolerance. In R. H. Brown & B. R. Kerry (Eds.), *Principles and Practice of Nematode Control in Crops* (pp. 179–231). New York: Academic Press.

Cook, C. G., Namken, L. N., & Robinson, A. F. (1997). Registration of N220-1-91, N222-1-01, N320-2-91, and N419-1-91 nematode-resistant cotton germplasm lines. *Crop Science, 37*, 1028–1029.

Cook, C. G., & Robinson, A. F. (2005). Registration of RN96425, RN96527, and RN96625-1 nematode-resistant cotton germplasm lines. *Crop Science, 45*, 1667–1668.

Cook, C. G., Robinson, A. F., Bridges, A. C., Percival, A.E., Prince, W. B., Bradford, J. M., et al. (2003). Field evaluation of cotton cultivar response to reniform nematodes. *Proceedings of the Beltwide Cotton Conferences,* Memphis, TN: National Cotton Council of America, 861–862.

Cook, C. G., Robinson, A. F., & Namken, L. N. (1997). Tolerance to *Rotylenchulus reniformis* and resistance to *Meloidogyne incognita* race 3 in high-yielding breeding lines of upland cotton. *Journal of Nematology, 29*, 322–328.

Creech, R. G., Jenkins, J. N., Tang, B., Lawrence, G. W., & McCarty, J. C. (1995). Cotton resistance to root-knot nematode: I. Penetration and reproduction. *Crop Science, 35*, 365–368.

Crow, W. T., Dickson, D. W., Weingartner, D. P., McSorley, R., & Miller, G. L. (2000a). Yield reduction and root damage to cotton induced by *Belonolaimus longicaudatus*. *Journal of Nematology, 32*, 205–209.

Crow, W. T., Weingartner, D. P., McSorley, R., & Dickson, D. W. (2000b). Population dynamics of *Belonolaimus longicaudatus* in a cotton production system. *Journal of Nematology, 32*, 210–214.

Dasgupta, D. R., & Seshadri, A. R. (1971a). Reproduction, hybridization and host adaptation in physiological races of the reniform nematode, *Rotylenchulus reniformis*. *Indian Journal of Nematology, 1*, 128–144.

Dasgupta, D. R., & Seshadri, A. R. (1971b). Races of the reniform nematode, *Rotylenchulus reniformis* Lindford and Oliveira, 1940. *Indian Journal of Nematology, 1*, 21–24.

Da Silva, R. A., Serrano, M. A. S., Gomes, A. C., Borges, D. C., De Souza, A. A., Asmus, G. L., et al. (2004). Ocorrência de *Pratylenchus brachyurus* e *Meloidogyne incognita* na cultura do algodoneiro no estado do Mato Grosso. *Fitopatologia Brasileiro, 29*, 337.

Da Ponte, J. J., Jilho, J. S., Lordello, R. R. A., & Lordello, A. I. L. (1998). Sinopse da literatura brasileira sobre *Meloidogyne* em algodão. *Summa Phytopathologica, 24*, 101–104.

Davis R. F., Koenning, S. R., Kemerait, R. C., Cummings, T. D., & Shurley, W. D. (2003). *Rotylenchulus reniformis* management in cotton with crop rotation. *Journal of Nematology, 35*, 58–64.

Davis, R. F., & May, O. L. (2003). Relationship between tolerance and resistance to *Meloidogyne incognita* in cotton. *Journal of Nematology, 35*, 411–416.

Davis, R. F., & May, O. L. (2005). Relationship between yield potential and percentage yield suppression caused by the southern root-knot nematode in cotton. *Crop Science, 45*, 2312–2317.

Davis, R. F., & Noe, J. P. (2000). Extracting *Hoplolaimus columbus* from soil and roots: Implications for treatment comparisons. *Journal of Cotton Science, 4*, 105–111.

Davis, R. F., & Timper, P. (2000). Resistance in selected corn hybrids to *Meloidogyne arenaria* and *M. incognita*. *Journal of Nematology, 32*, 633–640.

Davis, R. F., & Webster, T. M. (2005). Relative host status of selected weeds and crops for *Meloidogyne incognita* and *Rotylenchulus reniformis*. *Journal of Cotton Science, 9*, 41–46.

DeVay, J. E., Gutierrez, A. P., Pullman, G. S., Wakeman, R. J., Garber, R. H., Jeffers, G. P., et al. (1997). Inoculum densities of *Fusarium oxysporum* f. sp. *vasinfectum* and *Meloidogyne incognita* in relation to the development of wilt and the phenology of cotton plants (*Gossypium hirsutum*). *Phytopathology, 87*, 341–346.

Diez, A., Lawrence, G. W., & Lawrence, K. W. (2003). Competition of *Meloidogyne incognita* and *Rotylenchulus reniformis* on cotton following separate and concomitant inoculations. *Journal of Nematology, 35*, 422–429.

Dighe, N., Bell, A. A., Robinson, A. F., Menz, M. A., Cantrell, R. C., & Stelly, D. M. (2007). Tagging and mapping of the reniform nematode resistance gene introgressed from wild diploid, *G. longicalyx*, into upland cotton, *G. hirsutum*. *Crop Science,* 47, in print.

Dismukes, A., Lawrence, K. S., Price, A. J., Lawrence, G. W., & Akridge, R. (2006). Host status of noxious weed plants associated with *Gossypium hirsutum* – *Zea mays* rotation systems to *Rotylenchulus reniformis*. *Proceedings of the Beltwide Cotton Conferences,* Memphis, TN: National Cotton Council of America, 7–11.

Elgawad, A. M. M., Ismail, A. E., & El-Metwally, E. A. (1997). Evaluation of cotton rotation for production in a nematode-infested field in Egypt. *International Journal of Nematology, 7*, 103–106.

Esau, K. (1977). Anatomy of seed plants. New York: John Wiley & Sons.

Esser, R. P. (1976). *Sting nematodes, devastating parasites of Florida crops: Nematology circular no. 18.* Gainesville, FL: Florida Department of Agriculture and Consumer Services.

Faske, T. R., & Starr, J. R. (2006). Sensitivity of *Meloidogyne incognita* and *Rotylenchulus reniformis* to abamectin. *Journal of Nematology, 38*, 240–244.

Farias, P. R. S., Barbosa, J. C., Vieira, S. R., Sánchez-Vila, X., & Ferraz, L. C. C. B. (2002). Geostatistical analysis of the spatial distribution of *Rotylenchulus reniformis* on cotton cultivated under crop rotation. *Russian Journal of Nematology, 10*, 1–9.

French, N. M., Kirkpatrick, T. L., Colyer, P. D., Starr, J. L., Lawrence, K. S., Rich, J. R., et al. (2006). Influence of N-Hibit and ProAct on nematodes in field cotton. *Proceedings of the Beltwide Cotton Conferences,* Memphis, TN: National Cotton Council of America, 137–142.

Garber, R. H., DeVay, J. E., Goodel, P. B., & Roberts, P. A. (1996). Cotton diseases and nematodes. In S. J. Hake, T. A. Kerby, & K. D. Hake (Eds.), *Cotton production manual: Publication 3352* (pp. 150–174). Oakland, CA: University of California Division of Agriculture and Natural Resources.

Gaur, H. S., & Perry, R. N. (1991). The biology and control of plant-parasitic nematode *Rotylenchulus reniformis*. *Agricultural Zoology Reviews, 4*, 177–212.

Gazaway, W. S., Akridge, J. R., & McLean, K. (2000). Impact of various crop rotations and various winter cover crops on reniform nematode in cotton. *Proceedings of the Beltwide Cotton Conferences,* Memphis, TN: National Cotton Council of America, 162–163.

Gazaway, W. S., Akridge, J. R., & McLean, K. (2001). Impact of nematicides on cotton production in reniform infested fields. *Proceedings of the Beltwide Cotton Conferences,* Memphis, TN: National Cotton Council of America, 128.

Gazaway, W. S., Akridge, J. R., & Rodriguez-Kabana, R. (1998). Management of reniform nematode in cotton using various rotation schemes. *Proceedings of the Beltwide Cotton Conferences,* Memphis, TN: National Cotton Council of America, 141–142.

Gazaway, W. S., & McLean, K. S. (2003). A survey of plant-parasitic nematodes associated with cotton in Alabama. *Journal of Cotton Science, 7*, 1–7.

Gilman, D. F., Jones, J. E., Williams, C., & Birchfield, W. (1978). Cotton soybean rotation for control of reniform nematodes. *Louisiana Agriculture, 21*, 10–11.

Gilman, D. F., Marshall, J. G., Rabb, J. G., Lawrence, J. L., Boquet, D. J, & Bartleson, J. L. (1979). Performance of soybean varieties in Louisiana, 1976–78. *Louisiana Agriculture, 22*, 22.

Gonçalves de Oliveira, C. M., Kubo, R. K., Siloto, R. C., & Raga, A. (1999). Eficiência de carbofuran e terbufos sobre nematoides e pragas incias na cultura algodoneira. *Revista de Agricultura, 74,* 325–344.

Goodell, P. B., McClure, M. A., Roberts, P. A., & Thomas, S. H. (1996). Nematodes. In P. B. Goodell, L. D. Godfrey, & R. N. Vargas (Eds.), *Integrated pest management for cotton in the western region of the United States: Publication 3305* (pp. 103–110). Oakland, CA: University of California Division of Agriculture and Natural Resources.

Graham, T. W., & Holdeman, Q. L. (1953). The sting nematode, *Belonolaimus gracilis* Steiner: A parasite of cotton and other crops in South Carolina. *Phytopathology, 43*, 434–439.

Hallmann, J., Rodríguez-Kábana, R., & Kloeper, J. W. (1999). Chitin-mediated change in bacterial communities of the soil, rhizosphere and within roots of cotton in relation to nematode control. *Soil Biology and Biochemistry,* 31, 551–560.
Hartwig, E. E., & Epps, J. M. (1977). Registration of Centennial soybeans (Reg. No. 114). *Crop Science,* 17, 979.
Harville, B. G. (1985). Genetic resistance to reniform nematodes in soybeans. *Plant Disease, 69,* 587–589.
Heald, C. M. (1975). Pathogenicity and histopathology of *Rotylenchulus reniformnis* infecting cantaloupe. *Journal of Nematology, 7,* 149–152.
Heald, C. M., & Heilman, M. D. (1971). Interaction of *Rotylenchulus reniformis*, soil salinity and cotton. *Journal of Nematology, 3,* 179–182.
Heald C. M., & Inserra, R. N. (1988). Effect of temperature on infection and survival of *Rotylenchulus reniformis*. *Journal of Nematology, 20,* 356–361.
Heald, C. M., & Orr, C. C. (1984). Nematode parasites of cotton. In W. R. Nickle (Ed.), *Plant and insect nematodes* (pp. 147–166). New York: Marcel Dekker, Inc.
Heald, C. M., & Robinson, A. F. (1987). Effects of soil solarization on *Rotylenchulus reniformis* in the Lower Rio Grande Valley of Texas. *Journal of Nematology, 19,* 93–103.
Heald, C. M., & Robinson, A. F. (1990). Survey of current distribution of *Rotylenchulus reiniformis* in the United States. *Journal of Nematology, 22,* 695–699.
Hillocks, R. J., & Kibani, T. H. M. (2002). Factors affecting the distribution, incidence, and spread of Fusarium wilt of cotton in Tanzania. *Experimental Agriculture, 38,* 13–27.
Hinchliffe, D. J., Lu, Y., Potenza, C., Segupta-Gopalan, C., Cantrell, R. G., & Zhang, J. (2005). Resistance gene markers are mapped to homologous chromosomes in cultivated tetraploid cotton. *Theoretical and Applied Genetics, 110,* 1074–1085.
Inserra, R. N., Dunn, R. A., McSorley, R., Langdon, K. R., & Richmer, A. Y. (1989). *Weed hosts of Rotylenchulus reniformis in ornamental nurseries of southern Florida: Nematology circular no. 171.* Gainesville, FL: Florida Department of Agriculture and Consumer Services.
Jaskaran, R. K., Singh, J., & Vats, R. (2000). Avoidable yield losses in cotton due to root-knot nematode (*Meloidogyne incognita*) race 4. *Indian Journal of Nematology, 30,* 86–110.
Jayakuma, J., Ramakrishnan, S., & Rajendran, G. (2002). Bio-efficacy of *Paecilomyces lilacinus* against reniform nematode, *Rotylenchulus reniformis* infesting cotton. *Current Nematology, 13,* 19–21.
Jayakumar, J., Ramakrishnan, S., & Rajendran, G. (2003). Bio-efficacy of fluorescent pseudomonad isolates against reniform nematode, *Rotylenchulus reniformis* infecting cotton. *Indian Journal of Nematology, 33,* 13–15.
Jenkins, J. N., Creech, R. G., Tang, B., Lawrence, G. W., & McCarty, J. C. (1995). Cotton resistance to root-knot nematode: II. Post-penetration development. *Crop Science, 35,* 369–373.
Jepson, S. B. (1987). Identification of root-knot nematodes (*Meloidogyne* species). Wallingford, UK: CAB International.
Jones, J. E., Beasley, J. P., Dickson, J. I., & Caldwell, W. D. (1988). Registration of four cotton germplasm lines with resistance to reniform and root-knot nematodes. *Crop Science, 28,* 199–200.
Jones, J. E., & Birchfield, W. (1967). Resistance of the experimental cotton variety, Bayou, and related strains to root knot nematode and Fusarium wilt. *Phytopathology, 57,* 1327–1331.
Jones, J. E., Dickson, J. E., Aguillard, W., Caldwell, W. D., Moore, S. H., Hutchinson, R. L., et al. (1991). Registration of 'LA 887' cotton. *Crop Science, 31,* 1701.
Jones, J. E., Wright, S. L., & Newsom, L. D. (1958). Sources of tolerance and inheritance of resistance to root-knot nematode in cotton. *Proceedings of the 11th Annual Cotton Improvement Conference,* Memphis, TN: National Cotton Council of America, 34–39.
Jones, J. E., Newsom, L. O., & Finley, E. L. (1959). Effect of the reniform nematode on yield, fiber properties of upland cotton. *Agronomy Journal, 51,* 353–356.
Jones, J. G. W., & Nortcote, D. H. (1972). Nematode-induced syncytium – a multinucleate transfer cell. *Journal of Cell Science, 10,* 789–809.
Katsantonis, D., Hillocks, R. J., & Gowen, S. (2003). Comparative effect of root-knot nematode on severity of Verticillium and Fusarium wilt of cotton. *Phytoparasitica, 31,* 1–9.

Kemerait, R. C., Jost, P. H., Davis, R. F., Brown, S. N., Green, T. W., Mitchell, B. R., et al. (2006). Assessment of seed treatments for management of nematodes in Georgia. *Proceedings of the Beltwide Cotton Conferences,* Memphis, TN: National Cotton Council of America, 144–149.

Khadr, A. S., Salem, A. A., & Oteifa, B. A. (1972). Varietal susceptibility and significance of the reniform nematode, *Rotylenchulus reniformis*, in Fusarium wilt of cotton. *Plant Disease Reporter, 56,* 1040–1042.

Khalilian, A., Sullivan, M. J., Mueller, J. D., Shiralipour, A., Wolak, F. J., Williamson, R. E., et al. (2002). Effects of surface application of MSW compost on cotton production – Soil properties, plant responses and nematode management. *Compost Science and Utilization, 10,* 270–279.

Kinloch, R. A., & Rich, J. R. (2001). Management of root-knot and reniform nematode in ultra-narrow row cotton with 1,3-dichloropropene. *Supplement to Journal of Nematology, 33,* 311–313.

Koenning, S. R. (2002). Economics and ecology put to use – Action thresholds. *Proceedings of the Beltwide Cotton Conferences,* Memphis, TN: National Cotton Council of America, 143–147.

Koenning, S. R., & Barker, K. R. (2004). Influence of poultry litter applications on nematode communities in cotton agroecosystems. *Supplement to Journal of Nematology, 36,* 524–533.

Koenning, S. R., Barker, K. R., & Bowman, D. T. (2000). Tolerance of selected cotton lines to *Rotylenchulus reniformis. Supplement to Journal of Nematology, 32,* 519–523.

Koenning, S. R., Barker, K. R., & Bowman, D. T. (2001). Resistance as a tactic for management of *Meloidogyne incognita* on cotton in North Carolina. *Journal of Nematology, 33,* 126–131.

Koenning, S. R., & Bowman, D. T. (2005). Cotton tolerance to *Hoplolaimus columbus* and impact on population densities. *Plant Disease, 89,* 649–653.

Koenning, S. R., Edmisten, K. L., Barker, K. R., & Morrison, D. E. (2003). Impact of cotton production systems on management of *Hoplolaimus columbus. Journal of Nematology, 35,* 73–77.

Koenning, S. R., Kirkpatrick, T. L., Starr, J. L., Wrather, J. A., Walker, N. R., & Mueller, J. D. (2004). Plant parasitic nematodes attacking cotton in the United States: Old and emerging production challenges. *Plant Disease, 88,* 100–113.

Koenning, S. R., Walters, S. A., & Barker, K. R. (1996). Impact of soil texture on the reproductive and damage potentials of *Rotylenchulus reniformis* and *Meloidogyne incognita* on cotton. *Journal of Nematology, 28,* 527–536.

Komar, S. J., Wiley, P. D., Kermerait, R. C., & Shurley, W. D. (2003). Nematicide treatment effects on reniform nematodes in cotton. *Proceedings of the Beltwide Cotton Conferences,* Memphis, TN: National Cotton Council of America, 272.

Lal, A., Yadav, B. S., & Nandwana, R. P. (1976). A record of some new and known weed hosts of *Rotylenchulus reniformis* Linford & Oliveira, 1940 from Rajasthan. *Indian Journal of Nematology, 6,* 94–116.

Lawrence, G. W., & McLean, K. S. (1996). Reniform nematode and cotton production in Mississippi. *Proceedings of the Beltwide Cotton Conferences,* Memphis, TN: National Cotton Council of America, 251–253.

Lawrence, G. W., & McLean, K. S. (2000). Effect of foliar applications of oxamyl with aldicarb for the management of *Rotylenchulus reniformis* on cotton. *Supplement to Journal of Nematology, 32,* 542–549.

Lawrence, G. W., & McLean, K. S. (2001). Reniform nematodes. In T. L. Kirkpatrick & C. S. Rothrock (Eds.), *Compendium of cotton diseases* (pp. 42–44). St. Paul, MN: APS Press.

Lawrence, G. W., McLean, K. S., Batson, W. E., Miller, D., & Borbon, J. C. (1990). Response of *Rotylenchulus reniformis* to nematicide applications on cotton. *Supplement to Journal of Nematology, 22,* 707–711.

Lawrence, K. S., Lawrence, G. W., & VanSantan, E. (2005a). Effect of cold storage on recovery of *Rotylenchulus reniformis* from naturally infested soil. *Journal of Nematology, 37,* 272–275.

Lawrence, K. S., Usery, S. R., Burmester, C. H., & Lawrence, G. W. (2005b). *Evaluation of seed treatment nematicides for reniform nematode management in cotton in North Alabama.* In *2004 cotton research report no. 26.* Auburn, AL: Alabama Agricultural Experiment Station.

Linford, M. B., & Oliveira, J. M. (1940). *Rotylenchulus reniformis*, nov. gen. n. sp., a nematode parasite of roots. *Proceedings of the Helminthological Society of Washington, 7,* 35–42.

Linford, M. B., & Yap, F. (1940). Some host plants of the reniform nematode in Hawaii. *Proceedings of the Helminthological Society of Washington, 7*, 42–44.

Louw, I. W. (1982). Nematode pests of cotton. In D. P. Keetch & J. Heyns (Eds.), *Nematology in Southern Africa* (pp. 73–39). Pretoria: Republic of South Africa Department of Agriculture and Fisheries.

Maggenti, A. (1981). *General nematology*. New York: Springer-Verlag.

Martin, W. J., Newsom, L. D., & Jones, J. E. (1956). Relationship of nematodes to the development of Fusarium wilt in cotton. *Phytopathology, 46*, 285–289.

May, O. L., Davis, R. F., & Baker, S. H. (2001). Registration of GA 161 cotton. *Crop Science, 41*, 995–1996.

May, O. L., Davis, R. F., & Baker, S. H. (2004). Registration of GA96-211 upland cotton germplasm line. *Crop Science, 44*, 700–701.

McBride, R. G., Mikkelsen, R. L., & Barker, K. R. (1999). Survival and infection of root-knot nematodes added to soil amended with rye at different stages of decomposition and cropped with cotton. *Applied Soil Ecology, 13*, 231–235.

McClure, M. A., & Robertson, J. (1973). Infection of cotton seedlings by *Meloidogyne incognita* and a method of producing uniformly infected root segments. *Nematologica, 19*, 428–434.

McGawley, E. C., & Overstreet, C. (1995). Reproduction and pathological variation in populations of *Rotylenchulus reniformis*. *Journal of Nematology, 27*, 508 [Abstract].

McLean, K. S., & Lawrence, G. W. (2003). Efficacy of aldicarb to *Rotylenchulus reniformis* and biodegradation in cotton field soils. *Journal of Nematology, 35*, 373–375.

McLean, K. S., Lawrence, G. W., Overstreet, C., & Young, L. D. (2003). Efficacy of anhydrous ammonia on reniform nematode in cotton. *Proceedings of the Beltwide Cotton Conferences,* Memphis, TN: National Cotton Council of America, 282–283.

Mehta, U. K., & Sundara, P. R. 1989. Reniform nematode, *Rotylenchulus reniformis* Linford and Oliveira, 1940, on roots of sugarcane. *FAO Plant Protection Bulletin, 37*, 83–86.

Moresco, E., Morgan, E., Ripple, K. W., Smith, C. W., & Starr, J. L. (2004). Resistance to *Rotylenchulus reniformis* in interspecific *Gossypium* hybrids. *Journal of Nematology, 36*, 335 [Abstract].

Mueller, J. D. (1993). Lance nematodes. *Proceedings of the Beltwide Cotton Conferences,* Memphis, TN: National Cotton Council of America, 176–177.

Mueller, J. D., & Lewis, S. A. (2001). Lance nematodes. In T. L. Kirkpatrick & C. S. Rothrock (Eds.), *Compendium of cotton diseases* (pp. 44–45). St. Paul, MN: APS Press.

Mueller, J. D., & Sullivan, M. J. (1988). Response of cotton to infection by *Hoplolaimus columbus*. *Supplement to Journal of Nematology, 20*, 86–89.

Nakasono, K. (1983). Studies on morphological and physio-ecological variation of the reniform nematode, *Rotylenchulus reniformis* Linford and Oliveira, 1940 with an emphasis on differential geographical distribution of amphimictic and parthenogenetic populations in Japan. *Bulletin of the National Institute of Agricultural Sciences of Japan, 38*, 63–67.

Neal, D. C. (1954). The reniform nematode and its relationship to the incidence of Fusarium wilt of cotton at Baton Rouge, Louisiana. *Phytopathology, 44*, 447–450.

Newman, M. A., & Stebbins, T. C. (2002). Recovery of reniform nematodes at various soil depths in cotton. *Proceedings of the Beltwide Cotton Conferences,* Memphis, TN: National Cotton Council of America, 254–255.

Ogallo, J. M., Goodell, P. B., Eckert, J. W., & Roberts, P. A. 1999. Management of root-knot nematodes with resistant cotton cv. NemX. *Crop Science, 39*, 418–421.

Orr, C. C., & Brashears, A. D. (1977). Aldicarb and DBCP for root-knot nematode control of cotton. *Plant Disease Reporter, 62*,623–624.

Overstreet, C. (2001). *Survey of reniform nematodes in 2001: Update no. 3.* Baton Rouge, LA: Louisiana Cooperative Agricultural Extension Service.

Overstreet, C., & Erwin, T. L. (2003). The use of Telone in cotton production in Louisiana. *Proceedings of the Beltwide Cotton Conferences,* Memphis, TN: National Cotton Council of America, 277–278.

Overstreet, C., & McGawley. E. C. (2000). Geographical dispersion of reniform nematode in Louisiana. *Proceedings of the Beltwide Cotton Conferences,* Memphis, TN: National Cotton Council of America, 168–171.

Overstreet, C., & McGawley. E. C. (2001). Sting nematodes. In T. L. Kirkpatrick & C. S. Rothrock (Eds.), *Compendium of cotton diseases* (pp. 45–46). St. Paul, MN: APS Press.

Page, S. L. J. 1985. *Meloidogyne acronea*: Set 8, no. 114. In *CIH descriptions of plant-parasitic nematodes*. Wallingford, UK: CAB International.

Palanisamy, S., & Balasubramanian, M. (1983). Assessment of avoidable yield loss in cotton (*Gossypium barbadense L.*) by fumigation with metam sodium. *Nematologia Mediterranea, 11*, 201–202.

Palmatee, A. J., Lawrence, K. S., VanSanten, E., & Morgan-Jones, G. (2004). Interaction of *Rotylenchulus reniformis* with seedling disease pathogens of cotton. *Journal of Nematology, 36*, 160–166.

Pate, J. S., & Gunning, B. E. S. (1972). Transfer cells. *Annual Review of Plant Physiology, 23*, 173–196.

Patel, R. R., Patel, B. A., & Thakar, N. A. (2003). Organic amendments in management of *Rotylenchulus reniformis* on cotton. *Indian Journal of Nematology, 33*, 146–148.

Patel, R. R., Patel, B. A., & Thakar, N.A. (2004). Pathogenicity of reniform nematode, *Rotylenchulus reniformis* on cotton. *Indian Journal of Nematology, 34*, 106–107.

Percival, A. E., Wendel, J. F., & Stewart, J. M. (1999). *Taxonomy and germplasm resources.* In C. W. Smith & J. T. Cothren (Ed.), *Cotton: Origin, history, technology, and production* (pp. 33–63). New York: John Wiley & Sons, Inc.

Pettigrew, W. T., Meredith, W. R. Jr., & Young, L. D. (2005). Potassium fertilization effects on cotton lint yield, yield components, and reniform nematode populations. *Agronomy Journal, 97*, 1245–1251.

Prasad, K. S., & Padeganur, G. M. (1980). Observations on the association of *Rotylenchulus reniformis* with Verticillium wilt of cotton. *Indian Journal of Nematology, 10*, 91–92.

Quenéhérvé, P., Drob, F., & Topart, P. (1995). Host status of some weeds to *Meloidogyne* spp., *Pratylenchus* spp., *Helicotylenchus* spp., and *Rotylenchulus reniformis* associated with vegetables cultivated in polytunnels in Martinique. *Nematropica, 25*, 149–157.

Rebois, R. V. (1973). Effect of soil temperature on infectivity and development of *Rotylenchulus reniformis* on resistant and susceptible soybeans, *Glycine max. Journal of Nematology, 5*, 10–13.

Rebois, R. V., Madden, P. A., & Eldridge, B. J. (1975). Some ultrastructural changes induced in resistant and susceptible soybean roots following infection by *Rotylenchulus reniformis*. *Journal of Nematology, 7*, 122–139.

Riegel, C., & Noe, J. P. (2000). Chicken litter amendment effects on soil-borne microbes and *Meloidogyne incognita* on cotton. *Plant Disease, 84*, 1275–1281.

Robbins, R. T., & Barker, K. R. (1974). The effects of soil type, particle size, temperature, and moisture on reproduction of *Belonolaimus longicaudatus* from North Carolina and Georgia. *Journal of Nematology, 6*, 1–6.

Robbins, R. T., McNeely, V. M., & Lorenz, G. M. III. (1998). The lance nematode, *Hoplolaimus magnistylus,* in cotton in Arkansas. *Journal of Nematology, 30*, 590–591.

Robbins, R. T., Shipe, E. R., Rakes, L., Jackson, L. E., Gbur, E. E., & Dombek, D. G. (2001). Host suitability in soybean cultivars for the reniform nematode, 2000 tests. *Supplement to Journal of Nematology, 33*, 314–317.

Roberts, P. A., & Matthews, W. C. (1984). Cotton growth responses to nematode infection. *Proceedings of the Beltwide Cotton Conferences,* Memphis, TN: National Cotton Council of America, 17–19.

Robinson, A. F. (1989). Thermotactic adaptation in two foliar and two root-parasitic nematodes. *Revue de Nematologie, 12*, 125–131.

Robinson, A. F. (1994). Movement of five nematode species through sand subjected to natural temperature gradient fluctuations. *Journal of Nematology, 26*, 46–58.

Robinson, A. F. (2002). *Reniform nematodes:* Rotylenchulus *species.* In J. L Starr, R. Cook, & J. Bridge (Eds.), *Plant resistance to parasitic nematodes* (pp. 153–174). Wallingford, UK: CABI Publishing.

Robinson, A. F. (2007). Reniform in U.S. cotton: When where why and some remedies. *Annual Review of Phytopathology,* 45: 263–288.

Robinson, A. F., Akridge, J. R., Bradford, J. B., Cook, C. G., Gazaway, W. S., McGawley, E. C., et al. (2006). Suppression of *Rotylenchulus reniformis* 122 cm deep endorses resistance introgression in *Gossypium. Journal of Nematology, 38*, 195–209.

Robinson, A. F., Akridge, J. R., Bradford, J. M., Cook, C. G., Gazaway, W. S., Overstreet, C., et al. (2005). Vertical distribution of *Rotylenchulus reniformis* in cotton fields. *Journal of Nematology, 37*, 265–271.

Robinson, A. F., Bell, A. A., Dighe, N., Menz, M. A., Nichols, R. L., & Stelly, D. M. (2007). Introgression of resistance to nematode *Rotylenchulus reniformis* into upland cotton (*Gossypium hirsutum*) from *G. longicalyx*. *Crop Science,* 47: 1865–1877.

Robinson, A. F., Bowman, D. J., Colyer, P. D., Cook, C. G., Creech, R. G., Gannaway, J. R., et al. (2001). Nematode resistance. In T. L. Kirkpatrick & C. S. Rothrock (Eds.), *Compendium of cotton diseases* (pp. 68–72). St. Paul, MN: APS Press.

Robinson, A. F., Bridges, A. C., & Percival, A. E. (2004). New sources of resistance to the reniform (*Rotylenchulus reniformis* Linford and Oliveira) and root-knot (*Meloidogyne incognita* (Kofoid & White) Chitwood) nematode in upland (*Gossypium hirsutum* L.) and sea island (*G. barbadense* L.) cotton. *Journal of Cotton Science, 8*, 191–197.

Robinson, A. F., & Cook, C. G. (2001). Root-knot and reniform nematode reproduction on kenaf and sunn hemp compared with that on nematode resistant and susceptible cotton. *Industrial Crops and Products, 13*, 249–264.

Robinson, A. F., Cook, C. G., & Percival, A. E. (1999). Resistance to *Rotylenchulus reniformis* and *Meloidogyne incognita* race 3 in the major cotton cultivars planted since 1950. *Crop Science, 39*, 850–858.

Robinson, A. F., Cook, C. G., Westphal, A., & Bradford, J. M. (2005a). *Rotylenculus reniformis* below plow depth suppresses cotton yield and root growth. *Journal of Nematology, 37*, 285–291.

Robinson, A. F., Gutierrez, O. A., LaFoe, J. M., McCarty, Jr. J. C., & Jenkins, J. N. (2005b). Detection of reniform nematode resistance in primitive *Gossypium hirsutum* and *G. barbadesnse* during a survey of the U.S. National Cotton Collection and initiation of research to incorporate resistance into agronomic cotton. *Proceedings of the Beltwide Cotton Conferences,* Memphis, TN: National Cotton Council of America, 934.

Robinson, A. F., & Heald, C. M. (1989). Accelerated movement of nematodes from soil in Baermann funnels with temperature gradients. *Journal of Nematology, 21*, 370–378.

Robinson, A. F., & Heald, C. M. (1991). Carbon dioxide and temperature gradients in Baermann funnel extraction of *Rotylenchulus reniformis. Journal of Nematology, 23*, 28–38.

Robinson, A. F., & Heald, C. M. (1993). Movement of *Rotylenchulus reniformis* through sand and agar in response to temperature, and some obeservations on vertical descent. *Nematologica, 39*, 92–103.

Robinson, A. F., Heald, C. M., Flanagan, S. L., Thames, W. H., & Amador, J. (1987). Geographical distributions of *Rotylenchulus reniformis*, *Meloidogyne incognita*, and *Tylenchulus semipenetrans* in the Lower Rio Grande Valley as related to soil texture and land use. *Annals of Applied Nematology, 1*, 20–25.

Robinson, A. F., Inserra, R. N., Caswell-Chen, E. P., Vovlas, N., & Troccoli, A. (1997). *Rotylenchulus* species: Identification, distribution, host ranges, and crop plant resistance. *Nematropica, 27*, 127–180.

Robinson, A. F., & Jaffee, B. A. (1996). Repulsion of *Meloidogyhne incognita* by alginate pellets containing hyphae of *Monacrosporium cionopagum, M. ellipsosporum,* or *Hirsutella rhossiliensis. Journal of Nematology, 28*, 133–147.

Robinson, M., Jenkins, J. N., & McCarty, J. C. Jr. (1997). Root-knot nematode resistance of F_2 cotton hybrids from crosses of resistant germplasm and commercial cultivars. *Crop Science, 37*, 1041–1046.

Robinson A. F., & Orr, C. C. (1980). Histopathology of *Rotylenchulus reniformis* on sunflower. *Journal of Nematology, 12,* 84–85.

Robinson A. F., & Percival, A. E. (1997). Resistance to *Meloidogyne incognita* race 3 and *Rotylenchulus reniformis* in wild accessions of *Gosspium hirsutum* and *G. barbadense* from Mexico. *Journal of Nematology, 29*, 746–755.

Rodríguez-Fuentes, E. (1980). Supervivencia de *Rotylenchulus reniformis* Linford and Oliveira, 1940, sin la influencia del cultivo. *Revista Científica de la Facultad de Ciencias Agrícolas, Universidad Central de Cuba, 7*, 7–74.

Rodríguez-Fuentes, E., & Añorga-Morles, J. (1977). Estudios sobre la reproducción partenogenética y el desarrollo post-embrional de *Rotylenchulus reniformis* Linford y Oliveira, 1940. *Revista Científica de la Facultad de Ciencias Agrícolas, Universidad Central de Cuba, 4*, 49–55.

Roman, J. (1964). Immunity of sugarcane to the reniform nematode. *Journal of the Agricultural University of Puerto Rico, 48*, 162–163.

Sanaralingham, A., & McGawley, E. C. (1994). Interrelationships of *Rotylenchulus reniformis* with *Rhizoctonia solani* on cotton. *Journal of Nematology, 26*, 475–485.

Sasser J. (1972). *Nematode diseases of cotton.* In J. M. Webster (Ed.), *Economic nematology* (pp. 187–214). London: Academic Press.

Sasser, J. N., & Freckman, D. W. (1987). A world perspective on nematology: The role of the society. In J. A. Veech & D. W. Dickson (Eds.), *Vistas on nematology: A commemoration of the twenty-fifth anniversary of the Society of Nematologists* (pp. 7–14). Hyattsville, MD: Society of Nematologists.

Schwarz, M., Graham, C., & Kleyla, C. (2006). *New seed-applied nematicides from Bayer Cropscience. Proceedings of the Beltwide Cotton Conference,* Memphis, TN; National Cotton Council of America, 2284.

Sciumbato, G. L., Blessitt, J. A., & Blasingame, D. (2004). Mississippi cotton nematode survey: Results of an eight county survey. *Proceedings of the Beltwide Cotton Conferences,* Memphis, TN: National Cotton Council of America, 451.

Sehgal, M., & Gaur, H. S. (1988). Survival and infectivity of the reniform nematode, *Rotylenchulus reniformis*, in relation to moisture stress in soil without a host. *Indian Journal of Nematology, 18*, 49–54.

Sehgal, M., & Gaur, H. S. (1989). Effect of the rate of soil moisture loss on the survival, infectivity and development of *Rotylenchulus reniformis*, the reniform nematode. *Pakistan Journal of Nematology, 7*, 83–90.

Shen, X., Van Becelaere, G., Kumar, P., Davis, R. F., May, O. L., & Chee, P. 2006. QTL mapping for resistance to root-knot nematodes in the M-120 RNR upland cotton line (*Gossypium hirsutum* L.) of the Auburn 623 RNR source. *Theoretical and Applied Genetics, 113,* 1539–1549.

Shepherd, R. L. (1974a). Transgressive segregation for root-knot nematode resistance in cotton. *Crop Science, 14*, 872–875.

Shepherd, R. L. (1974b). Registration of Auburn 623 RNR cotton germplasm (Teg. No. GP20). *Crop Science, 14*, 911.

Shepherd, R. L. (1982). Registration of three germplasm lines of cotton (Reg. Nos. GP 164 to GP 166. *Crop Science, 22*, 692–293.

Shepherd, R. L. (1983). New sources of resistance to root-knot nematodes among primitive cottons. *Crop Science, 23*, 999–1002.

Shepherd, R. L. (1986a). Cotton resistance to the root-knot/Fusarium wilt complex. II. Relation to root-knot resistance and its implications on breeding for resistance. *Crop Science, 26*, 233–237.

Shepherd, R. L. (1986b). Genetic analysis of root-knot nematode resistance in cotton. *Proceedings of the Beltwide Cotton Conferences,* Memphis, TN: National Cotton Council of America, 502.

Shepherd, R. L., & Huck. M. G. (1989). Progression of root-knot nematode symptoms and infection on resistant and susceptible cottons. *Journal of Nematology, 21*, 245–241.

Shepherd, R. L., & Kappelman, A. J. (1986). Cotton resistance to the root-knot/Fusarium wilt complex. I. Relation to Fusarium wilt resistance and its implications on breeding for resistance. *Crop Science, 26*, 238–232.

Shepherd, R. L. (1988). Registration of twelve nonphotoperiodic lines with root-knot nematode resistant primitive cotton germplasm. *Crop Science, 28*, 868–869.

Shepherd, R. L. (1989). Registration of nine cotton germplasm lines resistant to root-knot nematode. *Crop Science, 36*, 820.

Sikora, R. A. (1979). Predisposition to *Meloidogyne* infection by the endotropic mycorrhizal fungus *Glomus mosseae*. In F. Lamberti & C. E. Taylor (Eds.), *Root-knot nematodes (*Meloidogyne *species) systematics, biology and control* (pp. 399–404). New York: Academic Press.

Silvey, D. T., Ripple, K. R., Smith, C. W., & Starr, J. L. (2003). Identification of RFLP loci linked to resistance to *Meloidogyne incognita* and *Rotylenchulus reniformis*. *Proceedings of the Beltwide Cotton Conferences,* Memphis, TN: National Cotton Council of America, 270.

Sivakumar, C. V., & Seshadri, A. R. (1971). Life history of the reniform nematode *Rotylenchulus reniformis* Linford and Oliveira, 1940. *Indian Journal of Nematology, 1*, 7–20.

Starr, J. L. (1998). *Cotton.* In K. R. Barker, G. A. Pederson, & G. L. Windham (Eds.), *Plant and nematode interactions: No. 36 in the series agronomy* (pp. 359–379). Madison, WI: American Society of Agronomy, Crop Science Society of America, Soil Science Society of America Publishers.

Starr, J. L., Bridge, J., & Cook, R. (2002). *Resistance to plant parasitic nematodes, history, current use, and future potential.* In J. L. Starr, R. Cook, & J. Bridge (Eds.), *Plant resistance to parasitic nematodes* (pp. 1–22). Wallingford, UK: CABI Publishing.

Starr J. L., Heald, C. M., Robinson, A. F., Smith, R. G., & Krause, J. P. (1993). *Meloidogyne incognita* and *Rotylenchulus reniformis* and associated soil textures from some cotton production areas of Texas. *Journal of Nematology, 25*, 252–256.

Starr, J. L., & Jeger, M. J. 1985. Dynamics of winter survival of eggs and juveniles of *Meloidogyne incognita* and *M. arenaria. Journal of Nematology, 17*, 252–256.

Starr, J. L., Jeger, M. J., Martyn, R. D., & Schilling, K. (1989). Effects of *Meloidogyne incognita* and *Fusarium oxysporum* f. sp. *vasinfectum* on plant mortality and yield of cotton. *Phytopathology, 79*, 640–646.

Starr, J. L., & Martyn, R. D. (1991). Reaction of cotton cultivars to Fusarium wilt and root-knot nematodes. *Nematropica, 21*, 51–58.

Starr, J. L., & Page, S. L. J. (1990). Nematode parasites of cotton and other tropical fibre crops. In M. Luc, R. A. Sikora, & J. Bridge (Eds.), *Plant parasitic nematodes in subtropical and tropical agriculture* (pp. 539–556). Wallingford, UK: CABI Publishing.

Starr, J. L., & Veech, J. A. (1986). Susceptibility to root-knot nematode in cotton lines resistant to the Fusarium wilt/root-knot complex. *Crop Science, 26*, 543–546.

Starr J. L., Wheeler, T. A., & Walker, N. R. (2001). Nematode fungal interactions. In T. L. Kirkpatrick & C. S. Rothrock (Eds.), *Compendium of cotton diseases* (pp. 46–48). St. Paul, MN: APS Press.

Stewart, J. M., & Robbins, R. T. (1995). Evaluation of Asiatic cottons for resistance to reniform nematode. In D. M. Oosterhiuis (Ed.), *Proceedings of the 1994 cotton research meeting: Special report 166,* Fayetteville, AR: Arkansas Agricultural Experiment Station, 165–168.

Stewart, J. M., & Robbins, R. T. (1996). Identification and enhancement of resistance to reniform nematode in cotton germplasm. *Proceedings of the Beltwide Cotton Conferences,* Memphis, TN: National Cotton Council of America, 255.

Sud, U. C., Varaprasad, K. J., Seshadri, A. R., & Kher, K. K. (1984). Relationship between initial densities of *Rotylenchulus reniformis* and damage to cotton with fits to Seinhorst curves. *Indian Journal of Nematology, 14*, 148–151.

Taylor, A. L., & Sasser, J. N. (1978). *Biology, identification and control of root-knot nematodes (*Meloidogyne *species).* Raleigh, NC: North Carolina State Unviersity Graphics.

Thames, W. H., & Heald, C. M. (1974). Chemical and cultural control of *Rotylenchulus reniformis* on cotton. *Plant Disease Reporter, 58*, 337–341.

Thomas, S H., & Kirkpatrick, T. L. (2001). Root-knot nematodes. In T. L. Kirkpatrick & C. S. Rothrock (Eds.), *Compendium of cotton diseases* (pp. 40–42). St. Paul, MN: APS Press.

Thorne, G. (1961). *Principles of nematology*. New York: McGraw-Hill.

Timper, P., Davis, R. F., & Tillman, P. G. (2006). Reproduction of *Meloidogyne incognita* on winter cover crops used in cotton production. *Journal of Nematology, 38,* 83–89.

Tsai, B. Y., & Apt, W. F. (1979). Anhydrobiosis of the reniform nematode: Survival and coiling. *Journal of Nematology, 11*, 316 [Abstract].

Veech, J. A. (1978). An apparent relationship between methoxy-substituted terpenoid adehydes and the resistance of cotton to *Meloidogyne incognita. Journal of Nematology, 9*, 222–229.

Veech, J. A. (1979). Histochemical localization and nematoxicity of terpenoid aldehydes in cotton. *Journal of Nematology, 11*, 240–246.

Veech, J. A. (1984). Nematodes. In R. J. Kohel & C. F. Lewis (Eds.), *Cotton* (pp. 309–330). Madison, WI: American Society of Agronomy, Crop Science Society of America, Soil Science Society of America.

Veech, J. A., & McClure, M. S. 1977. Terpenoid aldehydes in cotton roots susceptible and resistant to root-knot nematode, *Meloidogyne incognita. Journal of Nematology, 9*, 222–229.

Walker, N. R., Kirkpatrick, T. L., & Rothrock, C. S. (1998). Interaction between *Meloidogyne incognita* and *Thelaviopsis basicola* on cotton (*Gossypium hirsutum*). *Journal of Nematology, 30*, 415–422.

Walker, N. R., Kirkpatrick, T. L., & Rothrock, C. S. (1999). Effect of temperature on the interaction between *Meloidogyne incognita* and *Thelaviopsis basicola* on cotton. *Phytopathology, 89*, 613–617.

Wang, C. Matthews, W. C., & Roberts, P. A. (2006). Phenotypic expression of *rkn1*-mediated *Meloidogyne incognita* resistance in *Gossypium hirsutum* populations. *Journal of Nematology, 38*, 250–257.

Wang, K., Riggs, R. D., & Crippen, D. (2004). Suppression of *Rotylenchulus reniformis* on cotton by the nematophagous fungus ARF. *Journal of Nematology, 36*, 186–191.

Wang, K., Riggs, R. D., & Crippen, D. (2005). Isolation, selection, and efficacy of *Pochonia chlamydosporia* for control of *Rotylenchulus reniformis* on cotton. *Phytopathology, 95*, 890–893.

Westphal, A., Robinson, A. F., Scott, Jr. A. W., & Santini, J. B. (2004). Depth distribution of *Rotylenchulus reniformis* under crops of different host status and after fumigation. *Nematology, 6*, 97–108.

Westphal, A., & Scott Jr., A. W. (2005). Implementation of soybean in cotton cropping sequences for management of reniform nematode in South Texas. *Crop Science, 45*, 233–239.

Wheeler, T. A., Kaufman, H. W., Baugh, B., Kidd, P., Schuster, G., & Siders, K. (1999). Comparison of variable and single-rate applications of aldicarb on cotton yield in fields infested with *Meloidogyne incognita. Supplement to Journal of Nematology, 31*, 700–709.

Wheeler, T. A., Baugh, B., Kaufman, H., Schuster, G., & Siders, K. (2000). Variability in time and space of *Meloidogyne incognita* fall populations density in cotton fields. *Journal of Nematology, 32*, 258–264.

Windham, G. L., & Lawrence, G. W. (1992). Host status of commercial maize hybrids to *Rotylenchulus reniformis*. *Journal of Nematology, 24*, 745–748.

Wolcott, M., Overstreet, C., Burris, E., Cook, D., Sullivan, D., Padgett, G. B., et al. (2005). Evaluating cotton nematicide response across soil electrical conductivity zones using remote sensing. *Proceedings of the Beltwide Cotton Conferences,* Memphis, TN: National Cotton Council of America, 215–220.

Womersley, C., & Ching, C. (1989). Natural dehydration regimes as a prerequisite for the successful induction of anhydrobiosis in the nematode *Rotylenchulus reniformis*. *Journal of Experimental Biology, 143*, 359–372.

Wrather, J. A., Stevens, W. E., Kirkpatrick, T. L., & Kitchen, N. R. (2002). Effects of site-specific application of aldicarb on cotton in a *Meloidogyne incognita*-infested field. *Journal of Nematology, 34*, 115–119.

Yik, C. P., & Birchfield, W. (1984). Resistant germplasm in *Gossypium* species and related plants to *Rotylenchulus reniformis*. *Journal of Nematology, 16*, 146–153.

Zhang, X.-D., Callahan, F. E., Jenkins, J. N., Ma, D.-P., Karaca, M., Saha, S., et al. (2002). A novel root-specific gene, *MIC-3*, with increased expression in nematode-resistant cotton (*Gossypium hirsutum* L.) after root-knot nematode infection. *Biochimica et Biophysica Acta, 1576*, 214–218.

Zhou, E., & Starr, J. L. (2003). A comparison of damage function, root galling, and reproduction of *Meloidogyne incognita* on resistant and susceptible cotton cultivars. *Journal Cotton Science, 7*, 224–230.

Zhou, E., Wheeler, T. A., & Starr, J. L. (2000). Root galling and reproduction of *Meloidogyne incognita* isolates from Texas on resistant cotton genotypes. *Supplement to Journal of Nematology, 32*, 513–518.

Zimet, D. J., Rich, J. R., LaColla, A., & Kinloch, R. A. (1999). Economic analysis of Telone II (1,3-D) and Temik 15G (aldicarb) to manage reniform nematode (*Rotylenchulus reniformis*). *Proceedings of the Beltwide Cotton Conferences,* Memphis, TN: National Cotton Council of America, 111–112.

Zimet, D. J., Smith, J. L., Kinloch, R. A., & Rich, J. R. (2002). Improving returns using nematicides in northwestern Florida fields infested with root-knot nematodes. *Journal of Cotton Science, 6*, 28–33.

Section 3

Technological Advances in Sustainable Management

9

CATHERINE J. LILLEY*, WAYNE L. CHARLTON*, MANJULA BAKHETIA AND PETER E. URWIN*

THE POTENTIAL OF RNA INTERFERENCE FOR THE MANAGEMENT OF PHYTOPARASITIC NEMATODES

Centre for Plant Sciences, University of Leeds, Leeds, LS2 9JT, UK

Abstract. RNA interference (RNAi) is a natural cellular phenomenon in which double stranded RNA (dsRNA) is recognised as foreign by virtue of its conformation and thus sets in motion a chain of events in which the dsRNA and its mRNA homologue are degraded. This leads to silencing of the targeted gene. First described for the microbivorous nematode *Caenorhabditis elegans*, RNAi has emerged as a powerful tool for investigating gene function in a range of organisms. Practical applications proposed for RNAi include the genetic improvement of crop plants to create novel resistance to plant pathogens. Recent studies have described the successful application of RNAi to plant parasitic nematodes. Key developments in the last year have demonstrated that *in planta* expression of a double-stranded RNA can target a gene of a feeding plant parasitic nematode, inducing a silencing effect. When the targeted gene has an essential function this leads to a level of nematode resistance, paving the way for the potential use of RNAi technology to control plant parasitic nematodes.

1. INTRODUCTION

Plant parasitic nematodes represent one of the major biotic constraints in world agriculture causing global yield losses estimated to be around US$70 billion in 1987 (Sasser & Freckman, 1987). Adjusting for inflation, this figure was revised to US$125 billion in 2003 (Chitwood, 2003). No recent, comprehensive surveys of nematode losses have been carried out and the real figures may be higher than this, as a lack of clear disease symptoms can lead some growers to underestimate yield loss. Yield reductions may also be wrongly attributed to the secondary diseases suffered by crop plants already weakened by nematode attack.

Nematodes of the order Tylenchida are responsible for the majority of crop damage. Agronomically important species include both migratory parasites such as *Radopholus* spp. and *Pratylenchus* spp. that feed sequentially from plant cells in a destructive manner, and the more specialised sedentary endoparasites. These nematodes each form a unique, biotrophic interaction with the host plant, modifying root cells to establish a permanent feeding site that provides a sustained

These authors contributed equally to this review

A. Ciancio & K. G. Mukerji (eds.), Integrated Management and Biocontrol of Vegetable and Grain Crops Nematodes, 185–203.

source of nutrition. As shown in other chapters of this volume, the most economically important are the many species of the *Meloidogyne* genus (root-knot nematodes) and the *Heterodera* and *Globodera* genera of cyst nematodes. Root-knot nematodes have a very wide host range that potentially covers most flowering plants. Consequently they are responsible for a large part of the global yield loss to nematodes with the most widespread species, *M. incognita*, being possibly the single most damaging crop pathogen worldwide (Trudgill & Blok, 2001). Each individual cyst nematode species has a much narrower host range. The cyst nematode species that attack potato (*Globodera pallida* and *G. rostochiensis*), sugar beet (*Heterodera schachtii*) and soybean (*H. glycines*) are of particular economic importance (Lilley, Atkinson, & Urwin, 2005a).

2. CURRENT CONTROL MEASURES

Effective control of many nematode species is problematic as each of the current measures has aspects that limit its utility. Consequently a number of strategies are usually deployed in an integrated approach. Nematicides are widely used, but chemical control is often limited by factors such as grower preference, economic constraints or legislation to restrict the use of harmful chemicals. Nematicides impose substantial costs and are often not economic for some crops. The compounds applied are amongst the most toxic and environmentally hazardous pesticides in widespread use. Some have already been withdrawn, due to their mammalian toxicity and environmental concerns.

Cultural systems of control are widely practised and the relatively narrow host ranges of cyst nematodes ensure that crop rotation is an effective means of partial control for these parasites. Long rotations are required, however, to decline populations of species such as *G. pallida* that can remain dormant for many years. (Phillips & Trudgill, 1998). Crop rotation is not a practical solution for controlling species with wide host ranges such as *Meloidogyne* spp. and *Rotylenchulus reniformis*.

The most sustainable method of nematode control, requiring no changes to existing cultural practices, is the use of resistant plants that suppress nematode reproduction. Although naturally resistant cultivars have been a commercial success for some crops, the approach is unable to control many nematode problems. No useful source of resistance has been identified for crop plants affected by a range of important nematode species, particularly the migratory ectoparasites (Starr, Bridge, & Cook, 2002). Resistance also tends to be highly specific to certain nematode species or even pathotypes and can be compromised by virulent nematode populations. For example, the natural resistance gene H1 present in potato cultivars such as Maris Piper is useful in controlling populations of *G. rostochiensis* in the United Kingdom (Atkinson, 1995). However, these cultivars do not control *G. pallida* which has now become the prevalent potato cyst nematode in the UK as a direct consequence of selection. The lack of a comparable single dominant natural resistance gene for *G. pallida* has resulted in emphasis on multi-trait quantitative resistance that is often overcome by virulent pathotypes. Only a limited number of crop plants have been identified with resistance to

Meloidogyne species and for many of these the resistance can be overcome by virulent biotypes (Hussey & Janssen, 2002).

2.1. Potential for Biotechnological Control of Plant Parasitic Nematodes

The limitations of conventional control measures provide an important opportunity for plant biotechnology to deliver effective and durable forms of nematode resistance. The approach has a number of advantages. Utilisation of the new crop varieties would not require other changes to agronomic practices and there would be a reduction in nematicide use and its associated toxicological and environmental risks. Resistant crops could be developed to control more than one nematode species, including pathotypes that display virulence to currently used sources of resistance.

Biotechnological approaches have already delivered useful levels of resistance for a range of plant parasitic nematode species. The most advanced strategy to date is the transgenic expression of plant proteinase inhibitors that act to impair function of nematode digestive enzymes. A gene encoding a rice cysteine proteinase inhibitor (cystatin), Oc-I, was engineered to have an enhanced inhibitory activity (Urwin, Atkinson, Waller, & McPherson, 1995). Expression of the engineered variant (Oc-IΔD86) in *Arabidopsis* plants was the first transgenic technology shown to work against both root-knot and cyst nematodes (Urwin, Lilley, McPherson, & Atkinson, 1997a). Using the same approach, resistance has also been demonstrated against *Rotylenchulus reniformis* (Urwin, Levesley, McPherson, & Atkinson, 2000), *Radopholus similis* infecting banana (Atkinson, Grimwood, Johnston, & Green, 2004), and the potato cyst nematode *Globodera*. For the latter species, the technology progressed to the stage of field trials where full resistance to *G. pallida* was achieved by stacking natural partial and transgenic resistance (Urwin, Green, & Atkinson, 2003).

Inhibitors of serine proteinases also have proven potential for the control of plant parasitic nematodes. The trypsin inhibitor sporamin inhibited development of *Heterodera schachtii* when expressed in sugar beet hairy roots (Cai et al., 2003) and a serine proteinase inhibitor expressed in wheat provided protection against the cereal cyst nematode *H. avenae* (Vishnudasan, Tripathi, Rao, & Khurana, 2005). In a different approach, peptides that disrupt chemoreception in cyst nematodes afforded protection from *G. pallida* when engineered to be secreted from potato roots (Liu et al., 2005).

RNA interference (RNAi) technology is now routinely used as an investigative tool for understanding gene function in a range of organisms including plant parasitic nematodes. This review will consider the utility of RNAi technology both to identify potential target genes for novel control strategies and to provide resistance to plant parasitic nematodes.

2.2. The RNAi Mechanism

RNAi is a fundamental mechanism for controlling the flow of genetic information first described, in work with *Caenorhabditis elegans,* by Andrew Fire and Craig Mello (Fire et al., 1998), who in 2006 won the Nobel prize in physiology or medicine. As the mechanism has been elucidated it has become clear that it shares mechanistic similarities to post transcriptional gene silencing in plants (Jorgensen, Cluster, English, Que, & Napoli, 1996; Waterhouse, Graham, & Wang, 1998). The RNAi process has a role in guarding against viral and transposable elements in the genome (Voinnet, 2001; Waterhouse, Wang, & Lough, 2001). The RNAi mechanism recognises double-stranded RNA (dsRNA) that subsequently causes a chain of events in which both the dsRNA and its mRNA homologue are degraded, leading to sequence-specific, homology-dependent gene silencing. Since the discovery of RNAi the effect has been described in mammals (e.g. Hannon, 2002; Silva, Chang, Hannon, & Rivas, 2004; Zamore, Tuschl, Sharp, & Bartel, 2000), insects (e.g. Kennerdell & Carthew, 2000) and amphibians (Dirks, Bouw, Huizen, Jansen, & Martens, 2003; Li & Rohrer, 2006). Additionally, the phenomenon of quelling in *Neurospora crassa* may also prove to be mechanistically similar to mammalian RNAi (Cogoni & Macino, 2000; Forrest, Cogoni, & Macino, 2004). It is now widely accepted that the process is also an integral part of normal gene regulation processes (Voinnet, 2002). Several reviews are available that concern themselves with the intricacies of RNAi in mammalian, insect and plant systems and describe them in some depth (e.g. Brodersen & Voinnet, 2006; Denli & Hannon, 2003; Hammond, 2005; Qi & Hannon, 2005; Sen & Blau, 2006; Tomari & Zamore, 2005).

3. THE MECHANISM OF RNAi IN NEMATODES

The basic mechanism of RNAi in eukaryotic organisms appears to be conserved, although there are differences in the systemic nature and heritability of the effect. RNAi is triggered by dsRNA molecules that have homology with an endogenous gene. In *C. elegans*, any dsRNA longer than 100bp can induce silencing, with fragments in the range 500–1500 bp commonly used (Kaletta & Hengartner, 2006).

The RNAi pathway is comprised of two basic steps (Sontheimer, 2005). In the initiation phase the Dicer complex recognises the exogenous dsRNA molecules and cleaves the dsRNA in a processive, ATP-dependent manner into short (21–23 bp) interfering RNA duplexes (siRNAs) by way of its multidomain RNaseIII enzyme activity. The resulting siRNAs have 5' phosphate groups and 2 bp overhangs at their 3' ends. *Caenorhabditis elegans* has only one Dicer enzyme that functions in a complex with the dsRNA binding protein Rde-4, the PAZ-PIWI domain family protein Rde-1 and the Dicer-related helicase Drh-1 (Tabara, Yigit, Siomi, & Mello, 2002). Following dsRNA cleavage, it has been postulated that Rde-1 may bind the siRNAs to direct them to the second, effector phase of the RNAi pathway (Fig. 1) (Grishok, 2005).

The multicomponent RNA-induced silencing complex (RISC) degrades specific mRNAs in a targeted manner. During assembly of the active RISC, the siRNA duplex is unwound, the strands separate and only the anti-sense (guide) strand is incorporated into the RISC. Base pairing between the single-stranded siRNA and the complementary mRNA results in the activation of RISC complex that then recognises the target transcript. Endonucleolytic cleavage of the mRNA involves a member of the Argonaute family of proteins (Hammond, 2005) some of which have endonuclease activity in their PIWI domain. The identity of this key

RISC component is unknown in *C. elegans* but around 25 candidate Argonaute homologues exist (Grishok, 2005). The cleaved mRNA is subsequently degraded by exonuclease activity, leading to a gene silencing effect.

Other components of the *C. elegans* RISC have been identified, including an RNA binding protein Vig-1, and Tsn-1, a protein with a Tudor domain and five staphylococcal nuclease domains that may be responsible for the exonuclease activity of RISC (Caudy et al., 2003).

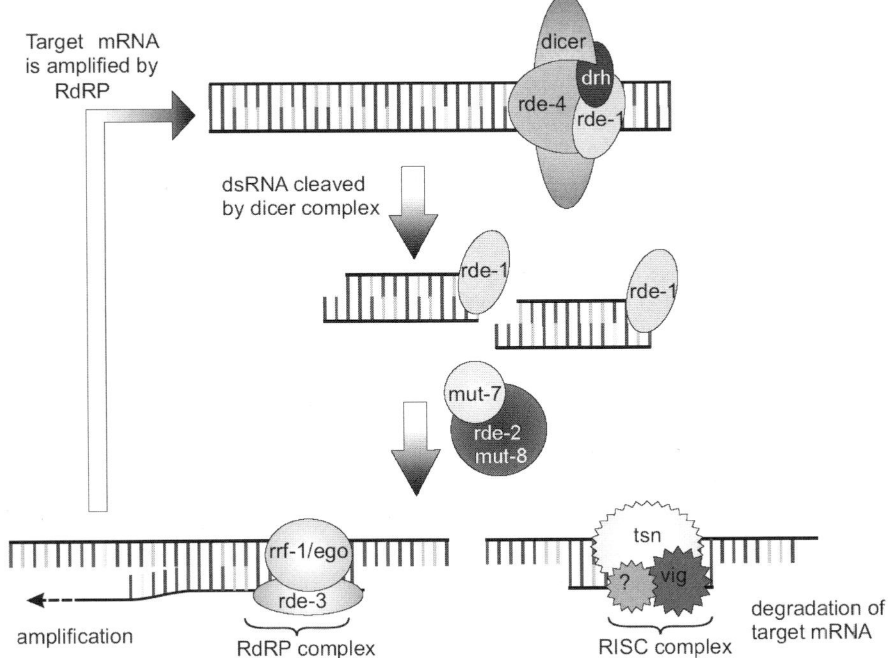

Figure 1. The RNAi gene silencing pathway in C. elegans. *In the initiation phase exogenous dsRNA is recognised by the Dicer complex and processed into siRNAs. The Dicer complex includes the multidomain RNaseIII Dicer enzyme itself, the dsRNA binding protein Rde-4, the PAZ-PIWI domain family protein Rde-1 and the Dicer-related helicase Drh-1. Following dsRNA cleavage, Rde-1 may bind the siRNAs to direct them to the second, effector phase of the RNAi pathway. Assembly of the active RISC complex may involve the helicase/exonuclease domain protein Mut-7 and Rde-2/Mut-8. The siRNA strands separate and the guide strand is incorporated into the active RISC. The activated RISC complex that also contains an unidentified Argonaute family protein, the RNA binding protein Vig-1 and the Tudor-SN protein Tsn-1, recognises and cleaves the target mRNA leading to a gene silencing effect. In a separate amplification step, siRNAs act as primers for the RNA-dependent RNA polymerases Ego-1 or Rrf-1 that may function in a complex with the Rde-3 polymerase. The dsRNA thus produced can enter the Dicer complex and trigger further, transitive, gene silencing.*

Silencing of abundant transcripts by only a few introduced molecules of dsRNA is achieved in *C. elegans* through an amplification step. The initial siRNAs produced by the Dicer complex can act as primers for an RNA-dependent RNA polymerase (RdRP), using the target mRNA as a template. The two RdRPs, EGO-1 and RRF-1, active in the germline and somatic cells respectively, have been implicated in this process. They may function in a complex with the RDE-3 polymerase (Chen et al., 2005a). The dsRNA made in this manner can enter the Dicer complex and trigger further, transitive, gene silencing.

For many species, including the model system Drosophila, dsRNA must be introduced directly into cells by microinjection or electroporation in order to elicit an RNAi response. Following reports of the microinjection technique (Fire et al., 1998) it was later shown that RNAi can be effectively induced in *C. elegans* by simple soaking (Tabara, Grishok, & Mello, 1998) or by feeding the worms with bacteria expressing dsRNA (Timmons & Fire, 1998). This facilitates large-scale functional genomic analysis of RNAi in *C. elegans*. The dsRNA moves systemically from the gut or injected tissue throughout most cells of the worm including the germline. This leads to induced gene-silencing in any cells that express the target mRNA. The systemic movement to the germline results in an RNAi phenotype that can be inherited, with progeny of the treated worms displaying a strong effect. The RNAi effect only extends beyond the F1 generation, however, when germline genes are targeted (Grishok, Tabara, & Mello, 2000).

The isolation of mutants defective in systemic RNAi has led to the identification of some of the genes involved in this process. The *sid-1* gene encodes a protein with 11 membrane spanning domains that localises to cells in contact with the environment, including some neurons (Winston, Molodowitch, & Hunter, 2002). Sid-1 promotes passive uptake of dsRNA with longer molecules transported more efficiently than siRNAs (Feinberg & Hunter, 2003). Three RNAi spreading defective (*rsd*) mutants are all deficient in transmission of the RNAi effect to the germline. A role in vesicle trafficking is predicted for RSD3, suggesting that dsRNA may be transported within endocytotic vesicles (Tijsterman, May, Simmer, Okihara, & Plasterk, 2004). The role of the endocytic pathway in uptake and translocation of dsRNA has recently been confirmed. Several *C. elegans* gene products with roles in intracellular vesicular transport and lipid modification were found to be essential for systemic RNAi (Saleh et al., 2006).

3.1. Plant Parasitic Nematodes RNAi

Reports in the literature provide evidence for the efficacy of RNAi in plant parasitic nematodes but the molecular detail of the RNAi process in plant parasitic nematodes has yet to be elucidated. A range of genes have been targeted for silencing in cyst and root-knot nematode species, and both the phenotypic and molecular effects were documented.

3.1.1. Uptake of dsRNA

The obligatory parasitic nature and small size of infective stages of plant parasitic nematodes makes them refractory to microinjection. Prior to infection of plant roots the non-feeding, pre-parasitic nematodes do not normally ingest. Octopamine was first used to stimulate oral ingestion by pre-parasitic 2nd stage juveniles (J2) of the cyst nematodes *G. pallida* and *H. glycines* leading to uptake of dsRNA from the soaking solution (Urwin, Lilley, & Atkinson, 2002). The same method has been used successfully to induce uptake of dsRNA by J2 of the root-knot nematode *M. incognita* (Bakhetia, Charlton, Atkinson, & McPherson, 2005; Shingles, Lilley, Atkinson, & Urwin, 2007). Resorcinol and serotonin also induce dsRNA uptake by J2 of *M. incognita* and may be more effective than octopamine for this nematode (Rosso, Dubrana, Cimbolini, Jaubert, & Abad, 2005). The addition of spermidine to the soaking buffer and an extended incubation time were reported to increase the efficiency of RNAi for the cyst nematode *G. rostochiensis* (Chen, Rehman, Smant, & Jones, 2005b).

Genes that are expressed in a range of different tissues and cell types have been targeted by RNAi. The ingested dsRNA can silence genes in the intestine (Urwin et al., 2002; Shingles et al., 2007); female reproductive system (Lilley et al., 2005b), subventral and dorsal oesophageal glands (Bakhetia, Urwin, & Atkinson, 2007; Chen et al., 2005b; Huang et al., 2006a; Rosso et al., 2005) and sperm (Urwin et al., 2002; Steeves, Todd, Essig, & Trick, 2006). In *C. elegans* the uptake of dsRNA from the gut has been shown to lead to systemic RNAi. When plant parasitic nematodes ingest dsRNA a systemic response is seen in other tissues. This suggests that plant parasitic nematodes share similar uptake and dispersal pathways with *C. elegans*.

There are other reports in the literature of alternate routes for uptake of dsRNA by plant parasitic nematodes. Soaking intact eggs of *M. artiellia* contained within their gelatinous matrix, in a solution containing dsRNA allowed successful targeting of the chitin synthase gene (Fanelli, Di Vito, Jones, & De Giorgi, 2005). The enzyme plays a role in the synthesis of the chitinous layer in the egg shell. The reduction of its transcript by RNAi led to a reduction in the amount of chitin in the eggshells and a subsequent delay in hatching of juveniles from treated eggs. The results imply that the eggs of this nematode are permeable to dsRNA.

RNAi of genes that are expressed in the neuronal system of *C. elegans* can be difficult to achieve (Kamath, Martinez-Campos, Zipperlen, Fraser, & Ahringer, 2000; Timmons, Court, & Fire, 2001) although RNAi effects in these cells can be enhanced by using a mutant strain defective in the RdRP *rrf-3* (Simmer et al., 2002, 2003). A recent study describes the silencing of five FMRFamide-like (*flp*) neuropeptide genes of *G. pallida*, each with a unique neuronal expression pattern (Kimber et al., 2007). Absence of transcript in treated worms and abnormal behavioural phenotypes were observed when the genes were targeted by RNAi, demonstrating the susceptibility of these neuronal genes to RNAi. The effect occurred for pre-parasitic J2 nematodes soaked only in water containing dsRNA. RNAi of intestinal and pharyngeal gland cell expressed genes has a requirement for stimulated oral uptake to achieve transcript knockout (Urwin et al., 2002; Rosso et al., 2005). RNAi-mediated silencing of the neuronal *flp* genes must therefore use an alternative route to take up dsRNA. The dsRNA may be gaining access via the secretory/excretory pore, the cuticle, or the amphids. Amphids are paired sense organs of nematodes at the anterior of the animal. The amphidial cavity has sensory neurons that are exposed to the external environment. These neurons demonstrate uptake of fluorescein isothiocyanate (FITC), a feature common to both *C. elegans* (Hedgecock, Culotti, Thomson, & Perkins, 1985) and cyst nematodes (Winter, McPherson, & Atkinson, 2002). Fluorescent dextran conjugates of 12 kDa but not 19.5 kDa are also taken up by the sensory neurons of *C. elegans*, suggesting a size constraint. It is postulated that the retrograde transport along cyst nematode chemosensory dendrites can provide a route for uptake of soluble compounds such as peptides from the external environment (Winter et al., 2002). The exposed nerve processes could also take up dsRNA molecules.

A gene encoding a secreted amphid protein of unknown function (*gr-ams-1*) has been targeted by RNAi of *G. rostochiensis* (Chen et al., 2005b). Although octopamine was included in the soaking solution on this occasion, *gr-ams-1* was more susceptible to an RNAi effect than a gland cell expressed endoglucanase, raising the possibility that neuronal retrograde transport offers more efficient dsRNA uptake that forced ingestion.

3.2. Comparative Observation of Reported Strategies

A range of techniques has been used both to deliver the RNAi effect and to determine the phenotype. The RNAi experiments described in the literature with plant parasitic nematodes have used a range of approaches. While the basic methodology is similar, adjustments are continually being made in order to maximise the strength of the observed phenotype. Comparison between experiments is difficult when the methodology differed, but some observations regarding the persistence and duration of RNAi in these nematodes can be made.

RNAi effects have been observed following exposure of preparasitic nematodes to dsRNA for time periods ranging from 4hr to 7 days. While a 4 hr incubation leads to effective RNAi for some cyst nematode genes (Lilley et al., 2005b; Urwin et al., 2002) increasing the incubation period generally results in increased silencing and stronger phenotypic effects for cyst nematodes (Chen et al., 2005b; Kimber et al., 2007). An incubation of 24 hr was found to be critical for RNAi-induced silencing in *G. rostochiensis* (Chen et al., 2005b). In *G. pallida,* incubation periods in excess of 18 hr were critical when targeting the *flp* genes to produce an aberrant phenotype. The severity of the effect was greatest after 7 days incubation (Kimber et al., 2007). Efficient silencing has been consistently observed in *Meloidogyne* spp. following 4 hr incubation of J2 in dsRNA (Huang et al., 2006a; Rosso et al., 2005; Shingles et al., 2007).

Successful RNAi has been observed in both cyst and root-knot nematodes treated with double stranded RNA molecules ranging in size from 42 bp to 1300 bp. A 309 bp dsRNA targeting a β-1,4-endoglucanase of *G. rostochiensis* induced weaker silencing than 244 bp dsRNA targeting an amphid secreted protein in the same nematode (Chen et al., 2005b). It is difficult to draw any conclusions from such observations. A number of factors may influence this outcome: differing spatial expression patterns, level and turnover rate of the endogenous transcript in addition to length of the dsRNA. 88 bp, 227 bp and 316 bp of dsRNA have been used to target the same gene, *Gp-flp-6* in *G. pallida* (Kimber et al., 2007). The shortest length was insufficient to induce any silencing effect. Both the 277 bp and 316 bp dsRNAs silenced the target transcript and resulted in reduced motility, but the shorter molecule consistently induced stronger effects. Further studies are required to determine if this is a general phenomenon for RNAi of plant parasitic nematodes, or if the effects are gene specific. Both 42 bp and 271 bp dsRNAs covering the coding region or the full-length transcript of the oesophageal gland peptide 16D10, led to a 93–97% reduction in target transcript in *M. incognita* J2 (Huang et al., 2006a). This suggests that different nematode species and/or different genes may have dissimilar requirements for inducing dsRNA molecules.

The nature of soaking J2 nematodes in dsRNA limits the strength of the RNAi effect, as the RNAi-induced gene silencing is time-limited, once nematodes are removed from exposure to dsRNA. Calreticulin (*mi-crt*) and polygalacturonase (*mi-pg-1*) genes of *M. incognita* targeted by RNAi, in a 4 hr incubation of dsRNA, displayed maximum transcript repression after a further 20 hr and 44 hr respectively. After a 68 hr recovery period the transcript level of both genes had returned to normal (Rosso et al., 2005). Similar results have been observed for cyst nematodes: transcript repression of a β-1,4-endoglucancase was observed immediately following a 16 hr dsRNA treatment of *H. glycines* J2 and after a 5 day recovery period. Transcript abundance increased at 10 days and had returned to pre-treatment levels by 15 days after dsRNA exposure (Bakhetia et al., 2007). The persistence of *Gp-flp-12* gene silencing was monitored by the reduced motility of treated J2s. There was no recovery of phenotype after 24 hrs and a significant but not complete recovery after 6 days (Kimber et al., 2007). If dsRNA-treated juveniles are allowed to invade plants and continue development, the phenotypic consequences of RNAi can be evident after a number of weeks (Bakhetia et al., 2005; Bakhetia et al., 2007; Chen et al., 2005b; Huang et al., 2006a; Lilley et al., 2005b; Urwin et al., 2002). If nematodes are compromised during the early invasion of plants or during induction of the feeding cell, then subsequent development will be affected. Time-limitation is not an issue in *C. elegans* when RNAi is achieved by feeding because the nematodes are being continuously exposed to dsRNA. A similar situation would arise in plant parasitic nematodes if dsRNA was produced in the feeding cell. This would similarly prolong the effective exposure and maximise the RNAi effect.

3.3. Observation of Phenotype

Careful analysis of the phenotypic effect of any RNAi experiment is a key challenge. Due to the obligate parasitic lifecycle of these species, many RNAi phenotypes can only be revealed after the treated pre-parasitic juveniles have been allowed to infect host roots and develop to adulthood. Subtle phenotypic effects may be missed. RNAi can be a powerful approach for functional analysis of nematode-specific or species-specific genes, with no putative homologues in current databases. Study of such genes, that may be good targets for novel control strategies, could provide insight into unique aspects of the plant-nematode interaction. Experiments to date have commonly analysed impact on the numbers of nematodes able to invade roots and successfully initiate feeding sites, or the proportion of cyst nematodes that develop as either males or females. Effects on female size and fecundity can also be measured as can the size and shape of developing nematodes at a given time point post infection.

In the cyst nematode *Heterodera glycines* a range of genes have been targeted by soaking the infective juvenile animals in dsRNA. These include a cysteine proteinase, C-type lectin (Urwin et al., 2002), β-1,4-endoglucanase, pectate lyase, chorismate mutase, the secreted peptide SYV46 and a gland protein of unknown function (Bakhetia et al., 2007). The phenotypic outcome of all these experiments was a reduced parasitic burden when the treated nematodes were used to infect a host plant.

RNAi using the soaking technique in *Globodera* species has been used to determine the phenotypic effect of targeting a cysteine proteinase, β-1,4-endoglucanase, a secreted amphid protein and FMRFamide-like peptides (Urwin et al., 2002; Chen et al., 2005b; Kimber et al., 2007). Again these lead to a reduction in the parasite burden on the host plant with the exception of the latter that impaired motility. In the root knot nematode *M. incognita*, soaking has been used to target a cysteine proteinase, dual oxidase and a secreted peptide 16D10, the observed phenotypes of which all showed a detrimental effect on pathogenesis (Shingles et al., 2007; Bakhetia et al., 2005; Huang et al., 2006a). Soaking of egg masses has been carried out to target the *M. artiellia* chitin synthase gene resulting in delayed egg hatch (Fanelli et al., 2005).

Gross population analysis of this sort can define the importance of a gene, but not necessarily help elucidate its particular role. If the gene of interest has a putative function based on sequence homology, often corroborated with in situ hybridisation studies, then more directed phenotypic analysis can be carried out. The *flp* genes of *G. pallida* are expressed in the nervous system (Kimber et al., 2002) with a likely role in coordinating motor activities. Consequently, phenotypic effects of RNAi silencing were analysed using migration assays to detect impaired motility (Kimber et al., 2007). RNAi combined with detailed analysis of gene expression by quantitative PCR helped to elucidate stages in the infection process of *H. glycines* when particular oesophageal gland secreted proteins were required (Bakhetia et al., 2007). Molecular and biochemical characterisation of a cathepsin L cysteine proteinase of *M. incognita* targeted by RNAi was correlated with the effect on parasitism (Shingles et al., 2007).

4. *IN PLANTA* DELIVERY OF dsRNA TO TARGET GENES OF PLANT PARASITIC NEMATODES

Delivery of dsRNA from the feeding cell to target specific, essential nematode genes has been proposed as a novel means for plant parasitic nematode control since the first demonstration that RNAi is effective in these nematodes (Urwin et al., 2002; Atkinson, Urwin & McPherson, 2003; Lilley et al., 2005a). The mode of feeding, particularly of sedentary endoparasitic nematodes, is ideally suited to such an approach. The nematode feeds exclusively from one or a few plant cells and continues to feed throughout development to a mature male or egg-laying female. This ensures constant ingestion of plant cell derived molecules and potentially enables targeted expression of the dsRNA construct in the feeding cells.

RNAi is widely used in plants as a research tool for analysis of gene function. More recently there has been interest in using it to engineer novel traits (Kusaba, 2004; Mansoor, Amin, Hussain, Zafar, & Briddon, 2006) with a number of potential commercial applications already described (e.g. Ogita, Uefuji, Morimoto, & Sano, 2004; Byzova, Verduyn, De Brouwer, & De Block, 2004; Davuluri et al., 2005). RNAi has also been used in plants to confer resistance to viruses (Waterhouse et al., 1998; Pooggin, Shivaprasad, Veluthambi, & Mohn, 2003) and the bacterial pathogen *Agrobacterium tumefaciens* (Escobar, Civerolo, Summerfelt, & Dandekar, 2001).

A variety of vectors are now available for induction of RNAi in plants by production of a dsRNA molecule. The general approach is to clone both sense and anti-sense cDNA sequences of the target gene, separated by a spacer region or intron into a binary vector under the control of a strong plant promoter. The transcribed RNA then forms into a self complementary hairpin structure. Use of an intron sequence as the spacer increases the silencing efficiency (Smith et al., 2000; Wesley et al., 2001).

4.1. The Feeding Strategy of Sedentary Endoparasitic Nematodes

A nematode establishes a feeding site through the modification of root cells to create a specialised feeding cell. This process is well described, particularly for cyst and root-knot nematode species. Following invasion of a host root, the infective juvenile ($J2_i$) migrates either intracellularly (cyst species) or intercellularly (root-knot species) through cortical cells, towards the vascular cylinder where an initial feeding cell is selected. One or more plant cells are then modified to re-differentiate into a specialised feeding site. Nematode proteins from the three pharyngeal gland cells are secreted through the bore of the stylet into the initial feeding cell, and induction of the feeding site is triggered. This results in dramatic changes in gene expression and considerable re-programming of root cell development. Interestingly, although there are morphological similarities and a shared function, the nature of the transformations differs between the syncytia induced by cyst nematodes and the giant cells induced by root-knot nematodes (Davis et al., 2000).

Cyst nematodes initiate a syncytium from a single cell located at the periphery of the formed vasculature (Golinowski, Sobczak, Kurek, & Grymaszewska, 1997). The syncytium then expands by recruitment of up to 200 adjacent cells from the stele through cell wall dissolution. This seems to be a modification of a normal root morphogenesis process (Jones, 1981a; Golinowski et al., 1997). In contrast, the $J2_i$ of *Meloidogyne* spp. selects a small number of parenchymal cells in the differentiating stele and induces them to undergo repeated cycles of acytokinetic mitoses. This results in enlarged and multinucleate giant cells from which the parasite feeds in turn (Sijmons, Grundler, Von Mende, Burrows, & Wyss, 1991). The cell architecture of these two feeding sites differs considerably but some common features occur. Both have a reduced number of smaller secondary vacuoles and high metabolic activity with increased numbers of organelles and nuclei. There is also a low plasmodesmatal density in cell walls adjacent to unmodified cells and a large number of wall ingrowths into xylem vessels (Jones, 1981b, de Almeida Engler et al., 2004). These feeding sites, if continually stimulated by the nematode, function as sinks that supply the nematode with all its nutritional requirements during the parasitic stages of its life cycle. The sequestration of plant material by the nematode results in serious, deleterious consequences for the host plant.

The nematodes feed using a hollow, protrusible stylet that pierces the giant cell or syncytium to allow pharyngeal gland secretions to pass into the cells and cell contents to be removed. A semi-permeable blind-ended structure known as the feeding tube extends into the cytoplasm of the feeding cell from the stylet orifice. The feeding tube acts as a molecular sieve, permitting the uptake of certain molecules and excluding others. Nematode secretions, as yet unidentified, are probably involved in formation of this tube. This is a unique, self-assembling structure that is reformed each time the stylet is reinserted for a cycle of feeding. The feeding tubes of root-knot and cyst nematodes differ in their structure. The former have thick, electron dense, crystalline walls (Hussey & Mims, 1991) and the latter have a thinner, uneven wall (Sobczak, Golinowski, & Grundler, 1999). The dsRNA or siRNA produced by a plant must pass into the nematode gut if *in planta* RNAi is to be effective. The divergent feeding tube structures lead to differences in size exclusion limits between nematode species. The cyst nematode feeding tube of *H. schachtii* permits uptake of 20 kDa dextrans but not 40 kDa (Böckenhoff & Grundler, 1994) and proteins of 11 kDa but not 23 kDa and 28 kDa (Urwin, Møller, Lilley, McPherson, & Atkinson, 1997b; Urwin, McPherson, & Atkinson, 1998). Uptake of 28 kDa green fluorescent protein (GFP) by *G. rostochiensis* could only be visualised using sensitive detection techniques (Goverse et al., 1998) whereas it was readily ingested by *M. incognita* (Urwin et al., 1997b).

After production of a dsRNA molecule in the plant cell, the RNAi trigger could be available for uptake by the feeding tube in one of three conformations. The first is unprocessed full-length dsRNA. In this form uptake would be possible by those nematodes possessing feeding tubes with pore apertures greater than 26Å in diameter assuming the molecule is drawn lengthwise through the pores. As GFP is a barrel-shaped protein with a diameter of 30Å, any nematode capable of ingesting GFP should also ingest dsRNA. If ingested in this conformation, it is envisaged that the dsRNA would pass into the gut cells, be processed by the nematode Dicer complex and induce silencing via the RNAi pathway.

Alternatively, siRNAs processed by the plant RNAi machinery could be available for ingestion via the feeding tube. This would require the same pore size as full length dsRNA, however lengthwise entry into the pore would be favoured. After cleavage of dsRNA by the plant cell Dicer, problems with uptake may arise if the siRNAs are immediately complexed with the multi-component RISC. It would appear impossible for this large protein-nucleic acid complex to pass through the wall of the feeding tube.

4.2. In Planta RNAi as a Biotechnological Strategy

The year 2006 saw three publications that described the feasibility of silencing nematode genes using dsRNA produced in the host plant (Huang, Allen, Davis, Baum, & Hussey, 2006a; Steeves et al., 2006; Yadav, Veluthambi, & Subramaniam, 2006). Nevertheless, questions still remain concerning the precise mode of action and the form in which the RNAi trigger is taken up by the nematodes. Yadav et al. (2006) demonstrated silencing of *Meloidogyne* genes by RNAi delivered from host tobacco plants. Nematode splicing factor and integrase genes were targeted based on their RNAi phenotype in *C. elegans* and their presumed essential role in *Meloidogyne*. Plants expressing hairpin dsRNA for each of the two sequences displayed >95% resistance to *M. incognita*. The few nematodes that formed galls appeared developmentally compromised and lacked detectable transcript for the targeted genes (Yadav et al., 2006). Unfortunately, no evidence was presented for the presence of either dsRNA or siRNAs in the transgenic plants, so the route by which silencing occurred cannot be deduced.

The second report came from the group of R. S. Hussey. They targeted a 13 amino acid peptide (16D10) that is secreted from the subventral oesophageal gland cells of *M. incognita* (Huang et al., 2006a). The peptide is highly conserved among four *Meloidogyne* species (*M. incognita, M. javanica, M. arenaria* and *M. hapla*) and appears to mediate an early signalling event required for giant cell formation, possibly through interaction with a plant transcription factor (Huang et al., 2006b). *In vitro* delivery of 16D10 dsRNA to J2 *M. incognita* suppressed their subsequent development by up to 81% when inoculated onto *Arabidopsis* roots. Transgenic *Arabidopsis* plants expressing the 16D10 sequence as a hairpin construct were found to contain both intact dsRNA and approx. 21bp siRNAs corresponding to 16D10. Infection of these plants with all four *Meloidogyne* species revealed a 63–90% reduction in the number of galls that developed with a decrease in gall size and corresponding reduction in total egg production. This clearly demonstrates that uptake by the feeding nematode of either full length dsRNA or processed siRNAs can occur and is sufficient to induce an RNAi phenotype. These results also highlight the utility of targeting highly conserved, nematode specific sequences to protect against more than one nematode species.

Most recently, evidence emerged that cyst nematodes can also be targeted by expressing dsRNA molecules in plant roots (Steeves et al., 2006). This is an important finding given the differences in feeding tube structure of the cyst and root-knot genera. Transgenic soybean plants were shown to accumulate siRNAs arising from expression of a hairpin construct targeting the major sperm protein (MSP) gene of the soybean cyst nematode *H. glycines*. Nematodes infecting these plants displayed an overall 68% reduction in egg production. Remarkably, the progeny hatched from the eggs that did develop displayed an overall 75% reduction in egg production when allowed to infect wildtype susceptible soybean plants. These results suggest that RNAi can be transmitted to the F1 progeny in a similar manner to that documented for *C. elegans* (Grishok et al., 2000).

4.3. Future Prospects for RNAi-Based Control of Plant Parasitic Nematodes

RNAi silencing of a gene that plays a key role in the development of the nematode, either directly or indirectly can adversely affect the progression of pathogenesis. Good targets for this technology are likely to be those genes that are nematode specific and have sequence conservation with orthologues from related species to maximise the spectrum of resistance. Putative parasitism genes such as that targeted by Huang et al. (2006a) may be targets of choice but are likely to be selective for particular nematode genera. Cross-species RNAi in nematodes has however, recently been demonstrated with sequences from the animal parasite *Ascaris suum* inducing gene silencing of their *C. elegans* counterparts (Gao et al., 2006).

The limited data that are currently available suggest that levels of plant resistance from RNAi biotechnology are generally comparable to those observed with other transgenic strategies. Transgenically expressed cysteine proteinase inhibitors (cystatins) have typically delivered 70–80% resistance against a number of nematode species (Atkinson et al., 2003). Total protection however, is achievable by pyramiding cystatins with partial natural resistance (Urwin et al., 2003). Additive effects can be achieved by introducing a number of transgenes into a single plant (Urwin et al., 1998). RNAi biotechnology may also benefit from being stacked with other defences. Expressing hairpin constructs targeting more than one gene by RNAi may increase the level of resistance. Co-expressing a number of dsRNA sequences to target multiple genes is highly effective in *Drosophila* (Schmid, Schindelholz, & Zinn, 2002) and up to five genes can be silenced simultaneously in *C. elegans* (Geldhof, Molloy, & Knox, 2006) RNAi based strategies have the advantage that no novel protein is produced. This may ease the progress of this new technology from development to a commercial product.

The possibility of siRNAs inducing "off-target" gene silencing effects presents a concern with RNAi-based technologies. These occur in animal systems (Jackson et al., 2003; Scacheri et al., 2004) but have not yet been reported in plants, even when specifically sought (Kumar, Gustafsson, & Klessig, 2006). Those siRNAs that play a crucial role in homologous sequence gene silencing via the RNAi pathway are, in many respects, analagous to regulatory microRNAs (miRNAs) that are typically 19–24 nucleotides long. In animal cells these miRNAs can trigger translational repression by imperfect annealing to the 3' untranslated regions (UTRs) of genes. Most reported off-target effects are considered to result from siRNAs with partial homology to non-target gene 3' UTRs acting in a similar manner (Birmingham et al., 2006). Plant miRNAs, in contrast to animal miRNAs, have almost perfect complementarity to their target sequences. Plant miRNAs also differ from the animal counter-part by triggering local transcript cleavage rather than translational arrest (Du & Zamore, 2005). This increased sequence specificity of plant miRNA mechanisms should, with bioinformatic input, facilitate selection of nematode genes that minimise the risk of off-target effects.

REFERENCES

Atkinson, H. J. (1995). Plant nematode interactions: molecular and genetic basis. In K. Kohmoto, U. S. Singh, & R. P. Singh (Eds.), *Pathogenesis and host specificity in plant diseases: Eukaryotes* (Vol. II, pp. 355–370). Pergamon Press Oxford, UK.

Atkinson, H. J., Grimwood, S., Johnston, K., & Green, J. (2004). Prototype demonstration of transgenic resistance to the nematode *Radopholus similis* conferred on banana by a cystatin. *Transgenic Research, 13,* 135–142.

Atkinson, H. J., Urwin, P. E., & McPherson, M. J. (2003). Engineering plants for nematode resistance. *Annual Review of Phytopathology, 41,* 615–639.

Bakhetia, M., Charlton, W., Atkinson, H. J., & McPherson, M. J. (2005). RNA interference of dual oxidase in the plant nematode *Meloidogyne incognita. Molecular Plant-Microbe Interactions, 18,* 1099–1106.

Bakhetia, M., Urwin, P. E., & Atkinson, H. J. (2007). qPCR analysis and RNAi define pharyngeal gland cell-expressed genes of *Heterodera glycines* required for initial interactions with the host. *Molecular Plant-Microbe Interactions,* 20, 306–312.

Birmingham, A., Andersen, E. M., Reynolds, A., Ilsley-Tyree, D., Leake, D., Fedorov, Y., et al. (2006). 3'UTR seed matches, but not overall identity, are associated with RNAi off-targets. *Nature Methods, 3,* 199–204.

Böckenhoff, A., & Grundler, F. M. W. (1994). Studies on the nutrient uptake by the beet cyst nematode *Heterodera schachtii* by *in situ* microinjection of fluorescent probes into the feeding structures in *Arabidopsis thaliana. Parasitology, 109,* 249–254.

Brodersen, P., & Voinnet, O. (2006). The diversity of RNA silencing pathways in plants. *Trends in Genetics, 22,* 268–280.

Byzova, M., Verduyn, C., De Brouwer, D., & De Block, M. (2004). Transforming petals into sepaloid organs in *Arabidopsis* and oilseed rape: implementation of the hairpin RNA-mediated gene silencing technology in an organ-specific manner. *Planta, 218,* 379–387.

Caudy, A. A., Ketting, R. F., Hammond, S. M., Denli, A. M., Bathoorn, A. M. P., Tops, B. B. J., et al. (2003). A micrococcal nuclease homologue in RNAi effector complexes. *Nature, 425,* 411–414.

Cai, D., Thurau, T., Tian, Y., Lange, T., Yeh, K-W., & Jung, C. (2003). Sporamin-mediated resistance to beet cyst nematodes (*Heterodera schachtii* Schm.) is dependent on trypsin inhibitory activity in sugar beet (*Beta vulgaris* L.) hairy roots. *Plant Molecular Biology, 51,* 839–849.

Chen, C.-C. G., Simard, M. J., Tabara, H., Brownell, D. R., McCollough, J. A. & Mello C. C. (2005a). A member of the polymerase nucleotidyltransferase superfamily is required for RNA interference in *C. elegans. Current Biology*, 15, 378–383.

Chen, Q., Rehman, S., Smant, G., & Jones, J. T. (2005b). Functional analysis of pathogenicity proteins of the potato cyst nematode *Globodera rostochiensis* using RNAi. *Molecular Plant-Microbe Interactions, 18,* 621–625.

Chitwood, D. J. (2003). Research on plant-parasitic nematode biology conducted by the United States Department of Agriculture-Agricultural Research Service. *Pest Management Science, 59,* 748–753.

Cogoni, C., & G. Macino, G. (2000). Post-transcriptional gene silencing across kingdoms. *Current Opinion in Genetics & Development, 10,* 638–643.

Davis, E. L., Hussey, R. S., Baum, T. J., Bakker, J., Schots, A., Rosso, M. N., & Abad, P. (2000). Nematode parasitism genes. *Annual Review of Phytopathology 38,* 365–396.

Davuluri, G. R., van Tuinen, A., Fraser, P. D., Manfredonia, A., Newman, R., Burgess, D., et al. (2005). Fruit-specific RNAi-mediated suppression of DET1 enhances carotenoid and flavonoid content in tomatoes. *Nature Biotechnology, 23,* 890–895.

de Almeida Engler, J. D., Van Poucke, K., Karimi, M., De Groodt, R., Gheysen, G., & Engler, G. (2004). Dynamic cytoskeleton rearrangements in giant cells and syncytia of nematode-infected roots. *Plant Journal, 38,* 12–26.

Denli, A.M., & Hannon, G. J. (2003). RNAi: An ever-growing puzzle. *Trends in Biochemical Sciences, 28,* 196–201.

Dirks, R. P., Bouw, G. B., Huizen, R. R., Jansen, E. J., & Martens, J. M. (2003). Functional genomics in *Xenopus laevis*: Towards transgene-driven RNA interference and cell-specific transgene expression. *Current Genomics, 4*, 699–711.

Du, T., & Zamore, P.D. (2005). MicroPrimer: The biogenesis and function of microRNA. *Development, 132*, 4645–4652.

Escobar, M. A., Civerolo, E. L., Summerfelt, K. R., & Dandekar, A. M. (2001). RNAi-mediated oncogene silencing confers resistance to crown gall tumorigenesis. *Proceedings of the National Academy of Sciences of the USA, 98*, 13437–13442.

Fanelli, E., Di Vito, M., Jones, J. T., & De Giorgi, C. (2005). Analysis of chitin synthase function in a plant parasitic nematode, *Meloidogyne artiellia*, using RNAi. *Gene, 349*, 87–95.

Feinberg, E. H., & Hunter, C. P. (2003). Transport of dsRNA into cells by the transmembrane protein SID-1. *Science, 301*, 1545–1547.

Fire, A., Xu, S., Montgomery, M. K., Kostas, S. A., Driver, S. E., & Mello, C. C. (1998). Potent and specific genetic interference by double-stranded RNA in *Caenorhabditis elegans*. *Nature, 391*, 806–811.

Forrest, E. C., Cogoni, C., & Macino, G. (2004). The RNA-dependent RNA polymerase, QDE-1, is a rate-limiting factor in post-transcriptional gene silencing in *Neurospora crassa*. *Nucleic Acids Research, 32*, 2123–2128.

Gao, G., Raikar, S., Davenport, B., Mutapcic, L., Montgomery, R., Kuzmin, E., et al. (2006). Cross-species RNAi: Selected *Ascaris suum* dsRNAs can sterilize *Caenorhabditis elegans*. *Molecular and Biochemical Parasitology, 146*, 124–128.

Geldhof, P., Molloy, C., & Knox, D. P. (2006). Combinatorial RNAi on intestinal cathepsin B-like proteinases in *Caenorhabditis elegans* questions the perception of their role in nematode biology. *Molecular and Biochemical Parasitology, 145*, 128–132.

Golinowski, W., Sobczak, M., Kurek, W., & Grymaszewska, G. (1997). The structure of syncytia. In C. Fenoll, F. M. W. Grundler, & S. A. Ohl (Eds.), *Cellular and molecular aspects of plant-nematode interactions* (pp. 80–97). Dordrecht, Netherlands: Kluwer Academic Publishers.

Goverse, A., Biesheuvel, J., Wijers, G. J., Gommers, F. J., Bakker, J., Schots, A., et al. (1998). *In planta* monitoring of the activity of two constitutive promoters, CaMV 35S and TR2', in developing feeding cells induced by *Globodera rostochiensis* using green fluorescent protein in combination with confocal laser scanning microscopy. *Physiological and Molecular Plant Pathology, 52*, 275–284.

Grishok, A. (2005). RNAi mechanisms in *Caenorhabditis elegans*. *FEBS Letters, 579*, 5932–5939.

Grishok, A., Tabara, H., & Mello, C. C. (2000). Genetic requirements for inheritance of RNAi in *C. elegans*. *Science, 287*, 2494–2497.

Hammond, S. M. (2005). Dicing and slicing – The core machinery of the RNA interference pathway. *FEBS Lettsers, 579*, 5822–5829.

Hannon, G. J. (2002). RNA interference. *Nature, 418*, 244–251.

Hedgecock, E. M., Culotti, J. G., Thomson, J. N., & Perkins, L. A. (1985). Axonal guidance mutants of *Caenorhabditis elegans* identified by filling sensory neurons with fluorescein dyes. *Developmental Biology, 111*, 158–170.

Huang, G., Allen, R., Davis, E. L., Baum, T. J., & Hussey, R. S. (2006a). Engineering broad root-knot resistance in transgenic plants by RNAi silencing of a conserved and essential root-knot nematode parasitism gene. *Proceedings of the National Academy of Sciences of the USA, 103*, 14302–14306.

Huang, G., Dong, R., Allen, R., Davis, E. L., Baum, T. J., & Hussey, R. S. (2006b). A root-knot nematode secretory peptide functions as a ligand for a plant transcription factor. *Molecular Plant-Microbe Interactions, 19*, 463–470.

Hussey, R. S., & Janssen, G. J. W. (2002). Root-knot nematodes: *Meloidogyne* species. In J. L. Starr, R. Cook, & J. Bridge (Eds.), *Plant resistance to parasitic nematodes* (pp. 43–70). Oxford, UK: CAB International.

Hussey, R. S., & Mims, C. W. (1991). Ultrastructure of feeding tubes formed in giant-cells induced in plants by the root-knot nematode *Meloidogyne incognita*. *Protoplasma, 162*, 99–107.

Jackson, A. L., Bartz, S. R., Schelter, J., Kobayashi, S. V., Burchard, J., Mao, M., et al. (2003). Expression profiling reveals off-target gene regulation by RNAi. *Nature Biotechnology, 21*, 635–637.

Jones, M. G. K. (1981a). The development and function of plant cells modified by endoparasitic nematodes. In B. M. Zuckerman, W. F. Mai, & R. A. Rohde (Eds.), *Plant parasitic nematodes* (pp. 255–280). New York, USA: Academic Press.

Jones, M. G. K. (1981b). Host cell responses to endoparasitic nematode attack: structure and function of giant cells and syncytia. *Annals of Applied Biology, 97*, 353–372.

Jorgensen, R. A., Cluster, P. D., English, J., Que, Q., & Napoli, C. A. (1996). Chalcone synthase cosuppression phenotypes in petunia flowers: comparison of sense vs. antisense constructs and single-copy vs. complex T-DNA sequences. *Plant Molecular Biology, 31*, 957–973.

Kaletta, T., & Hengartner, M. O. (2006). Finding function in novel targets: *C. elegans* as a model organism. *Nature Reviews Drug Discovery, 5*, 387–398.

Kamath, R. S., Martinez-Campos, M., Zipperlen, P., Fraser, A. G., & Ahringer, J. (2000). Effectiveness of specific RNA-mediated interference through ingested double-stranded RNA in *Caenorhabditis elegans*. *Genome Biology, 2*(1), research0002.1–0002.10.

Kennerdell, J. R., & Carthew, R. W. (2000). Heritable gene silencing in *Drosophila* using double stranded RNA. *Nature Biotechnology, 18*, 896–898.

Kimber, M. J., Fleming, C. C., Prior, A., Jones, J. T., Halton, D. W., & Maule, A. G. (2002). Localisation of *Globodera pallida* FMRFamide-related peptide encoding genes using *in situ* hybridisation. *International Journal of Parasitology, 32*, 1095–1105.

Kimber, M. J., McKinney, S., McMaster, S., Day, T. A., Fleming, C. C., & Maule, A. G. (2007). *flp* gene disruption in a parasitic nematode reveals motor dysfunction and unusual neuronal sensitivity to RNA interference. *The FASEB Journal,* 21, 1233–1243.

Kumar, D., Gustafsson, C., & Klessig, D. F. (2006). Validation of RNAi silencing specificity using synthetic genes: salicylic acid-binding protein 2 is required for innate immunity in plants. *Plant Journal, 45*, 863–868.

Kusaba, M. (2004). RNA interference in crop plants. *Current Opinion in Biotechnology, 15*, 139–143.

Li, M., & Rohrer, B. (2006) Gene silencing in *Xenopus laevis* by DNA-vector based RNA interference and transgenesis. *Cell Research, 16*, 99–105.

Lilley, C. J., Atkinson, H. J., & Urwin, P. E. (2005a). Molecular aspects of cyst nematodes. *Molecular Plant Pathology, 6*, 577–588.

Lilley, C. J., Goodchild, S. A., Atkinson, H. J., & Urwin, P. E. (2005b). Cloning and characterisation of a *Heterodera glycines* aminopeptidase cDNA. *International Journal of Parasitology, 35*, 1577–1585.

Liu, B., Hibbard, J. K., Urwin, P. E., & Atkinson, H. J. (2005). The production of synthetic chemodisruptive peptides *in planta* disrupts the establishment of cyst nematodes *Plant Biotechnology Journal*, 3, 487–496.

Mansoor, S., Amin, I., Hussain, M., Zafar, Y., & Briddon, R. W. (2006). Engineering novel traits in plants through RNA interference. *Trends in Plant Science, 11*, 559–565.

Ogita, S., Uefuji, H., Morimoto, M., & Sano, H. (2004). Application of RNAi to confirm theobromine as the major intermediate for caffeine biosynthesis in coffee plants with potential for construction of decaffeinated varieties. *Plant Molecular Biology, 54*, 931–941.

Phillips, M. S., & Trudgill, D. L. (1998). Population modelling and integrated control options for potato cyst nematodes. In R. J. Marks & B. B. Brodie (Eds.), *Potato cyst nematodes biology, distribution and control* (pp. 153–163). Oxford, UK: CAB International.

Pooggin, M., Shivaprasad, P. V., Veluthambi, K., & Mohn, T. (2003). RNAi targeting of DNA virus in plants. *Nature Biotechnology, 21*, 131–132.

Qi, Y., & Hannon, G. J. (2005). Uncovering RNAi mechanisms in plants: Biochemistry enters the foray. *FEBS Letters, 579*, 5899–5903.

Rosso, M. N., Dubrana, M. P., Cimbolini, N., Jaubert, S., & Abad, P. (2005). Application of RNA interference to root-knot nematode genes encoding esophageal gland proteins. *Molecular Plant-Microbe Interactions, 18*, 615–620.

Saleh, M-C., van Rij, R. P., Hekele, A., Gillis, A., Foley, E., O'Farrell, P. H. et al. (2006). The endocytic pathway mediates cell entry of dsRNA to induce RNAi silencing. *Nature Cell Biology, 8*, 793–802.

Sasser, J. N., & Freckman, D. W. (1987). A world perspective on nematology: the role of the society. In J. A. Veech & D. W. Dickerson (Eds.), *Vistas on nematology* (pp. 7–14). Hyatsville, USA: Society of Nematologists.

Scacheri, P. C., Rozenblatt-Rosen, O., Caplen, N. J., Wolfsberg, T. G., Umayam, L., Lee, J. C. et al. (2004). Short interfering RNAs can induce unexpected and divergent changes in the levels of untargeted proteins in mammalian cells. *Proceedings of the National Academy of Sciences of the USA, 101*, 1892–1897.

Schmid, A., Schindelholz, B., & Zinn, K. (2002). Combinatorial RNAi: a method for evaluating the functions of gene families in Drosophila. *Trends in Neuroscience, 25*, 71–74.

Sen, G. L., & Blau, H. M. (2006). A brief history of RNAi: the silence of the genes. *The FASEB Journal, 20*, 1293–1299.

Shingles, J., Lilley, C. J., Atkinson, H. J., & Urwin, P. E. (2007). *Meloidogyne incognita*: molecular and biochemical characterisation of a cathepsin L cysteine proteinase and the effect on parasitism following RNAi. *Experimental Parasitology, 115*, 114–120.

Sijmons, P. C., Grundler, F. M. W., Von Mende, N., Burrows, P. R., & Wyss, U. (1991). *Arabidopsis thaliana* as a new model host for plant-parasitic nematodes. *Plant Journal, 1*, 245–254.

Silva, J., Chang, K., Hannon, G. J., & Rivas, F. V. (2004). RNA-interference-based functional genomics in mammalian cells: reverse genetics coming of age. *Oncogene, 23*, 8401–8409.

Simmer, F., Tijsterman, M., Parrish, S., Koushika, S. P., Nonet, M. L., Fire, A., et al. (2002). Loss of the putative RNA-directed RNA polymerase RRF-3 makes *C. elegans* hypersensitive to RNAi. *Current Biology, 12*, 1317–1319.

Simmer, F., Moorman, C., van der Linden, A. M., Kuijk, E., van den Berghe, P. V. E., Kamath, R. S., et al. (2003). Genome-wide RNAi of *C. elegans* using the hypersensitive *rrf-3* strain reveals novel gene functions. *PLoS Biology, 1*, e12, 077–084.

Smith, N. A., Singh, S. P., Wang, M.-B., Stoutjesdijk, P. A., Green, A. G., & Waterhouse, P. M. (2000). Total silencing by intron-spliced hairpin RNAs. *Nature, 407*, 319–320.

Sobczak, M., Golinowski, W., & Grundler, F. M. W. (1999). Ultrastructure of feeding plugs and feeding tubes formed by *Heterodera schachtii*. *Nematology, 1*, 363–374.

Sontheimer, E. J. (2005). Assembly and function of RNA silencing complexes. *Nature Reviews Molecular Cell Biology, 6*, 127–138.

Starr, J. L., Bridge, J., & Cook, R. (2002). Resistance to plant-parasitic nematodes: History, current use and future potential. In J. L. Starr, R. Cook, & J. Bridge (Eds.), *Plant resistance to parasitic nematodes* (pp. 1–22). Oxford, UK: CAB International.

Steeves, R. M., Todd, T. C., Essig, J. S., & Trick, H. N. (2006). Transgenic soybeans expressing siRNAs specific to a major sperm protein gene suppress *Heterodera glycines* reproduction. *Functional Plant Biology, 33*, 991–999.

Tabara, H., Grishok, A., & Mello, C. C. (1998). RNAi in *C. elegans*: soaking in the genome sequence. *Science, 282*, 430–431.

Tabara, H., Yigit, Y., Siomi, H., & Mello, C. C. (2002). The dsRNA binding protein RDE-4 interacts with RDE-1, DCR-1, and a DExH-box helicase to direct RNAi in *C. elegans*. *Cell, 109*, 861–871.

Tijsterman, M., May, R. C., Simmer, F., Okihara, K. L., & Plasterk, R. H. A. (2004). Genes required for systemic RNA interference in *Caenorhabditis elegans*. *Current Biology, 14*, 111–116.

Timmons, L., & Fire, A. (1998). Specific interference by ingested dsRNA. *Nature, 395*, 854.

Timmons, L., Court, D. L., & Fire, A. (2001). Ingestion of bacterially expressed dsRNAs can produce specific and potent genetic interference in *Caenorhabditis elegans*. *Gene, 263*, 103–112.

Tomari, Y., & Zamore, P. D. (2005). Perspective: machines for RNAi. *Genes & Development, 19*, 517–529.

Trudgill, D. L., & Blok, V. C. (2001). Apomictic, polyphagous root-knot nematodes: exceptionally successful and damaging biotrophic root pathogens. *Annual Review of Phytopathology, 39*, 53–77.

Urwin, P. E., Atkinson, H. J., Waller, D. A., & McPherson, M. J. (1995). Engineered oryzacystatin-1 expressed in transgenic hairy roots confers resistance to *Globodera pallida*. *Plant Journal, 8*, 121–131.

Urwin, P. E., Green, J., & Atkinson, H. J. (2003). Expression of a plant cystatin confers partial resistance to *Globodera*, full resistance is achieved by pyramiding a cystatin with natural resistance. *Molecular Breeding, 12*, 263–269.

Urwin, P. E., Levesley, A., McPherson, M. J., & Atkinson, H. J. (2000). Transgenic resistance to the nematode *Rotylenchulus reniformis* conferred by *A. thaliana* plants expressing proteinase inhibitors. *Molecular Breeding, 6,* 257–264.

Urwin, P. E., Lilley, C. J., & Atkinson, H. J. (2002). Ingestion of double-stranded RNA by pre parasitic juvenile cyst nematodes leads to RNA interference. *Molecular Plant-Microbe Interactions, 15,* 747–752.

Urwin, P. E., Lilley, C. J., McPherson, M. J., & Atkinson, H. J. (1997a). Resistance to both cyst and root-knot nematodes conferred by transgenic *Arabidopsis* expressing a modified plant cystatin. *Plant Journal, 12,* 455–461.

Urwin, P. E., McPherson, M. J., & Atkinson, H. J. (1998). Enhanced transgenic plant resistance to nematodes by dual protease inhibitor constructs. *Planta, 204,* 472–479.

Urwin, P. E., Møller, S. G., Lilley, C. J., McPherson, M. J., & Atkinson, H. J. (1997b). Continual green-fluorescent protein monitoring of cauliflower mosaic virus 35S promoter activity in nematode-induced feeding cells in *Arabidopsis thaliana. Molecular Plant-Microbe Interactions, 10,* 394–400.

Vishnudasan, D., Tripathi, M. N., Rao, U., & Khurana, P. (2005). Assessment of nematode resistance in wheat transgenic plants expressing potato proteinase inhibitor (*PIN2*) gene. *Transgenic Research, 14,* 665–675

Voinnet, O. (2001). RNA silencing as a plant immune system against viruses. *Trends in Genetics, 17,* 449–459.

Voinnet, O. (2002). RNA silencing: Small RNAs as ubiquitous regulators of gene expression. *Current Opinion in Plant Biology, 5,* 444–451.

Waterhouse, P. M., Graham, M. W., & Wang, M. B. (1998). Virus resistance and gene silencing in plants can be induced by simultaneous expression of sense and antisense RNA. *Proceedings of the National Academy of Sciences of the USA, 95,* 13959–13964.

Waterhouse, P. M., Wang, M. B., & Lough, T. (2001). Gene silencing as an adaptive defence against viruses. *Nature, 411,* 834–842.

Wesley, S. V., Helliwell, C. A., Smith, N. A., Wang, M.-B., Rouse, D. T., Liu, Q., et al. (2001). Construct design for efficient, effective and high-throughput gene silencing in plants. *Plant Journal, 27,* 581–590.

Winston, W. M., Molodowitch, C., & Hunter, C. P. (2002). Systemic RNAi in *C. elegans* requires the putative transmembrane protein SID-1. *Science, 295,* 2456–2459.

Winter, M. D., McPherson, M. J., & Atkinson, H. J. (2002). Neuronal uptake of pesticides disrupts chemosensory cells of nematodes. *Parasitology, 125,* 561–565.

Yadav, B. C., Veluthambi, K., & Subramaniam, K. (2006). Host-generated double stranded RNA induces RNAi in plant-parasitic nematodes and protects the host from infection. *Molecular and Biochemical Parasitology, 148,* 219–222.

Zamore, P. D., Tuschl, T., Sharp, P. A., & Bartel, D. P. (2000). RNAi: Double-stranded RNA directs the ATP-dependent cleavage of mRNA at 21 to 23 nucleotide intervals. *Cell, 101,* 25–33.

10

S. GOWEN[1], K. G. DAVIES[2] AND B. PEMBROKE[1]

POTENTIAL USE OF *PASTEURIA* SPP. IN THE MANAGEMENT OF PLANT PARASITIC NEMATODES

[1]*The University of Reading, Department of Agriculture, School of Agriculture, Policy and Development, Reading, RG6 6AR, UK*
[2] *Rothamsted Research, Harpenden, Hertfordshire, AL5 2JQ, UK*

Abstract. Potentials of *Pasteuria penetrans* and close bacterial nematode parasites are reviewed. Several aspects concerning the identification and recognition of *P. penetrans* are discussed, with description of the bacterium's life cycle, biology, host range and specificity. The application of traditional and molecular taxonomic methods for the identification of isolates and species as well as the available technologies for *in vivo* and *in vitro* mass culture are also reviewed.

1. INTRODUCTION

A general, satisfactory control strategy has not yet been developed for plant parasitic nematodes. Host plant resistance, chemicals, rotations and cultural practices can all have a role in nematodes management but for differing reasons they are not widely adopted. Biological control is an attractive practice in theory but, when compared to other available control means, it has not been extensively or sufficiently researched. The success of a biological control agent will depend on its ability to reduce the multiplication of the pest, but this has been often difficult to assess in the field.

Among nematode antagonists, the Gram+ bacteria of the *Pasteuria* group attracted, in the last decades, considerable interest due to several peculiarities of their parasitic behaviour. Research on *Pasteuria* was mainly focused on *Pasteuria penetrans* and its potential as a biological control agent of root-knot nematodes (*Meloidogyne* spp.) (Stirling, 1991).

Although *P. penetrans* is a naturally occurring parasite of root-knot nematodes, it rarely exerts a suppressive effect on populations, when detected by growers or agronomists. However, we consider that the opportunity for exploiting *P. penetrans* is worth the effort.

In this chapter we highlight the positive attributes of this bacterium showing how, through a thorough understanding of its biology, it might be manipulated (within other control strategies) to decrease nematode population densities to an extent resulting in measurable benefits to crop growth.

A. Ciancio & K. G. Mukerji (eds.), Integrated Management and Biocontrol of Vegetable and Grain Crops Nematodes, 205–219.

2. RECOGNISING *PASTEURIA*

Nematode parasites of the *Pasteuria* group are often overlooked because their presence on or within nematodes can only be seen under a microscope at more than 100× magnification. This may be an impediment to their recognition in samples taken to a laboratory. When soil samples are processed for nematode extraction, juvenile or vermiform stages of the nematode species present may be recovered. These may have *Pasteuria* endospores attached to them if the bacterium is present in that soil, but spore attachment will only be seen if nematodes are observed under high power magnification. If a number of root-knot nematode juveniles are endospore encumbered, they may appear to aggregate into clumps: this is often a useful characteristic, that can be noticed at lower magnifications.

Infected female root-knot nematodes can be found in root systems but where the incidence of *P. penetrans* is low, then the chance of detection is small (see section below). Infected females do not produce egg masses, they appear dense and cream coloured in contrast to healthy females which become partially translucent as they mature and produce egg masses.

In summary, we do not disguise the fact that the recognition of *P. penetrans* from the field requires some nematological expertise. This reinforces our consideration that the practical development of this bacterium as a biological control agent will be achieved only with progresses in our understanding of its biology, biochemistry and life cycle.

2.1. Life Cycle and Development

The first observation of a *Pasteuria* from plant parasitic nematodes (*Pratylenchus pratensis*) was provided by Thorne (1940), which considered the organism a microsporidian and named it *Dubosqia penetrans*. The life cycle of *P. penetrans* was first described and illustrated by Mankau (1975), Mankau and Imbriani (1975), Imbriani and Mankau (1977) and Sayre and Wergin (1977).

The initial stage of the life cycle of *P. penetrans* on root-knot nematodes is the chance contact of endospores to the second (infective) stage juvenile, which occurs in the soil as the juvenile seeks a suitable host root. Endospore attachment does not necessarily cause infection, implying that not all endospores may be viable. The extent of endospore viability is difficult to determine, as infection can proceed from the attachment by one single endospore to a juvenile (Trotter, Darban, Gowen, Bishop, & Pembroke, 2005). Greater than 15 endospores may disable the nematode in its movements, and invasion may not take place (Davies, Kerry, & Flynn, 1988). The optimal attachment level should be around 5–10 endospores per juvenile, as enough endospores will initiate infection without reducing the ability of the nematode to invade roots (Davies et al., 1988; Davies, Laird, & Kerry, 1991; Rao, Gowen, Pembroke, & Reddy, 1997). Even so, when plants are infected only with endospore encumbered juveniles, 100% *Pasteuria* infection is not certain (Pembroke, 2007 unpublished).

Once a spore-encumbered juvenile has invaded a root, it will establish a feeding site and apparently normal development will continue. Stirling (1991), quoting the life cycle as described by Imbriani and Mankau (1977), states that germination of the spore(s) and production of the germ tube does not occur until approximately 8 days after nematode invasion. Sayre and Wergin (1977) suggested that germination of the endospore is initiated by the onset of the nematode feeding activity, however endospores on second stage juveniles have been observed to germinate in the absence of a host plant (Davies, personal communication).

The germ tube (infection peg) emerges through the central opening of the basal ring of the endospore and penetrates the nematode cuticle entering the hypodermal tissue. No deformation of the nematode cuticle occurs, suggesting that there is no appreciable force exerted on the nematode during this process. Rhizomorphs are first observed close to the site of penetration. However, mycelial colonies (microcolonies) up-to 20 μm are ultimately found in the pseudocoelom (Sayre & Wergin, 1977).

The exponential growth phase of the bacterium in the infected host is not altogether clear, but recently rod-shaped bacillus-like cells have been observed (Davies et al., 2004). These are likely to be the vegetative growth stages of the bacterium, but their identity awaits confirmation.

Sporogenesis is triggered in a manner similar to other *Bacillus* spp. and involves a phosphorylation pathway (Kojetin et al., 2005). It results initially in the production of microcolonies that mature to contain fewer but larger cells. These fragments lead to quartets and then doublets, each eventually developing a single sporangium which gives rise to a single, true endospore (Stirling, 1991). The external development of the nematode remains unaltered, undergoing normal moults and only microscopic examination (higher than 200× magnification) would reveal the intensification of the hyperparasite within the nematode. The resultant swollen adult female is almost devoid of eggs and may contain greater than 2 million spores (Stirling, 1981; Stirling, 1991; Darban, Pembroke, & Gowen, 2004).

The developmental cycles of *Pasteuria* spp. can be different with respect to the duration and the nematode stages that are capable of being infected. Although detailed studies in this area have not been undertaken, preliminary observations show that the length of life cycles differs and that the development of *Pasteuria* within different life stages of the nematodes also differs. For example, the life cycle of *P. thornei* was shorter than that for *P. penetrans* and all forms of the life cycle were observed in one or other of the migratory stages of the nematode (Starr & Sayre, 1988). Similarly, all stages of the life cycle have been observed in the pseudocoelom of second stage juveniles of *Heterodera avenae* (Davies, Flynn, Laird, & Kerry, 1990) whereas this has not been seen in *Meloidogyne* spp.

2.2. Morphology

There are two major developmental structures that have been used to characterise the species: the sporangial shape and the structures and dimensions of the endospore, as viewed by brightfield scanning and electron microscopy (Sayre, Wergin, Schmidt, & Starr, 1991). Although the dimensions of mature endospores are robust features on which to characterise different isolates of the bacterium, developmental structures are more problematic to use, as these will often be age related, continually undergoing changes and therefore difficult to assess. Also, the early stages of development are often fleeting and difficult to observe in some specimens and therefore do not represent reliable structures useful for taxonomic purposes.

The mature endospore itself is probably the best morphological structure to use to characterise an isolate. Measurements of the width and height of the endospores and of the sporangium can be made with a high degree of accuracy in order to compare different populations. In undertaking such studies it should be born in mind that specimen fixation, staining and orientation can all play a part in generating the observable variation encountered and the only valid comparisons are those made between populations that have been processed following the same method.

2.3. Traditional and Molecular Taxonomy

Six species of *Pasteuria* have been identified to date. Five of these parasitize plant parasitic nematodes: *P. penetrans* (parasitic on *Meloidogyne* spp.); *P. thornei* (parasitic on *Pratylenchus penetrans)*; *P. nishizawae* (parasitic on *Heterodera* and *Globodera* spp.); Candidatus P. usage (parasitic on *Belonolaimus longicaudatus*), and *P. hartismeri* (parasitic on *M. ardenensis). Pasteuria ramosa* parasitic on the Cladoceran *Daphnia magna*, a water flea, was first described by Metchnikoff in 1888 (Metchnikoff, 1888). Identification and characterisation of *Pasteuria* spp. have been based on a number of features that include morphology, life cycle and development, host range and more recently DNA sequences (Sayre & Starr, 1985; Sayre, Starr, Golden, Wergin, & Endo, 1988; Sayre et al.,1991; Giblin-Davis et al., 2003; Bishop, Gowen, Pembroke, & Trotter, 2007).

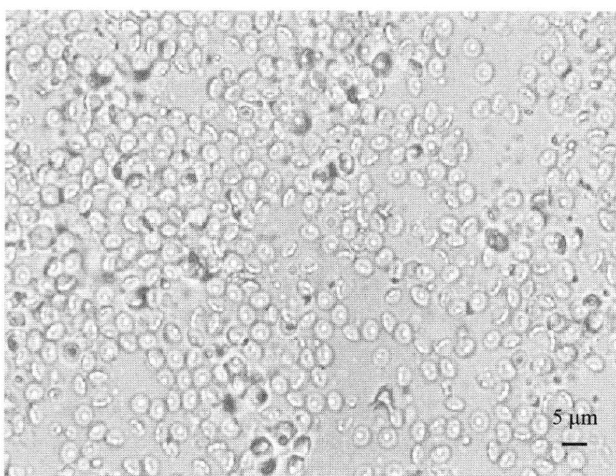

Figure 1. Endospores of Pasteuria penetrans *at life-cycle completion, 35 days after nematode infection (circa 650 degree days).*

Pasteuria endospores have an ellipsoidal characteristic shape, and with experience they should always be recognisable at 150–200× magnifications. The endospores are 3–5 μm wide, non-motile and have two distinct components: a central refractive core, surrounded by a peripheral matrix. These are readily recognised when an infected nematode is squashed and the endospores are released (Fig. 1).

Recognition of the early development stages after the endospore has germinated in a nematode is difficult. The identification of this stage requires careful examination of squashed nematodes at high magnification (Fig. 2). As the colonies develop into sporangia, the enlarging mycelia assume characteristic forms initially displaying branches of four (quartets) and then two (doublets) cells. The new endospore eventually develops from the terminal part of the sporangium.

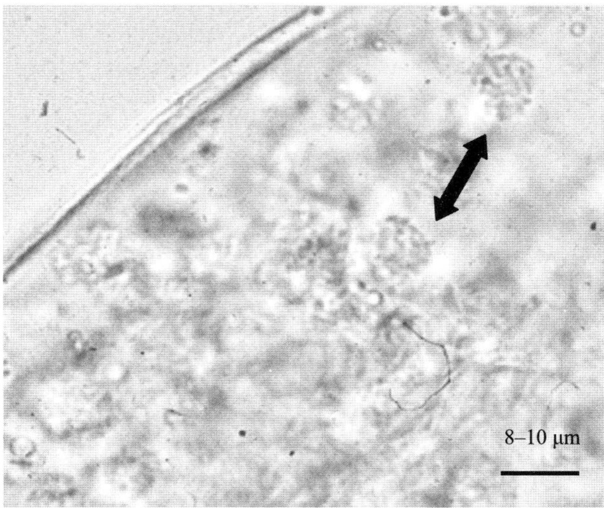

Figure 2. Microcolonies of Pasteuria penetrans *inside a developed second stage juvenile of root-knot nematode.*

2.4. DNA Approaches

As DNA replicates it undergoes mutations with a given likelihood. Therefore closely related organisms sharing a recent common ancestor will have less sequence divergence than organisms that have a very distant common ancestor. As a rule of thumb it is accepted that a DNA reassociation value higher than 70% is the threshold for delineating a bacterial species (Wayne et al., 1987). However, recent research data suggest that a DNA similarity lower than 97% can be regarded as the threshold for delineating a new bacterial species (Amann, Ludwing, & Schleifer, 1995; Hagström et al., 2002).

DNA based techniques are now being routinely applied to populations of *Pasteuria*: endospores are collected from infected nematodes and then bead beating is used to release the DNA. PCR is then employed with primers that recognise the 16S rDNA ribosomal subunit (Ebert, Rainey, Embley, & Scholz 1996; Anderson et al., 1999; Atibalentja, Noel, & Domier, 2000; Bekal, Borneman, Springer, Giblin-Davis, & Becker, 2001; Sturhan, Shutova, Akimov, & Subbotin, 2005; Bishop et al., 2007). There are at present 58 sequences from *Pasteuria* 16S rDNA genes submitted to GenBank (*http://www.ncbi.nlm.nih.gov/Genbank*), which can be used to characterise the different populations.

However, it has been recently suggested that even greater stringency should be applied to species definitions and that an average nucleotide identity higher than 99% should be applied, because several bacteria show minimum differences of their (well characterised) genes (Konstantinidis & Tiedje, 2005). There are few available studies to date which used genes other than the 16S rDNA to characterise *Pasteuria* spp. (Trotter & Bishop, 2003; Schmidt, Preston, Nong, Dickson, & Aldrich, 2004; Charles et al., 2005). Where these studies have been undertaken the results tended to be consistent, showing that *Pasteuria* lies deep within the *Bacillus-Clostridium* clade, with species designation being related to the host from which the bacterium had been isolated.

2.5. Host Range

Spores of *Pasteuria* spp. represent the resting stage and can remain viable for several years (Giannakou, Pembroke, Gowen, & Davies, 1997). These propagules are non-motile. They are responsible of transmission, since the host infection process starts when they passively adhere to the nematode cuticle on contact. The encounter with the nematode occurs as the host migrates through the soil in search of a plant root. Therefore, the endospore adhesion is a crucial step in the infection process.

Numerous studies have been carried out testing the ability of different populations to adhere to different nematode populations (Stirling, 1985; Davies et al., 1988; Channer & Gowen, 1992; Sharma & Davies, 1996a, 1996b; Español, Verdejo-Lucas, Davies, & Kerry, 1997; Mendoza de Gives, Davies, Morgan, & Behnke, 1999; Davies et al., 2001; Wishart, Blok, Phillips, & Davies, 2004). Data showed a range of variation, varying from endospore populations whose attachment is highly restricted to one population of nematodes but not to any others (either within a species or between species), to those having a much broader host range and adhere not only to the population from which they were originally isolated, but even to nematodes of a different genus. This aspect is particularly important for the use of *Pasteuria* spp. as biological control agents, because spores must be targeted to the nematode species that is occurring as a pest.

The endospores ability to adhere to different host life stages is also variable, with recent research showing that they can sometimes adhere to the cuticle of males, but sometimes they cannot (Carneiro, Randig, Freitas, & Dickson, 1999; Davies & Williamson, 2006). Endospores adhesion to the nematode cuticle cannot necessarily be interpreted as conducive to an infection, since not all adhering endospores will germinate and lead to the development of parasitism.

3. MASS PRODUCTION

3.1. *In vitro* Culture

The need for an environmentally benign method to control plant parasitic nematodes as an alternative for chemical pesticides, combined with the fact that *Pasteuria* is associated with nematode suppressive soils and has been shown to unequivocally act as a biological control agent (Stirling, 1984), has focused research on *in vitro* production methods. However, the obligate nature of the bacterium life cycle has proved difficult to overcome and initial attempts at *in vitro* mass production have proved illusive (Reise, Hackett, & Huettel, 1991; Bishop & Ellar, 1991). As early as 1992 a patent was submitted showing that endospores could be cultured *in vitro* in a media that contained explanted tissues from nematodes (Previc & Cox, 1992), however this achievement has not been developed further.

Media have also been developed whereby one was able to sustain very small amounts of vegetative growth, while another led to sporulation and the development of endospores (Bishop & Ellar, 1991), however exponential growth was not obtained. More recently research has focused on the possibility of other bacteria being necessary for the development of *Pasteuria* and experiments have been undertaken in which *P. penetrans* was co-cultured with another bacterium, *Enterobacter cloacae* (Dupponis, Ba, & Mateille, 1999).

Research by Pasteuria BioScience LLC in Florida, USA has focused on the possibility that *Pasteuria* was an acidophile (Gerber & White, 2001; Hewlett, Gerber, & Smith, 2004). Pasteuria BioScience LLC is clearly making progress and *in vitro* cultured endospores produced in a fermentation vessel have recently been tested in the field (Hewlett, Griswold, & Smith, 2006).

3.2. *In vivo* Culture

This technique is based on a system first described by Stirling and Wachtel (1980) in which a plant host is inoculated with spore-encumbered root-knot nematode juveniles. The root systems, containing infected (spore-filled) female nematodes, are harvested after an appropriate period of time and then dried for long term storage as a fine powder (Pembroke, Darban, Gowen, & Karanja, 2005).

To achieve the best possible mass production of *Pasteuria* spores, it is important to consider all the organisms involved in the system: host plant, nematodes and *Pasteuria* propagules, giving particular attention to the conditions under which they are grown. The overall objective should be to harvest root systems containing as many *Pasteuria*-infected females as possible. Temperature and the time of harvest are the most critical factors. The optimum temperature for *P. penetrans* development is around 28–30°C, and spores can be found in females after 35 days. Greater endospore numbers are obtained if the plants are left to grow for longer periods (Darban et al., 2004).

The host origins and species composition appear very important. If the host nematode is a field population proceeding from a tropical location, it could include a mixture of species e.g. *M. incognita, M. javanica* or *M. arenaria*, which may differ in their susceptibility to *P. penetrans*. Considering that the nematode life cycle from egg to egg is about 3 weeks at 28–30°C, the production system may appear unsuitable. The uninfected females, and the consequent egg production and secondary infections (due to uninfected nematodes) can increase the stress on the host plant and thus may reduce its potential to sustain the *Pasteuria*-infected females, thus affecting the numbers of endospores produced.

Pasteuria is a hyperparasite, dependant on the well being of its nematode host which in turn requires a thriving host plant. Tomato is often the preferred plant host because it is easy to grow and is highly susceptible to root-knot nematodes. However, tomato grows best in a diurnally fluctuating temperature regime. A constant temperature of 28–30°C may be suitable for nematode reproduction and *Pasteuria* development, but cannot result as an optimum for the plant host. The number of endospores in an infected female nematode will hence increase as long as the nematode is receiving sufficient nutrients from the plant.

In the fluctuating conditions of a glasshouse (20–32°C), infected female nematodes were observed to increase in weight for up to 88 days, contained 2.3 million endospores and numbers of endospores in each female had not peaked (Darban et al., 2004). Although Stirling (1981) showed that endospores can be found after 700-degree days at 30°C, his data also show that the highest number of endospores per female was achieved at 20°C.

Attention should be given to the growth habit of the variety used in an *in vivo* production system, if tomato is the chosen host plant. At temperatures around 30°C tomato plants senesce early, perhaps hastened by nematode induced stress. An indeterminate variety may live for longer and may produce more roots, thus providing more nutrients to the nematodes than a determinate variety. This may be an advantage for endospore production, if host vigour and longevity are important. No data from such studies have yet been reported.

Also the size of the host plant at inoculation is another important parameter when *in vivo* system is chosen for endospores mass production. In general, the larger the plant, the greater can be the initial inoculum of endospore encumbered nematodes. However, a compromise is required because of the expected longevity of the host plant, as described above. If the host plant is too old at inoculation then the maximum possible endospore production may not be achieved.

Finally, the longevity of a female infected by *P. penetrans* is unknown.

3.3. Distribution in Natural Systems

Records of *Pasteuria* endospore attachment on nematodes have been made from many countries and all continents, apart from the Antarctic. Most records are for *P. penetrans* on the major tropical root-knot species in warmer climates. There are relatively fewer records on root-knot nematodes from temperate regions (Bishop et al., 2007) although these areas show a greater number of *Pasteuria* endospore attachment records, on a diversity of nematodes (Sturhan, 1988; Subbotin, Sturhan, & Ryss, 1994). It cannot be excluded that this possibly occurs, because more nematologists (particularly taxonomists) observe specimens under high power magnification in these regions.

3.3.1. Finding Pasteuria

There is no universally established method for the direct recovery of *Pasteuria* endospores from soil. Spore attachment on nematodes has been the most frequent means of detection. However, recent studies looking at the interaction between plant parasitic nematodes in a natural sand dune system has successfully employed immunological techniques to quantify the number of endospores in this ecosystem (Costa, Kerry, Bardgett, & Davies, 2006).

In general, suggested methods required to find *Pasteuria* spp. rely on:

1. Collection of root-knot nematode infested plants, drying the roots and then re-hydrating root segments, carefully looking for females in the roots. *Pasteuria*-infected females are characterised by a porcelain colour with no transparent regions in the body (Fig. 3).
2. Grinding roots and making a suspension which is then probed with juveniles, leaving them in the suspension for 24 hrs and looking for endospore attachment. If endospores are present, encumbered juveniles may clump together.
3. Direct observations of endospores in suspensions may be made from ground roots at high power (higher than 200×). However, recognition of endospores in such suspensions is difficult, particularly if their concentations are low. The problem with the grinding roots method is that at low densities, the chances of missing infected females are high (Pembroke et al., 2005).
4. Collecting soil samples from around infested plants, extracting nematodes and looking for juveniles encumbered with spores. The examination of samples taken from crops where root-knot nematode damage is less than expected may be more effective in perennial crops. Stirling and White (1982), found that numbers of root-knot nematodes were lower in 25 year old vineyards where *P. penetrans* was widely distributed. Dabiré, Chotte, Fardoux, and Mateille, (2001) developed techniques for direct microscopic observation of spores, following their dispersion in soil aggregates.
5. Immunological techniques can be used for recognition and quantification of *Pasteuria* endospores in soil (Fould, Dieng, Davies, Normand, & Mateille, 2001; Costa et al., 2006).

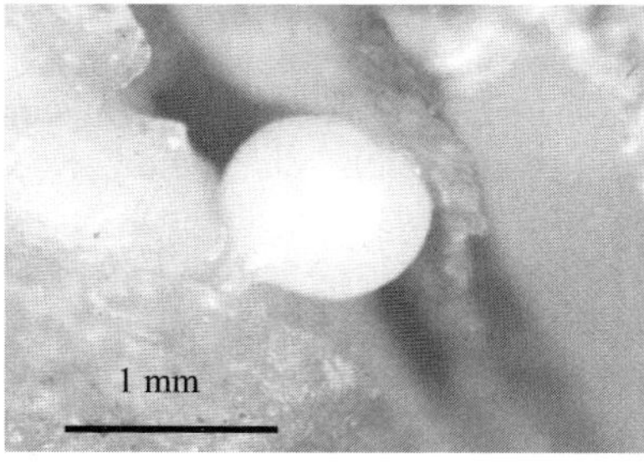

Figure 3. Mature root-knot nematode female infected with Pasteuria penetrans *dissected from re-hydrated roots.*

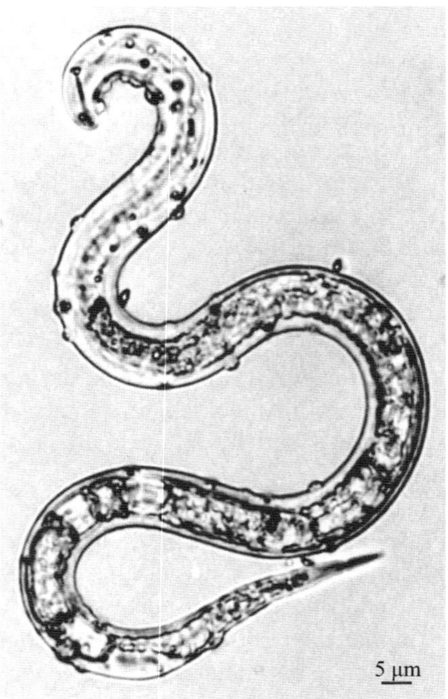

Figure 4. Root-knot nematode second stage juvenile encumbered with spores of Pasteuria penetrans.

4. *PASTEURIA* ASSOCIATION WITH NEMATODE SUPPRESSIVE SOILS

The goal of any biological management strategy is to develop suppressiveness of the pest population to a level which is less damaging to the host crop. Among nematode antagonists, *P. penetrans* has the attributes that would make it appropriate for such a strategy. However, there are few reports of instances where it established to the extent that root-knot nematode populations were suppressed. In intensively grown vegetable crops on light textured soils close to Dakar (Senegal), Mankau (1980) found that production was high and root-knot nematodes appeared not to be damaging. Upon examination, 80–98% of root-knot nematode juveniles were found to be encumbered with endospores of *P. penetrans*. Stirling and White (1982) found that numbers of root-knot nematodes were lower in vineyards more than 25 years old, where *P. penetrans* was widely distributed. Success was demonstrated in contriving an increase in *P. penetrans* endospore densities in field microplot experiments with *M. incognita/M. javanica* in Ecuador (Triviño & Gowen, 1996) and Tanzania (Trudgill et al., 2000). Similarly, positive results have been reported in a peanut-based cropping system in Florida where the pest was *M. arenaria* (Oostendorp, Dickson, & Mitchell, 1991; Chen, Dickson, Mc Sorley, Mitchell, & Hewlett, 1996) and in a 7-year tobacco monoculture where *M. incognita/M. javanica* were the principal species (Weibelzahl-Fulton, Dickson, & Whitty, 1996). Also, populations of *P. penetrans* have been shown to increase when nematode susceptible crops are grown continuously (Chen & Dickson, 1998; Cetintas & Dickson, 2004).

The texture of soil could be important in developing suppression: *P. penetrans* occurs more frequently in sand and sandy loam soils than in those with greater amounts of loam and clay (Spaull, 1984; Chen & Dickson, 1998). However, in the localities where suppression has been demonstrated, soils varied in texture from 94% sand in Florida (Chen & Dickson, 1998) to silty loam soil (50% silt, 39% clay, 11% sand) in Ecuador (Trudgill et al., 2000).

4.1. Biological and Ecological Features of Pasteuria penetrans

Endospores of *Pasteuria penetrans* are resistant to desiccation, as neither natural nor laboratory-induced drying affects the survival of the spores. Spores require a 3-day period of re-hydration (Brown & Smart, 1984) before maximum attachment ability is restored. Dried endospores can remain viable for long periods: Giannakou et al. (1997) showed that dried root powder that was stored in a laboratory drawer for 11 years contained viable spores. Though attachment did not differ from a freshly produced population, there was evidence to show that the pathogenicity may have declined. However, a criticism of this work is that the "new" population to which the original *Pasteuria penetrans* population was compared had been generated on a nematode population different from the original *Pasteuria*-infected one. Therefore it could be argued that the genetic make-up of the "new" population could have changed (Cook, personal communication), and that the two populations, though similar, could not categorically be described as identical.

There are no experimental data on the long-term survival of endospores in soils. Similarly, there is no published information on natural enemies that may parasitise or ingest endospores (Chen & Dickson, 1998). More information would be useful on the dynamics of populations of endospores in soil in the absence of hosts, if predictive studies on endospore densities and epidemiology are to be progressed.

5. CONCLUSIONS

For *Pasteuria* endospores to be applied inundatively, high quantities of propagules are needed. Mass production of endospores by *in vivo* techniques could under certain circumstances be sufficient for practical use, but there are few documented instances from field experiments where soil inundation has led to satisfactory root-knot nematode control. The main issues to consider concern the need for an immediate effect (in a nematicide-like manner) or if the endospores are applied with the expectation that their densities will increase over a number of crop cycles (Triviño & Gowen, 1996).

Like many biological control agents, it might be expected that greater success with *Pasteuria* will be achieved in the relatively smaller areas of protected crops, rather than in open fields (Pembroke, Gowen, & Giannakou, 1998). There are two reasons supporting this view: beds cultivated under glass or plastic are generally permanent and may be cropped with perennials (flowers) or annuals vegetables. In such intensively cropped systems the opportunities for integrating a soil applied microbial agent should be greatest. Also, the natural build-up of endospore densities in soil may be greater than in open fields. Finally in protected cropping systems with controlled irrigation, the movement of endospores to deeper soil layers may be less than in fields receiving natural rainfall.

In conclusion, the data produced since the rediscovery of these bacteria show that *P. penetrans* and other *Pasteuria* spp. have potentials for application in biological control of nematodes. However, several aspects of their biology and application remain yet to be investigated, with particular concern for the availability of low cost, mass production technologies and of isolates covering the broad range of host nematode diversity.

REFERENCES

Amann, R. I., Ludwing, W., & Schleifer, K. H. (1995). Phylogenetic identification and *in situ* detection of individual microbial cells without cultivation. *Microbiological Reviews, 59*, 143–169.

Anderson, J. M., Preston, J. F., Dickson, D. W., Hewlett, T. E., Williams, N. H., & Maruniak, J. E. (1999). Phylogenetic analysis of *Pasteuria penetrans* by 16s rDNA gene cloning and sequencing. *Journal of Nematology, 31*, 319–325.

Atibalentja, N., Noel, G. R., &. Domier, L. L. (2000). Phylogenetic position of the North American isolate of *Pasteuria penetrans* that parasitizes the soybean cyst nematode, *Heterodera glycines*, as inferred from the 16S rDNA sequence analysis. *International Journal for Systematic and Evolutionary Microbiology, 50*, 605–613.

Bekal, S., Borneman, J., Springer, M. S., Giblin-Davis, R. M., & Becker, J. O. (2001). Phenotypic and molecular analysis of a *Pasteuria* strain parasitic to the sting nematode. *Journal of Nematology, 33*, 110–115.

Bishop, A. H., Gowen, S. R., Pembroke, B., & Trotter, J. R. (2007). Morphological and molecular characteristics of a new species of *Pasteuria* parasitic on *Meloidogyne ardenensis*. *Journal of Invertebrate Pathology, 96*, 28–33.

Bishop, A. H., & Ellar, D. J. (1991). Attempts to culture *Pasteuria penetrans* in vitro. *Biocontrol Science and Technology, 1*, 101–114.

Brown, S. M., & Smart, G. C. (1984). Attachment of *Bacillus penetrans* to *Meloidogyne incognita*. *Nematropica, 14*, 171–172.

Carneiro, R. M. D. G., Randig, O., Freitas, L. G., & Dickson, D. W. (1999). Attachment of endospores of *Pasteuria penetrans* to males and juveniles of *Meloidogyne* spp. *Nematology, 1*, 267–271.

Cetintas, R., & Dickson, D. W. (2004). Persistence and suppressiveness of *Pasteuria penetrans* to *Meloidogyne arenaria* Race1. *Journal of Nematology, 36*, 540–549.

Channer, A. G. De R., & Gowen, S. R. (1992). Selection for increased host resistance and increased pathogen specificity in the *Meloidogyne-Pasteuria penetrans* interaction. *Fundamental and Applied Nematology, 15*, 331–339.

Charles, L., Carbone, I., Davies, K. G., Bird, D., Burke, M., Kerry, B. R. et al. (2005). Phylogenetic analysis of *Pasteuria penetrans* by use of multiple genetic loci. *Journal of Bacteriology, 187*, 5700–5708.

Chen, Z. X., & Dickson, D. W. (1998). Review of *Pasteuria penetrans*: biology, ecology and biological control potential. *Journal of Nematology, 30*, 313–340.

Chen, Z. X., Dickson, D. W., Mc Sorley, R., Mitchell, D. J., & Hewlett, T. E. (1996). Suppression of *Meloidogyne arenaria* race 1 by soil application of endopsores of *Pasteuria penetrans*. *Journal of Nematology, 28*, 159–168.

Costa, S., Kerry, B. R., Bardgett, R. D., & Davies, K. D. (2006). Exploitation of immunofluorescence for the quantification and characterisation of small numbers of *Pasteuria* endospores. *FEMS Microbiology Ecology, 58*, 593–600.

Darban, D. A., Pembroke, B., & Gowen, S. R. (2004). The relationship of time and temperature to body weight and numbers of endospores in *Pasteuria penetrans*-infected *Meloidogyne javanica* females. *Nematology, 6*, 33–36.

Dabiré, K. R., Chotte, J.-L., Fardoux, J., & Mateille, T. (2001). New developments in the estimation of spores of *Pasteuria penetrans*. *Biology and Fertility of Soils, 33*, 340–343.

Davies, K. G., Kerry, B. R., & Flynn, C. A. (1988). Observations on the pathogenicity of *Pasteuria penetrans*, a parasite of root-knot nematodes. *Annals of Applied Biology, 112*, 491–501.

Davies, K. G., Flynn, C. A., Laird, V., & Kerry, B. R. (1990). The life-cycle, population dynamics and host specificity of a parasite of *Heterodera avenae,* similar to *Pasteuria penetrans. Revue de Nématologie, 13*, 303–309.

Davies, K. G., Laird, V., & Kerry, B. R. (1991). The motility, development and infection of *Meloidogyne incognita* encumbered with spores of the obligate hyperparasite *Pasteuria penetrans*. *Révue de Nematologie, 14*, 611–618.

Davies, K. G., Fargette, M., Balla, G., Daudi, A. Duponnois, R., & Gowen, S. R. (2001). Cuticle heterogeneity as exhibited by *Pasteuria* spore attachment is not linked to the phylogeny of parthenogenetic root-knot nematodes (*Meloidogyne* spp.). *Parasitology, 122*, 11–120.

Davies, K. G., Waterman, J., Manzanilla-Lopez, R. & Opperman, C. H. (2004). The life-cycle of *Pasteuria penetrans* in excised root-knot nematodes, *Meloidogyne* spp.: a re-evaluation. XXVII ESN International Symposium, June 14-18, 2004, Rome, Italy, pp. 91–92.

Davies, K. G., & Williamson, V. M. (2006). Host specificity exhibited by populations of endospores of *Pasteuria penetrans* to the juvenile and male cuticles of *Meloidogyne hapla*. *Nematology, 8*, 475–476.

Dupponis, R., Ba, A. M., & Mateille, T. (1999). Beneficial effects of *Enterobacter cloacae* and *Pseudomonas mendocina* for biocontrol of *Meloidogyne incognita* with the endospore-forming bacterium *Pasteuria penetrans. Nematology, 1*, 95–101.

Ebert, D., Rainey, P., Embley, T. M., & Scholz, D. (1996). Development, life cycle, ultrastructure and phylogenetic position of *Pasteuria ramosa* Metchnikoff 1888: Rediscovery of an obligate endoparasite of *Daphnia magna* Straus. *Philosophical Transactions of the Royal Society of London* B, 351, 1689–1701.

Español, M., Verdejo-Lucas, S., Davies, K. G., & Kerry, B. R. (1997). Compatibility between *Pasteuria penetrans* and *Meloidogyne* populations from Spain. *Biocontrol Science and Technology, 7*, 219–230.

Fould, S., Dieng, A. L., Davies, K. G., Normand, P., & Mateille, T. (2001). Immunological quantification of the nematode parasitic bacterium *Pasteuria penetrans* in soil. *FEMS Microbiology Ecology, 37*, 187–195.

Gerber, J. F., & White, J. H. (2001). Materials and methods for the efficient production of *Pasteuria*. International patent application WO 01/11017 A2.

Giannakou, I. O., Pembroke, B., Gowen, S. R., & Davies, K. G. (1997). Effects of long-term storage and above normal temperatures on spore adhesion of *Pasteuria penetrans* and infection of the root-knot nematodes *Meloidogyne javanica*. *Nematologica, 43*, 185–192.

Giblin-Davis, R. M., Williams, D. S., Bekal, S., Dickson, D. W., Brito, J. A., Becker, J. O. & Preston, J. F. (2003). '*Candidatus* Pasteuria usgae' sp. nov., an obligate endoparasite of the phytoparasitic nematode *Belonolaimus longicaudatus*. *International Journal of Systematic and Evolutionary Microbiology*, 53, 197–200.

Hagström, Å., Pommier, T., Rohwer, F., Simu, K., Stolte, W., Svensson, D. et al. (2002). Use of 16S ribosomal DNA for delineation of marine bacterioplankton species. *Applied Environmental Microbiology, 68*, 3628–3633.

Hewlett, T. E., Gerber, J. F., & Smith, K. S. (2004). *In vitro* culture of *Pasteuria penetrans*. In *Nematology Monographs and Perspectives 2: Proceedings of the Fourth International Congress of Nematology* 8–13 June 2002, Tenerife Spain Eds R. Cook & D. J. Hunt. 175–185.

Hewlett, T. E., Griswold, S. T., & Smith, K. S. (2006). Biological control of *Meloidogyne incognita* using *in-vitro* produced *Pasteuria penetrans* in a microplot study. *Journal of Nematology, 38*, 274 [Abstract].

Imbriani, J. L., & Mankau, R. (1977). Ultrastructure of the nematode pathogen *Bacillus penetrans*. *Journal of Invertebrate Pathology, 30*, 337–347.

Kojetin, D., Thompson, R., Benson, L., Naylor, S., Waterman, J., & Davies, K. (2005). Structural analysis of divalent metals binding to the *Bacillus subtilis* response regulator Spo0F: The possibility for *in vitro* metalloregulation in the initiation of sporulation. *BioMetals, 18*, 449–466.

Konstantinidis, K. T., & Tiedje, J. M. (2005). Genomic insights that advance the species definition for prokaryotes. *Proceedings of the National Academy of Sciences, USA, 102*, 2567–2572.

Mankau, R. (1975). *Bacillus penetrans* n. comb. causing a virulent disease of plant-parasitic nematodes. *Journal of Invertebrate Pathology, 26*, 333–339.

Mankau, R. (1980). Biological control of *Meloidogyne* populations by *Bacillus penetrans* in West Africa. *Journal of Nematology, 12*, 230 [Abstract].

Mankau, R., & Imbriani, J. L. (1975). The life-cycle of an endoparasite in some tylenchid nematodes. *Nematologica, 21*, 89–94.

Mendoza de Gives, P., Davies, K. G., Morgan, M., & Behnke, J. M. (1999). Attachment tests of *Pasteuria penetrans* to the cuticle of plant and animal parasitic nematodes, free living nematodes and srf mutants of *Caenorhabditis elegans*. *Journal of Helminthology, 73*, 67–72.

Metchnikoff, E. (1888). *Pasteuria ramosa*, un representant des bacteries a division longitudinale. *Annales de l' Institut Pasteur, 2*, 165–170.

Oostendorp, M., Dickson, D. W., & Mitchell, D. J. (1991). Population development of *Pasteuria penetrans* on *Meloidogyne arenaria*. *Journal of Nematology, 23*, 58–64.

Pembroke, B., Gowen, S. R., & Giannakou, I. (1998). Advancement of the ideas for the use of *Pasteuria penetrans* for the biological control of root-knot nematodes (*Meloidogyne* spp.). *The Brighton Conference-Pests and Diseases, 3*, 555–560.

Pembroke, B., Darban, D. A., Gowen, S. R., & Karanja, D. (2005). Factors affecting quantification and deployment of root powder inoculum of *Pasteuria penetrans* for biological control of root-knot nematodes. *International Journal of Nematology, 15*, 17–20.

Previc, E. P., & Cox, R. J. (1992). Production of endospores from *Pasteuria* by culturing with explanted tissues from nematodes. U.S. Patent Number 5, 094, 954. U.S. Patent Office, U.S. Department of Commerce, Washington, DC.

Rao, M. S., Gowen, S. R., Pembroke, B., & Reddy, P. P. (1997). Relationship of *Pasteuria penetrans* spore encumberance on juveniles of *Meloidogyne incognita* and their infection in adults. *Nematologia Mediterranea, 25*, 129–131.

Reise, R. W., Hackett, K. L., & Huettel, R. N. (1991). Limited *in vitro* cultivation of *Pasteuria nishizawae*. *Journal of Nematology, 23*, 547 [Abstract].

Sayre, R. M., & Starr, M. P. (1985). *Pasteuria penetrans* (ex Thorne, 1940) nom. rev., comb. n., sp. n., a mycelial and endospore-forming bacterium parasitic in plant-parasitic nematodes. *Proceedings of the Helminthological Society of Washington, 52*, 149–165.

Sayre, R. M., & Wergin, W. P. (1977). Bacterial parasite of a plant nematode: morphology and ultrastructure. *Journal of Bacteriology, 129*, 1091–1101.

Sayre, R. M., Starr, M. P., Golden, A. M., Wergin, W. P., & Endo, B. Y. (1988). Comparison of *Pasteuria penetrans* from *Meloidogyne incognita* with a related mycelial and endospore-forming bacterial parasite from *Pratylenchus brachyurus*. *Proceedings of the Helminthological Society Washington, 55*, 28–49.

Sayre, R. M., Wergin, W. P., Schmidt, J. M., & Starr, M. P. (1991). *Pasteuria nishizawae* sp. nov., a mycelial and endospore-forming bacterium parasitic on cyst nematodes of genera *Heterodera* and *Globodera*. *Research in Microbiology, 142*, 551–564.

Schmidt, L. M., Preston, J. F., Nong, G., Dickson, D. W., & Aldrich, H. C. (2004). Detection of *Pasteuria penetrans* race 1 in planta by polymerase chain reaction. *FEMS Microbiology Ecology, 48*, 457–464.

Sharma, S. B., & Davies, K. G. (1996a). Characterisation of *Pasteuria* isolated from *Heterodera cajani* using morphology, pathology and serology of endospores. *Systematic and Applied Microbiology, 19*, 106–112.

Sharma, S. B., & Davies, K. G. (1996b). A comparison of two sympatric species of *Pasteuria* isolated from a tropical vertisol soil. *World Journal of Microbiology and Biotechnology, 12*, 361–366.

Spaull, V. W. (1984). Observations on *Bacillus penetrans* infecting *Meloidogyne* in sugarcane fields in South Africa. *Revue de Nematologie, 7*, 277–282.

Starr, M. P., & Sayre, R. M. (1988). *Pasteuria thornei* sp. nov. and *Pasteuria penetrans* sensu stricto emend., mycelial and endospores-forming bacteria parasitic, respectively, on plant parasitic nematodes of the genera *Pratylenchus* and *Meloidogyne*. *Annales de l'Institut Pasteur/Microbiologie, 139*, 11–31.

Stirling, G. R. (1981). Effect of temperature on infection of *Meloidogyne javanica* by *Bacillus penetrans*. *Nematologica, 27*, 458–462.

Stirling, G. R. (1984). Biological control of *Meloidogyne javanica* with *Bacillus penetrans*. *Phytopathology, 74*, 55–60.

Stirling, G. R. (1985). Host specificity of *Pasteuria penetrans* within the genus *Meloidogyne*. *Nematologica, 31*, 203–209.

Stirling, G. R. (1991). Biological control of plant-parasitic nematodes. Progress, problems and prospects (282 pp.). Wallingford, UK: CAB International.

Stirling, G. R., & Wachtel, M. F. (1980). Mass production of *Bacillus penetrans* for the biological control of root-knot nematodes. *Nematologica, 26*, 308–312.

Stirling, G. R., & White, A. M. (1982). Distribution of a parasite of root-knot nematodes in South Australian vineyards. *Plant Disease, 66*, 52–53.

Sturhan, D. (1988). New host and geographical records of nematode-parasitic bacteria of the *Pasteuria penetrans* group. *Nematologica, 34*, 350–356.

Sturhan, D., Shutova, T. S., Akimov, V. N., & Subbotin, S. A. (2005). Occurrence, hosts, morphology, and molecular characterization of *Pasteuria* bacteria parasitic to nematodes in the family Plectidae. *Journal of Invertebrate Pathology, 88*, 17–26.

Subbotin, S. A., Sturhan, D., & Ryss, A. Y. (1994). Ocurrence of nematode-parasitic bacteria of the genus *Pasteuria* in the former USSR. *Russian Journal of Nematology, 2*, 61–64.

Triviño, C. G., & Gowen, S. R. (1996). Deployment of *Pasteuria penetrans* for control of root-knot nematodes in Ecuador. *Brighton Crop Protection Conference, Pests and Diseases, 4*, 389–392.

Trotter, J. R., & Bishop, A. H. (2003). Phylogenetic analysis and confirmation of the endospore-forming nature of *Pasteuria penetrans* based on the spo0A gene. *FEMS Microbiology Letters, 225*, 249–256.

Trotter, J. R., Darban, D. A., Gowen, S. R., Bishop, A. H., & Pembroke, B. (2005). The isolation of a single spore isolate of *Pasteuria penetrans* and its pathogenicity on *Meloidogyne javanica*. *Nematology, 6*, 463–471.

Trudgill, D. L., Bala, G., Blok, V. C., Daudi, A., Davies, K. G., & Gowen, S. R. (2000). The importance of tropical root-knot nematodes (*Meloidogyne* spp.) and factors affecting the utility of *Pasteuria penetrans* as a biocontrol agent. *Nematology, 2*, 823–845.

Thorne, G. (1940). *Dubosqia penetrans* n. sp. (Sporozoa, Microsporidia, Nosematidae) a parasite of the nematode *Pratylenchus pratensis* (De Man) Filipjev. *Proceeding of the Helminthological Society of Washington, 7*, 51–53.

Wishart, J., Blok, V. C., Phillips, M. S., & Davies, K. G. (2004). *Pasteuria penetrans* and *P. nishizawae* attachment to *Meloidogyne chitwoodi*, *M. fallax* and *M. hapla*. *Nematology, 6*, 507–510.

Weibelzahl-Fulton, E., Dickson, D. W., & Whitty, E. B. (1996). Suppression of *Meloidogyne incognita* and *M. javanica* by *Pasteuria penetrans* in field soil. *Journal of Nematology, 28*, 43–49.

Wayne, L. G., Brenner, D. J. , Colwell, R. R., Grimont, P. A. D., Kandler, O., & Krichevsky, M. I. (1987). Report of the ad hoc committee on reconciliation of approaches to bacterial systematics. *International Journal of Systematic Bacteriology, 37*, 463–464.

11

GIOVANNA CURTO

SUSTAINABLE METHODS FOR MANAGEMENT OF CYST NEMATODES

Servizio Fitosanitario, Regione Emilia-Romagna, Laboratorio di Nematologia, 40128 Bologna, Italy

Abstract. The cyst nematode, *Heterodera schachtii* Schmidt, is the most dangerous sugar beet pest. It causes serious stands and yield decreases wherever sugar beet is grown. The adoption of wide crop rotations and the cultivation of *Brassicaceae* nematicidal plants and sugar beet tolerant varieties, concur to maintain good yields in infested soils. The history in the last 25 years regarding the progress in applied researches on agronomical, biological and genetic cyst nematode control, and the recommended practical techniques for the North-Italian farmers are reported.

1. INTRODUCTION

Sugar beet cyst nematode *Heterodera schachtii* Schmidt, one of the most dangerous and widespread pest of sugar beet (*Beta vulgaris* L. ssp. *saccharifera*), causes several plant damages. Changes in the absorbent cells, with subsequent nutritional imbalance and reduction of the root weight, may induce yield losses higher than 50% with an infestation of 300–400 eggs-2nd stage juveniles (J2) in 100 g of dry soil (Tacconi, 1987b).

Even if the recent restructuring of sacchariferous industrial sector caused a drastic reduction in the sugar beet crop surface in Italy, the cyst nematode infestations continue to represent a serious problem, since sugar beet crops are localized in areas close to the sugar refineries, with the aim of reducing the costs of taproot transportation. As a consequence, the choice of inserting sugar beet crops in a medium-long rotation scheme, results from the factory distance and not from the level of soil infestation.

Heterodera schachtii is widespread in the European sugar beet areas. In Italy, the infestations may reduce especially the weight of sugar beet roots and are strictly related to local climatic conditions: soil temperatures higher than 10°C for a long time increase the pest generation number and consequently the larval infestation level in soil. In addition, high temperatures stress infested sugar beet plants, reducing the roots ability to accumulate sucrose reserves.

A. Ciancio & K. G. Mukerji (eds.), Integrated Management and Biocontrol of Vegetable and Grain Crops Nematodes, 221–237.

The nematode activity stops in autumn and winter, after the sugar beet harvest and with soil temperature lower than 8–10°C, to start again in spring. The completion of one *H. schachtii* generation is reached at the thermal sum of 465°C, that is the sum of the daily mean temperatures higher than 10°C. Therefore, while in Central-Northern Europe *H. schachtii* completes 3 generations per year, currently in Northern Italy it may complete at least 3–4 generations (BETA, 2006), because of the general temperature increase. Consequently since the '80s, the damage threshold has been fixed as 100 eggs-J2 in 100 g of dry soil, significantly lower than in Central-Northern Europe.

Table 1. Field trial results on Heterodera schachtii *chemical control on sugar beet during 1974–1975 (Tacconi & Saretto, 1975).*

Active ingredient	*Living cysts before treatment (IP)*	*Living cysts at harvest (FP)*	*FP/IP*	*Root weight (ton/ha)**	*Polarization (%)*	*Sucrose (ton/ha)**
Methylisotiocyanate	8.25	7.75	0.93	61.83 a	14.67	9.07 a
(1.2-Dichloropropane + 1.3-Dichloropropene) 80% + Methylisotiocyanate 20%	8.75	6.75	0.77	47.53 b	14.50	6.91 b
Aldicarb 10%	8.50	6.75	0.79	43.72 bc	14.31	6.28 bc
Oxamyl 10%	7.75	12.25	1.58	38.25 bc	14.06	5.40 bc
Phenamiphos 10%	14.00	11.00	0.79	36.07 bc	14.30	5.20 bc
Phorate 10%	8.50	10.50	1.24	40.16 bc	13.63	5.45 bc
Carbofuran 5%	6.25	7.25	1.16	40.78 bc	13.51	5.51 bc
Untreated Control	7.75	9.50	1.23	31.57 c	13.87	4.47 c

**Values significantly different for P = 0.05 (Duncan Test).*

2. SUGAR BEET CYST NEMATODE IN NORTHERN ITALY

The distribution of *H. schachtii in* Italy was ascertained by surveys carried out in 1990 and in 2004. In 1990 the most affected Italian regions were those with largest sugar beet crop surfaces and sugar refinery densities: Emilia-Romagna, Veneto, Lombardy and Apulia (Tacconi, 1993a). In 2004, the *H. schachtii* most damaged areas, in Northern Italy, were: Emilia-Romagna (48% of sugar beet crop surface), Piedmont (14%), Lombardy (11%) and Veneto (11%), with the most of infestations between light (less than 100 eggs-J2 in 100 g of dry soil) and medium (100–200 eggs-J2) levels. The highest infested areas were identified both in the eastern part of Emilia-Romagna and in the provinces of Rovigo (Veneto), Pavia (Lombardy), Alexandria and Asti (both in Piedmont) (Beltrami, Zavanella, & Curto, 2006b) (Fig. 1).

Currently, infested areas exceed 10% of the sugar beet Italian surface in Abruzzo and Emilia-Romagna, and are lower than 5% in Piedmont, Lombardy, Veneto, Tuscany, Apulia and 1% in Latium and Basilicata (source Cooperative Sugar beet Producers – Co.PRO.B.). Therefore, pest control is crucial for maintaining the crop productivity and ensuring an adequate income to farmers.

Today, farmers may choose among a series of agronomical techniques, which may warrant a success in controlling cyst nematodes, if correctly and punctually applied.

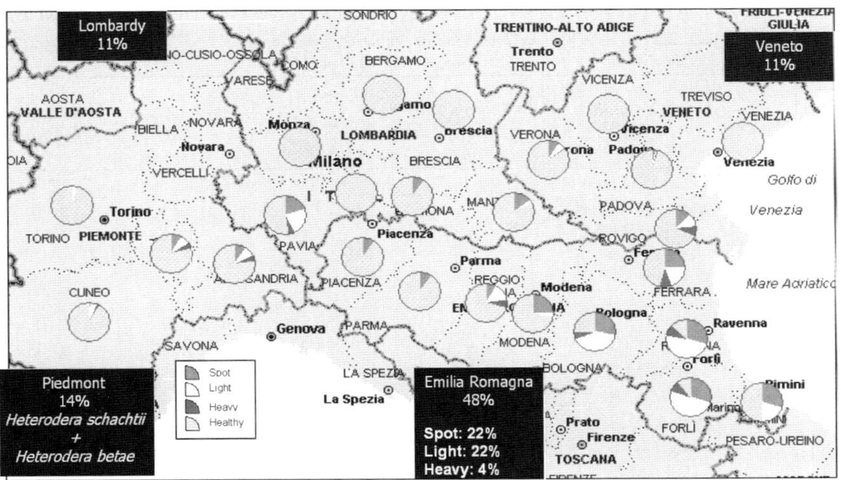

Figure 1. Spreading of Heterodera schachtii *in Northern Italy showing prevalence classes on sugar beet crop surface (from Beltrami* et al.*, 2006b).*

3. *HETERODERA SCHACHTII* BIOCONTROL IN NORTHERN ITALY

3.1. Chemical and Agronomic Control

From the '60s to the '70s, chemical control of cyst nematodes was investigated in several sugar beet field trials in Northern Italy, particularly in Emilia-Romagna and Veneto areas. The results were clear: chemical nematicidal applications were in most cases ineffective (Table 1), both in increasing sugar beet yields (Tacconi & Grasselli, 1978; Tacconi & Olimpieri, 1981; Bongiovanni, 1963; Tacconi & Ugolini, 1967; Greco, Lamberti, De Marinis, & Brandonisio, 1978) and in controlling the nematodes population (Zambelli & De Leonardis, 1974; Tacconi & Saretto, 1975). Furthermore, they appeared very expensive and toxic for the environment.

About 25 years ago, *H. schachtii* life cycle was investigated in greenhouse studies and in field trials (Tacconi, 1979, 1982), fixing the economic damage threshold of 100 eggs-J2 in 100 g of dry soil (Tacconi & Trentini, 1978; Tacconi &

Casarini, 1978; Greco, Brandonisio, & De Marinis, 1982a; Greco, Brandonisio, & De Marinis,1982b; Tacconi, 1987a). Researches effectively addressed the definition of appropriate four-years or six-years crop rotations, including *H. schachtii* non host crops, since these methods appeared more suitable for the environment, climate, soil and crops of Northern Italy plains (Table 2). Results showed that wide rotations always decreased the nematode population below the damage threshold, in moderately infested soils, and increased root yields (Tacconi & Olimpieri, 1985; Tacconi & Santi, 1991; Tacconi & Venturi, 1991).

Table 2. Effect of sugar beet crop rotations with non host crops of Heterodera schachtii *(from Tacconi & Venturi,1991)*

Rotations	*Crops in rotations **	*Nematode stages · g^{-1} before last crop*	*Root weight (ton/ha)*	*Polarization (%)*	*Sucrose (ton/ha)**
1981		Egg-J2			
Biennial	(B-O) + (B-W) + B	2.69	27.30	14.31	3.90
Quadriennal	(B-O-M-W) + B	1.34	34.60	13.95	4.81
1983		Cysts			
Biennial	(B-O) + (B-W) + (B-W) + B	15.75	52.70	13.23	7.01
Sexennial	(B-M-M-M-M-W) + B	5.00	73.30	13.79	10.05
Sexennial	(B-A-A-A-M-W) + B	3.25	77.60	12.35	9.59
Sexennial	(B-O-M-W-M-W) + B	6.50	77.10	14.14	10.86
1986		Egg-J2			
Biennial	(B-S) + (B-S) + B	6.80	10.60	13.15	1.38
Quadriennal	(B-S-W-M) + B	1.85	43.40	13.53	5.81
1988		Egg-J2			
Triennial	(B-S-W) + (B-S-W) + B	8.25	50.40	12.57	6.34
Sexennial	(B-S-W-M-M-W) + B	0.28	61.60	11.47	7.01

**B = sugar beet; W = wheat; M = maize; S = soybean; A = alfa-alfa; O = oats.*

In Northern Italy, in the same years, the susceptibility of some cultivated plant species towards indigenous populations of *H. schachtii* was screened in bioassays, in order to define the best rotations for agronomic control. They were: sunflower (*Helianthus annuus* L.), soybean (*Glycine max* L.), broad bean (*Vicia faba* L.), white clover (*Trifolium repens* L.), alfa-alfa (*Medicago sativa* L.), wheat (*Triticum* spp.), barley (*Hordeum vulgare* L.), maize (*Zea mays* L.), sorghum (*Sorghum vulgare* Pers.), potato (*Solanum tuberosum* L.) and tobacco (*Nicotiana tabacum* L.), which were classified as non host crops; reversed clover (*Trifolium resupinatum* L.), red clover (*Trifolium pratense* L.), eggplant (*Solanum melongena* L.), chickpea (*Cicer arietinum* L.), sweet pea (*Lathyrus odoratus* L.), hairy vetch (*Vicia villosa*

Roth.), classified as poor hosts (less than 1 adult female on the root); tomato (*Solanum lycopersicum* L.) classified as light host (1–4 females on the root); bean (*Phaseolus vulgaris* L.) classified as host (4.1–7 females on the root); carnation (*Dianthus caryophyllus* L.), pea (*Pisum sativum* L.) cv. Perfection, red radish (*Raphanus sativus* L. ssp. *major*), rape (*Brassica napus* L. var. *oleifera*), bird rape (*Brassica campestris* L. var. *oleifera*), cabbage (*Brassica oleracea* L.), white mustard (*Sinapis alba* L.), charlock mustard (*Sinapis arvensis* L.), spinach (*Spinacia oleracea* L.) and common buchwheat (*Polygonum fagopyrum* L.) classified as very good hosts (more than 10 adult females on the root) (Tacconi, 1993b, 1996, 1997).

Agronomic control represents even today one of the most effective methods for cyst nematodes management, together with a correct weeds management during rotation, since most widespread weeds are hosts of *H. schachtii* too. They are: redroot amaranth (*Amaranthus retroflexus* L.), bishop's weed (*Ammi majus* L.), scarlet pimpernel (*Anagallis arvensis* L.), shepherd's purse (*Capsella bursa pastoris* (L.) Medic.), fat-hen (*Chenopodium album* L.), black bind weed (*Fallopia convolvolus* (L.) A. Löve), willow weed (*Polygonum persicaria* L.), purslane (*Portulaca oleracea* L.), wild radish (*Raphanus raphanistrum* L.), sheep's sorrel (*Rumex acetosella* L.), black nightshade (*Solanum nigrum* L.) and common chickweed (*Stellaria media* L.) (Tacconi & Santi, 1981), while velvetleaf (*Abutilon theophrasti* Medic.) is non host of *H. schachtii* (Tacconi & De Vincentis, 1996).

Other cultural practices, which can help farmers to control cyst nematodes, are: efficient hydraulic layout; clean equipment; earlier sowing. The latter procedure aims at staggering both the sugar beet and the cyst nematode cycles and obtaining sturdier plants, able to resist to the nematode infestation. Further procedures include harvesting of susceptible varieties within August, in order to avoid the damage increase and the parasite development (BETA, 2006).

3.2. Biological Control

3.2.1. Brassicaceae *Nematicidal Intercrops*

The quality improvement of sugar beet crop, decreasing nematode infestations below the threshold value and increasing both taproot weight and sucrose, was effectively achieved through the study of rotations effects, including intercrops of *Brassicaceae* species, selected for high glucosinolate content. The cells of these plants, in fact, contain the glucosinolate-myrosinase system, which, following cell lesions and enzymatic hydrolysis, produces a number of biologically active compounds including isothiocyanates, nitriles, epithionitriles and thiocyanates (Fahey, Zalcmann., & Talalay, 2001).

Nematicidal *Brassicaceae* can accumulate the majority of glucosinolates either in the root system (catch effect) or in the stems and leaves (biofumigant effect). The first process is the most suitable to control cyst nematodes.

Brassicaceae catch crops attract the juvenile stages of endoparasitic nematodes working as a trap, since these, after root penetration, are poisoned by hydrolysis products and are not successful in completing their developmental cycle in 10–12 weeks, that is the intercropping time. Consequently, the nematode population in soil progressively decreases. At full flowering the plants are chopped and immediately incorporated at around 20 cm depth by means of a stalk cutter and a miller, working at some meters distance from each other. A light irrigation sprinkled after incorporation in soil, aims at promoting the glucosinolate hydrolysis and the subsequent isothiocyanate release (Lazzeri, Leoni, Bernardi, Malaguti, & Cinti, 2004b).

The nematicidal effect of a catch crop is produced during the whole cultivation time, while its incorporation as green manure shows an overall ammendant effect, increasing the organic matter amount and improving soil fertility, being the biofumigant effect during incorporation only secondary.

3.2.2. Application of Heterodera schachtii *Biocontrol in Northern Italy*

In Italy, the first researches regarding the control of cyst nematodes by means of nematicidal plants, go back to 1983 and continued with high impulse for all the '90s. These studies concerned the life cycle of *H. schachtii* in the roots of either cultivated or biocidal plants, through *in vitro* and *in vivo* experiments carrid out both in laboratory and in glasshouse, with the purpose of achieving the most effective rotation schemes for sugar beet crops, including nematicidal intercrops.

The first *in vitro* tests were performed in 5 cm diameter Petri dishes, soaking *H. schachtii* J2 in a *Brassicaceae* glucosinolate solution, at different concentrations, after glucosinolate hydrolysis by means of myrosinase. The nematodes were observed after 24, 48, 72 and 96 hrs, screening the percent mortality of J2. The allyl isothiocyanate, resulting by the hydrolysis of sinigrin, showed the highest J2 mortality after 24 hrs at an initial glucosinolate concentration of 0.5%, while at the same concentration other rapeseed glucosinolates (gluconapin, glucotropeolin, dehydroerucin) caused the J2 death after 48 hrs (Lazzeri, Tacconi, & Palmieri, 1993).

The *in vivo* studies were developed in subsequent steps, at first in glasshouse in either 5 l pots each containing 7–8 plants (Tacconi, Mambelli, Menichetti, & Pola, 1989) or 54 ml plastic microcells (units) each containing 1 plant (Tacconi & Pola, 1996). All the biocidal selections were cultivated in sterilised soil and inoculated with a known number of *H. schachtii* J2.

Results were checked in semifield conditions, in 1 m^2 plots each containing 1 m^3 of infested soil (Tacconi, De Vincentis, Lazzeri, & Malaguti, 1998; Tacconi, Lazzeri, & Palmieri, 2000) and in field trials, concerning the study of rotation schemes including either non host or biocidal catch crops (Tacconi & Olimpieri, 1983; Tacconi, Biancardi, & Olimpieri, 1990; Tacconi & Regazzi, 1990; Tacconi, Mambelli, & Venturi, 1991; Tacconi, Biancardi, & Olimpieri, 1995; Tacconi et al., 2000).

In glasshouse experiments, the development of juveniles (J3 and J4) and adults (males and females) in roots was examined after root homogenisation (Stemerding, 1964) in periodical checks. These studies showed the biocidal plant ability in interrupting the *H. schachtii* life cycle to J3 or J4 female, without any formation of adult female and cysts. On the contrary, juvenile males developed to adults, changing the sex-ratio of the nematode population. The effectiveness of biocidal plants in reducing *H. schachtii* population was described either by the biotest (Behringer, Heinicke, Von Kries., Müller, & Schmidt, 1984) as percent ratio between the number of adult white females on biocidal plant roots and on sugar beet ones, or by the reproduction factor (R) (Ferris et al., 1993), that is the ratio between nematode population in soil after the catch crop incorporation (FP) and before the catch crop sowing (IP), (R = FP/IP).

In 2006 *in vitro* tests were performed according to the method described in Lazzeri, Curto, Leoni, and Dallavalle (2004a). The *in vitro* experiments were carried out in glass cavity blocks, soaking the J2 in a glucosinolate solution and adding myrosinase which reacted directly in the block. The blocks were sealed to preserve the volatile compounds, and the nematicidal and nematistatic effects were observed either after 24 or 48 hrs. Gluconasturtiin, glucoerucin and sinigrin were tested at different concentrations for the definition of LC50 towards *H. schachtii* J2 (Table 3).

Table 3. Glucosinolate concentrations checked in vitro *bioessays towards* Heterodera schachtii *second stage juveniles.*

Glucosinolate	*Concentration (mM)*								
Gluconasturtiin	0.013	0.026	0.05	0.1	0.125	0.15	0.25	0.5	1
Glucoerucin	0.0625	0.0125	0.15	0.2	0.25	0.5			
Sinigrin	1								

The J2 mortality was the same in gluconasturtiin either after 24 or 48 hrs (0.125<LC50<0.25): the nematicidal action was very fast (already after 24 h) while the immobilisation effect resulted poor. In glucoerucin, the nematicidal action resulted slower than in gluconasturtiin: the highest J2 mortality was reached after 48 hrs (0.15<LC50<0.20 mM), while a strong immobilisation effect was observed after 24 hrs (0.20<LC50<0.25 mM). In general, the toxic effect towards *H. schachtii* J2 is achieved by highest glucosinolate concentrations (Lazzeri et al., 2004a).

3.2.3. *Nematicidal Plant Species in* Heterodera schachtii *Control*

Main glucosinolates effective against *H. schachtii* in Northern Italy agronomic conditions derive either from radish (*Raphanus sativus* L. ssp. *oleiformis*) or white mustard (*Sinapis alba* L.) varieties. *Raphanus sativus* ssp. *oleiformis* cv. Nemex and cv. Pegletta, *S. alba* cv. Maxi and other varieties with high nematicidal power were at first tested as catch crops and used as intercrops in quadriennial rotation schemes (Tables 4, 5) (Tacconi et al., 1989; Tacconi & Venturi, 1991).

Table 4. Reproduction factor of Heterodera schachtii *population, between rotation end and beginning, with and without nematicidal catch crops (Tacconi & Venturi, 1991).*

*Rotations (1983–1989)**	*FP/IP without catch crop*	*FP/IP with catch crop*
Biennial (B-Ba) + (B-W) + (B-W) + B	3.78	2.38
Triennial (B-Ba-W) + (B-S-W) + B	3.17	1.58
Quadriennal (O-M-W-B)	0.11	0.07
Quadriennal + Biennial (B-Ba-S-W) + (B-W) + B	2.25	2.35

**B = sugar beet; Ba = barley; M = maize; O = oats; S = soybean; W = wheat.*

In studies carried out in the '90th, some selections of other plant genera such as *Cleome spinosa* Jacq. (family *Capparaceae), Eruca sativa* Mill. cv. Prisca and *Reseda luteola* L. (family *Resedaceae*) resulted effective against *H. schachtii* (Table 6) (Tacconi et al., 1998), and recently also against the southern root-knot nematode, *Meloidogyne incognita* (Curto, Dallavalle, & Lazzeri, 2005).

Since 2004 both the main *Brassicaceae* varieties, marketed as nematicidal plants for control of *H. schachtii*, and other selections previously evaluated as effective in control of *M. incognita* (Curto et al., 2005; Curto, Lazzeri, Dallavalle, Santi, & Malaguti, 2006a; Curto, Lazzeri, Santi, & Dallavalle, 2006b), were tested in Northern Italy (Emilia-Romagna), checking their effectiveness on the indigenous population of *H. schachtii* in the local, environmental conditions.

Results showed a good genetic stability of the old varieties and generally a satisfying effectiveness in the newest selections, with a decrease in nematode population higher than 80% (Beltrami, Curto, & Zavanella, 2006a; Beltrami et al., 2006b). Only a brassica blend between white mustard (*S. alba* L.) and oriental mustard (*Brassica juncea* L.) allowed *H. schachtii* to multiply more than on sugar beet, while *R. sativus* L. ssp. *oleiformis* cv. Carlos, did not keep its performance in time, decreasing in two following years its ability in interrupting the cyst nematode cycle (R>1).

Eruca sativa cv. Nemat, very efficient as catch crop against *M. incognita*, did not confirm its biocidal effects on *H. schachtii* (Table 7). The *H. schachtii* life cycle in the roots was interrupted generally at the J3 stage, but several male adults were observed (Beltrami et al., 2006b). The green matter released in soil by biocidal varieties was always conspicuous, varying from 5 to 10.1 kg/m^2 (Beltrami, Zavanella, & Curto, 2007).

Table 5. Host status of fodder radish and white mustard biocidal selections vs. sugar beet cyst nematode Heterodera schachtii, *in a field test in Northern Italy (Tacconi* et al., *1989)*

Plant species	*Female specimens/10 g roots*				*Male specimens/10 g roots*		*Host* status*
	J2	*J3–J4*	*Adult*	*Cyst*	*J3–J4*	*Adult*	
Beta vulgaris L. ssp. *saccharifera* cv. Sigma	6.5	12.0	14.0	2.6	3.0	2.0	5
Raphanus sativus L. ssp. *oleiformis* cv. Sereno	1.9	1.8	0.7	0.1	2.8	4.0	3
R. sativus L. ssp. *oleiformis* cv. Pegletta	3.2	1.4	0.0	0.0	3.9	2.9	1
R. sativus L. ssp. *oleiformis* cv. Levana	5.4	5.5	0.8	1.1	6.4	4.1	3
R. sativus L. ssp. *oleiformis* cv. Nemex	2.7	0.8	0.0	0.0	2.5	1.9	1
Sinapis alba L. cv. Emergo	1.7	1.5	0.2	0.0	2.4	2.3	2
S. alba L. cv. Maxi	1.9	0.9	0.0	0.0	0.1	0.6	1

**Based on BIOTEST (Behringer* et al., *1984): a catch crop shows nematicidal effects, at a host status included from 1 to 3.*

3.2.4. Management of Nematicidal Intercrops in Northern Italy

In Northern Italy, two periods are recommended for the cultivation of nematicidal intercrops: a spring time on set-aside fields and a summer period, after the harvest of winter cereals. Currently, the spring intercropping is the most practised because of set-aside spreading, highest effectiveness in cyst nematode control and very low costs. In this case, the *Brassicaceae* catch crops must be kept far from red radish seed crops (*Raphanus sativus* L. ssp. *major*), which could be polluted by unwished crosses with nematicidal plants, since both crops flower at the same time.

Management of spring intercrops includes (BETA, 2006): a glyphosate-based herbicide treatment, 3–4 days before sowing; sowing of nematicidal varieties on unbroken soil at the end of March; either mowing or plant cutting and incorporation in soil at full flowering, and a deep ploughing in August, to prepare soil for the sugar beet crop in the following spring.

Table 6. Host status of biocidal selections vs. sugar beet cyst nematode Heterodera schachtii *(Tacconi* et al., *1998).*

Plant species		*Female/g roots*			*Males/g roots*		*Host* status*
	J2	*J3–J4*	*Adult*	*Cyst*	*J3–J4*	*Adult*	
Beta vulgaris L. ssp. *saccharifera* cv. Dima	2.6	10.4	19.9	0.8	11.5	8.3	5
Raphanus sativus L. ssp. *oleiformis* cv. Pegletta	5.6	3.3	0.0	0.0	3.1	0.3	1
Cleome spinosa Jacq. Italian ecotype	3.7	3.5	0.2	0.0	1.3	1.1	1
Eruca sativa Mill. cv. Prisca	4.4	1.6	0.1	0.0	1.6	0.6	1
Reseda luteola L.	0.1	0.0	0.0	0.1	0.0	0.0	0
Sinapis arvensis L. Sri Lanka ecotype	1.4	1.5	3.2	0.2	2.8	1.3	5

* *Based on BIOTEST (Behringer* et al., *1984): a catch crop shows nematicidal effects, at a host status included from 1 to 3.*

Management of a summer intercrop in quadriennal rotations requires more inputs than the spring one: at the end of August the biocidal variety must be sown on unbroken soil after the cereal harvesting. Its cultivation time lasts from September to November, and could necessitate an irrigation aid and an insecticidal application against *Altica* sp. At the end of November, the nematicidal intercrop must be dried up by glyphosate, then the soil tilled in winter and sown with maize, sorghum or soybean, in the following March–April.

The fall cultivation was initially studied to allow small farms to grow biocidal intercrops, but the results were inconsistent and in 35% of cases either indifference or increase in cyst nematode infestations were recorded. Late sowing delays the biocidal crop cycle, while the decrease in soil temperatures reduces the glucosinolate store into their roots. The thermal sum in soil remains below the nematode optimum, with a progressive cyst dormancy (biological minimum at 8–10°C). Therefore, in the autumnal cultivation some cover crop effects, such as the supply of organic matter and the limitation of nitrate leaching, become predominant.

Table 7. Host status of biocidal selections vs. a North Italian population of Heterodera schachtii *(Curto* et al., *unpublished data).*

Plant specie	*Variety*	R^a
Raphanus sativus L. ssp. *oleiformis*	Terranova*	0.00
	Comet**	0.05
	Corporal*	0.06
	Adios**	0.06
	Regresso*	0.08
	Diabolo***	0.14
	Arena***	0.15
	Remonta**	0.18
	Colonel***	0.25
	Pegletta**	0.38
	Karakter**	0.61
	Carlos**	1.61
Sinapis alba L. + *Brassica juncea* L.	Terraprotect*	2.85
Sinapis alba L.	Accent***	0.24
	Concerta*	0.34
Sinapis arvensis L.	*	12.30
Rapistrum rugosum L.	*	0.00
Eruca sativa Mill.	Nemat**	1.02
Sorghum vulgare	Triumph*	0.04
Crotolaria juncea L.	*	0.09
Beta vulgaris L. ssp. *saccharifera*	Orion*	1.50
	Gea**	3.50

R^a = eggs/J2 ratio in 100 g of dry soil at the beginning and the end of each cycle.
= varieties checked only one year; **= mean of two years; *= mean of three years.*

3.2.5. Promotion of Heterodera schachtii *Biocontrol in Northern Italy*

Since the '80th end, the biological management of *H. schachtii* with *Brassicaceae* nematicidal intercrops was effectively promoted both by sugar companies and sugar beet farmer associations (Co.PRO.B.), spreading this technique to the most of Northern Italy (Emilia-Romagna, Lombardy, Veneto) with innovative mind.

Emilia-Romagna regional administration and sugar beet farmer national associations supported the insertion of nematicidal intercrops in the rotation schemes, with the objective of reclaiming heavily infested soils. Grants to sugar beet farmers were warranted, both for purchasing biocidal radish seeds and getting technical assistance in the rotation planning and the biocidal crop cultivation.

From 1994 to 2001 more than 10,000 ha (on a total surface of 76,000 ha of sugar beet crops), were sown in Emilia-Romagna with biocidal *Brassicaceae* intercrops, within the "Sugar beet Cyst Nematode Project" promoted by Co.PRO.B. in the sugar beet districts of Bologna and Ferrara provinces, with an annual trend in continuous development. Regarding the sowing time, most of biocidal intercrops were cultivated in spring, on set-aside fields. Sowings within April 30th were 72%

in 1999 and 70% in 2000, the residual 30% being represented by September sowings and only the lowest part by Summer ones, after harvesting of either cereals or other crops (i.e. onion).

Currently in Italy, the contraction of sugar beet surface and the closure of sugar refineries induced farmers to abandon most infested fields, moving the sugar beet cultivation towards areas with low *H. schachtii* infestation, or closer to the sugar refineries. In the last years the insertion of a biocidal intercrop in the rotation schemes was considered as a possible way to increase the efficacy of sugar beet tolerant varieties, when cyst nematode infestation are higher than 400 egg-J2 in 100 g of dry soil.

3.2.6. Resistance and Tolerane

The selection of sugar beet genotypes tolerant to *H. schachtii,* achieved only recently interesting productive performances. The new genotypes derive from crosses between cultivated selections of sugar beet (*Beta vulgaris* L. ssp. *saccharifera*) and spontaneous species, such as *Beta maritima* and *Beta procumbens*, both carriers of resistance genes to the cyst nematode.

In 2003 the Italian National Technical Commission (CTN) performed the first trials concerning some new tolerant lines and in 2004 the commercialisation of resistant cv Paulina and tolerant cv Pauletta (both by KWS) started.

The definition of either resistant or tolerant sugar beet variety was recently described. A resistant variety is able to limit the nematode reproduction, while a tolerant variety is able to decrease the productive losses, if compared with a susceptible one (Plantard et al., 2006). On the contrary, results obtained in Italy showed that the tolerant variety Pauletta, grown on *H. schachtii* infested soil, had much higher yields than the resistant one and was equally able to limit the nematode reproduction. Currently, it is the only one variety marketed in a consistent number of unities in Italy. Trials carried out in Emilia-Romagna both in full field and pots demonstrated that the productivity of the resistant cv Paulina was lower than susceptible control (cv. Gea) with poor yields, in sugar and root weights, either in healthy or infested soils (Beltrami et al., 2006b). For this reason, it was no more commercialised in Italy, since 2006.

Other new varieties defined as tolerant both to rhizomania (Beet necrotic yellow vein virus) and sugar beet cyst nematode, were introduced on the Italian market in the last three years: Fenice and Flex (Delitzsch), Colorado and Florida (Betaseeds), Piera (KWS). Results (Table 8) of several trials (Beltrami et al., 2007) showed no relevant differences in root yields between the susceptible variety cv. Gea and the tolerant ones (cvs. Pauletta, Colorado, Fenice, Piera and Flex) when grown in healthy soil. However, a higher root yield was recorded in tolerant varieties when they were cultivated either in lightly infested soil (<100 eggs-J2) with a 20% increase, or in infested ones with a 50% increase, compared with the *H. schachtii* susceptible sugar beet cultivars.

Regarding polarization values, the susceptible variety always evidenced a heavy decrease in its polarization, coinciding with an increase in *H. schachtii* population density, while in the tolerant varieties and particularly in Piera and Flex, this reduction was lower. High levels of both thick juice and invert sugar reveal a poor quality of sugar beets, stressed by the cyst nematode. These unfavourable values were sensibly higher in the susceptible variety than in the cvs. Pauletta, Colorado, Fenice, Piera, Flex and Florida.

The Gross Sealable Production (GSP) in infested soil was much higher in tolerant varieties (cvs. Pauletta, Colorado, Fenice, Piera and Flex) than either susceptible beet cultivars or the "resistant" cv. Paulina. Both Piera and Flex, because of their higher polarimetric title, evidenced a higher GSP compared with Pauletta, Colorado and Fenice. In healthy soil, the results of the susceptible variety (cv. Gea) did not differ statistically from the effectiveness of the tolerant ones. But other traditional varieties, also susceptible to *H. schachtii* but more productive than cv. Gea, could be appropriately cultivated in soil where the cyst nematode was not recorded. It is worth to nota that all the current tolerant varieties are not tolerant to sugar beet leaf spots (*Cercospora baeticola* L.).

The ability of sugar beet tolerant varieties to lower the cyst nematode population was checked in Northern Italy fields and in pots. Results show R values between 2 and 4 in the tolerant varieties and between 16 and 20 in the susceptible ones (Beltrami et al., 2006b). In a soil with a *H. schachtii* infestation of 100 eggs-J2 in 100 g of dry soil, the cultivation of a tolerant variety allowed a nematode population density of 200–400 eggs-J2 100 g^{-1} of dry soil, whereas the susceptible variety reached a nematode population of 1600–2000 eggs-J2 100 g^{-1} of dry soil.

On the basis of further observations, tolerant sugar beet varieties seem to decrease their ability to control the *H. schachtii* population when the initial infestation is higher than 400 eggs-J2·100 g^{-1} of dry soil (Beltrami et al., 2007).

4. OUTLOOK OF BIOCONTROL IN NORTHERN ITALY

Even if the Italian sugar beet crop surface was sensibly reduced in these last years, the crop productivity is still threatened by *H. schachtii*. Therefore, sowing of tolerant sugar beet varieties, even in soils with a very low cyst nematode infestation, is strongly advised. The most recommended tolerant cultivars are Pauletta, Colorado, Fenice, Piera, Flex and Florida, the last three ones being more suitable for fall harvestings than the others.

In soils free from *H. schachtii*, the use of best traditional sugar beet varieties (both rhizomania and sugar beet leaf spots tolerant) is strongly suggested, because they allow the best productive results.

In soil characterised by very heavy cyst nematode infestations (more than 300 eggs-J2·100 g^{-1} of dry soil) the cultivation of nematicidal *Brassicaceae* intercrops is always recommended, since are able to quickly and effectively improve soil, releasing large amounts of organic matter.

Table 8. Productive results of tolerant sugar beet varieties either on healthy or lightly infested soil (normalized with average data 2005–2006) (Beltrami et al., *2007).*

	Healthy soil				
Variety	*Root*	*Polarization*	*Sucrose*	*Thick juice*	*GSP*
Gea**	100.0	100.0	100.0	100.0	100.0
Flex	94.6	102.9	97.5	100.3	98.4
Piera	96.4	101.4	97.5	100.1	97.5
Fenice	107.2	92.9	99.6	99.0	96.3
Colorado	105.3	93.0	98.0	99.0	94.8
Pauletta	103.4	93.9	97.2	99.1	94.3
Paulina	96.4	92.3	88.8	97.3	85.5
DMS 0.05	8.5	2.5	7.7	0.5	8.0
	*Lightly infested soil**				
	Root	*Polarization*	*Sucrose*	*Thick juice*	*GSP*
Gea**	100.0	100.0	100.0	100.0	100.0
Flex	117.0	104.8	123.1	100.1	125.5
Piera	117.7	106.0	125.4	100.1	128.9
Fenice	125.8	97.0	122.1	98.4	120.0
Colorado	132.8	95.1	126.7	98.1	122.9
Pauletta	130.2	95.9	124.5	98.1	120.5
Paulina	112.7	94.4	106.4	96.7	103.4
DMS 0.05	8.6	2.6	9.1	0.8	9.8

* *<100 eggs-J2 in 100 g of dried soil*
** *Commercial standard*

In Northern Italy, technical services are organised at the regional level, with the aim of supporting farmers in sugar beet crop decisions, according to regional guidelines of integrated crop management. These guidelines are updated every year on the basis of the results achieved by private and public research institutes and companies. One of the main investigation company in sugar beet is BETA ITALIA S.c.a.r.l., whose partners are Finbieticola, gathering the main sugar beet farmer associations (ANB, CNB, and ABI) and Assozucchero, which includes the whole sugar industry compartment (Italia Zuccheri, Eridania-Sadam, SFIR, COPROB and Zuccherificio del Molise). Public institutes involved in sugar beet research are the Research Institute for Industrial Crops – Council for Research in Agriculture (CRA-ISCI) Rovigo section, the Phytosanitary Service of Emilia-Romagna Region in Bologna and some Italian Universities.

The effective control of *H. schachtii*, linked to high productive levels in sugar beet crops, are currently achieved by the integration of agronomical and biological strategies. Either the soil health or the cyst infestation level, ascertained through nematological analysis, represent the factors for choosing the most suitable strategy. Most productive varieties, susceptible to *H. schachtii*, must be grown on healthy soil, while the tolerant ones, suitable for early or late harvests, must be cultivated on infested soil. Moreover, nematological analyses represent the only method suitable to reveal heavy nematode infestations with more than 300–400 eggs-J2, corresponding to the threshold excluding the sugar beet cultivation and recommends the sowing of biocidal *Brassicaceae* intercrops for soil recovery.

Anyway, even if few farmers of some sugar beet districts still follow short rotations, the technical support service recommends four year rotations in healthy soil and five year rotations with ascertained nematode infestation, as crucial cultural care for achieving best effectiveness of whatever pest and disease control strategy.

REFERENCES

Behringer, P., Heinicke, D., Von Kries, A., Müller, J., & Schmidt, J. (1984). Resistenz gegen Rübennematoden bei Zwischenfrüchten. *Nachrichtenblatt Deutschen Pflanzenschutz., Braunschweig, 36*, 125–126.

Beltrami, G., Curto, G., & Zavanella, M. (2006a). Lotta biologica e selezione varietale per la resistenza al nematode *Heterodera schachtii. Informatore fitopatologico, 10*, 16–20.

Beltrami, G., Zavanella, M., & Curto, G. (2006b). Combattere il nematode bieticolo in tre mosse. *L'Informatore Agrario*, *62*(3), 44–48.

Beltrami, G., Zavanella, M., & Curto, G. (2007). Piante biocide e varietà tolleranti contro il nematode della bietola. *L'Informatore Agrario, 2*, 57–61.

BETA S. c. a. r. l. (2006). Guida alla coltivazione della barbabietola da zucchero. MDM S.p.A. Forlì, Italy, p. 184.

Bongiovanni, G. C. (1963). Un biennio di prove di campo con nematocidi sperimentali contro "*Heterodera schachtii*" Schm (Vol. 1, pp. 221–226). Bologna: Atti Giornate Fitopatologiche.

Curto, G., Dallavalle, E., & Lazzeri, L. (2005). Life cycle duration of *Meloidogyne incognita* and host status of Brassicaceae and Capparaceae selected for glucosinolate content. *Nematology, 7*, 203–212.

Curto, G., Lazzeri, L., Dallavalle, E., Santi, R., & Malaguti, L. (2006a, June 25–29). Management of biocidal green manure in the control of southern root-knot nematode, *Meloidogyne incognita* (Kofoid *et* White) Chitw., in Northern Italy. *Proceedings of the second international biofumigation symposium*, Moscow, Idaho.

Curto, G., Lazzeri, L., Santi, R., & Dallavalle, E. (2006b, June 25–29). Evaluation of alternative strategies for the control of southern root-knot nematode *Meloidogyne incognita* (Kofoid *et* White) Chitw., on tomato crop, in plastic greenhouse. *Proceedings of the second international biofumigation symposium*, Moscow, Idaho.

Fahey, J., Zalcmann, A., & Talalay, P. (2001). The chemical diversity and distribution of glucosinolates and isothiocyanates among plants. *Phytochemistry, 56*, 5–51.

Ferris, H., Carlson, H. L., Viglierchio, D. R., Westerdhal, B. B., Wu, F. W., Anderson, C. E., et al. (1993). Host status of selected crops to *Meloidogyne chitwoodi. Journal of Nematology, 25* (Supplement), 849–857.

Greco, N., Brandonisio, A., & De Marinis, G. (1982a). Investigation on the biology of *Heterodera schachtii* in Italy. *Nematologia mediterranea, 10*, 201–214.

Greco, N., Brandonisio, A., & De Marinis, G. (1982b). Tolerance limit of the sugarbeet to *Heterodera schachtii. Journal of Nematology, 14*, 199–202.

Greco, N., Lamberti, F., De Marinis, G., & Brandonisio, A. (1978). Prove di lotta chimica contro *Heterodera schachtii* su barbabietola da zucchero nel Fucino. *Atti Giornate Fitopatologiche, 1*, 374–379.

Lazzeri, L., Curto, G., Leoni, O., & Dallavalle, E. (2004a). Effects of glucosinolates and their enzymatic hydrolysis products via myrosinase on the root-knot nematode *Meloidogyne incognita* (Kofoid *et* White) Chitw. *Journal of Agricultural and Food Chemistry, 52*, 6703–6707.

Lazzeri, L., Leoni, O., Bernardi, R., Malaguti, L., & Cinti, S. (2004b). Plants, techniques and products for optimising biofumigation in full field. *Agroindustria, 3*, 281–288.

Lazzeri, L., Tacconi, R., & Palmieri, S. (1993). *In vitro* activity of some glucosinolates and their reaction products toward a population of the nematode *Heterodera schachtii. Journal of Agricultural and Food Chemistry, 41*, 825–829.

Plantard, O., Porte, C., Denise, M., Muchembled, C., Richard-Molard, M., & Baril, C. (2006). Management of resistant and tolerant sugar beet cultivars for a durable control of the cyst nematode *Heterodera schachtii. Proceedings of the 69th IIRB Congress, Bruxelles.*

Stemerding, S. (1964). Een mixer wattenfilter methode om vrijbeweeglijke endoparasitaire nematoden uit wortels te verzamelen. *Verslagen En Mededelingen Plantenziektenkundige. Dienst Wageningen 141 (Jaarboek, 1963)*, 170–175.

Tacconi, R. (1979). Osservazioni in campo sul ciclo biologico dell'*Heterodera schachtii. Informatore fitopatologico, 6*, 13–18.

Tacconi, R. (1982). Osservazioni sul ciclo biologico di *Heterodera schachtii* Schmidt, 1871 su barbabietola da zucchero. *Informatore fitopatologico, 7–8*, 55–59.

Tacconi, R. (1987a). Problemi nematologici in Emilia-Romagna. *Informatore fitopatologico, 4*, 21–27.

Tacconi, R. (1987b). Il punto sul nematode a cisti (*Heterodera schachtii* Schmidt) della barbabietola da zucchero. *Informatore fitopatologico, 7/8*, 31–38.

Tacconi, R. (1993a). Il nematode *Heterodera schachtii* in Italia. *Informatore Fitopatologico, 2*, 22–23.

Tacconi, R. (1993b). Riproduzione di *Heterodera schachtii* su alcune piante coltivate. *Nematologia mediterranea, 21*, 9–12.

Tacconi, R. (1996). Riproduzione di *Heterodera schachtii* su colza, ravanello da seme e grano saraceno. *L'Informatore Agrario, 1*, 65–67.

Tacconi, R. (1997). Riproduzione di *Heterodera schachtii* su piante coltivate (II contributo). *Nematologia mediterranea, 25*, 93–97.

Tacconi, R., & Casarini, B. (1978). Rapporti fra livelli di infestazione di *Heterodera schachtii* e produttività della barbabietola da zucchero. *Informatore fitopatologico, 3*, 15–17.

Tacconi, R., & De Vincentis, F. (1996). Comportamento di *Heterodera schachtii* su piante infestanti di barbabietola. *Informatore Fitopatologico, 12*, 15–16.

Tacconi, R., & Grasselli, A. (1978). Prove di lotta contro il nematode *Heterodera schachtii* con geodisinfestanti (II contributo). *Atti Giornate Fitopatologiche* (Vol. 1, pp. 381–388). Catania.

Tacconi, R., & Olimpieri, R. (1981). Prove di lotta contro *Heterodera schachtii* con geodisinfestanti (III contributo). *Informatore Fitopatologico*, *3*, 3–9.

Tacconi, R., & Olimpieri, R. (1983). Effetto di avvicendamenti colturali e di colture intercalari su *Heterodera schachtii. Informatore Fitopatologico, 12*, 33–40.

Tacconi, R., & Olimpieri, R. (1985). Effetto di avvicendamenti colturali su *Heterodera schachtii. Informatore Fitopatologico, 3*, 39–45.

Tacconi, R., & Pola, R. (1996). Resistenza a *Heterodera schachtii* di rafano oleifero, senape bianca e kenaf. *L'Informatore Agrario, 26*, 76–77.

Tacconi, R., & Regazzi, D. (1990). Costi di coltivazione di colture intercalari resistenti a *Heterodera schachtii. Informatore Fitopatologico*, *9*, 47–48.

Tacconi, R., & Santi, R. (1981). Piante ospiti di *Heterodera schachtii* e rotazioni agrarie. *Informatore Fitopatologico, 5*, 21–24.

Tacconi, R., & Santi, R. (1991). Effetto di avvicendamenti colturali su *Heterodera schachtii* (II contributo). *Informatore Fitopatologico, 1*, 57–59.

Tacconi, R., & Saretto, L. (1975). Prove di lotta contro *Heterodera schachtii*, *Chaetocnema tibialis* e *Temnorrhinus mendicus* con geodisinfestanti di recente formulazione. *Informatore fitopatologico, 2–3*, 5–13.

Tacconi, R., & Trentini, L. (1978). Valutazioni preliminari di danni provocati da diversi livelli di infestazione di *Heterodera schachtii* su barbabietola. *Informatore Fitopatologico, 6*, 17–20.

Tacconi, R., & Ugolini, A. (1967). Prova sperimentale di lotta contro l'*Heterodera schachtii* Schm. *Atti Giornate Fitopatologiche*, Bologna, *1*, 443–448.

Tacconi, R., & Venturi, G. (1991). Mezzi agronomici di lotta contro il nematode *Heterodera schachtii* in Italia. *L'Informatore Agrario, 47*, 62–66.

Tacconi, R., Biancardi, E., & Olimpieri, R. (1990). Effetto di avvicendamenti colturali e di colture intercalari su *Heterodera schachtii* (II contributo). *Informatore Fitopatologico, 5*, 47–51.

Tacconi, R., Biancardi, E., & Olimpieri, R. (1995). Effetto di avvicendamenti colturali e di colture intercalari di piante-esca resistenti su *Heterodera schachtii*. *Nematologia Mediterranea,* Supplement, *23*, 113–120.

Tacconi, R., De Vincentis, F., Lazzeri, L., & Malaguti, L. (1998). Riproduzione su piante oleaginose di *Heterodera schachtii*. *L'Informatore Agrario, 4*, 57–59.

Tacconi, R., Lazzeri, L., & Palmieri, S. (2000). Effetto del sistema glucosinolati-mirosinasi contenuto nelle radici di *Raphanus sativus* sp. *oleiformis* su *Heterodera schachtii*. *Nematologia Mediterranea, 28* (Supplement), 55–63.

Tacconi, R., Mambelli, S., & Venturi, G. (1991). Effetto di piante-esca resistenti su *Heterodera schachtii*. *L'Informatore Agrario, 46*, 63–68.

Tacconi, R., Mambelli, S., Menichetti, P., & Pola, R. (1989). Osservazioni sul ciclo di *Heterodera schachtii* su piante resistenti. *Nematologia Mediterranea, 17*, 21–25.

Zambelli, N., & De Leonardis, A. (1974). Triennio di prove di lotta contro alcuni fitofagi della barbabietola da zucchero. *Informatore Fitopatologico, 5*, 13–18.

12

ANTOON PLOEG

BIOFUMIGATION TO MANAGE PLANT-PARASITIC NEMATODES

Department of Nematology, University of California, 1463 Boyce Hall, Riverside, CA 92521, USA

Abstract. Biofumigation is a sustainable strategy to manage soil-borne pathogens, nematodes, insects, and weeds. Initially it was defined as the pest suppressive action of decomposing *Brassica* tissues, but it was later expanded to include animal and plant residues. Most data on the efficacy of biofumigation are from *in vitro* studies using fungal pathogens. Biofumigation also attracted the interest of nematologists, and research on the potential of this method to manage plant-parasitic nematodes is reviewed.

1. INTRODUCTION

Annual yield losses on a worldwide scale that are attributed to plant-parasitic nematode are estimated to range between 5% and 12% (Sasser & Freckman, 1987). Depending on climate, crops grown, nematode species and their density levels, and economic factors, a number of tactics can be employed to minimize nematode damage. Preventing the introduction of nematodes with planting material, seeds, or soil, including non-host crops in rotation schemes, using nematode resistant varieties or rootstocks, and lowering nematode populations through nematicides are some of the most frequently used strategies.

However, concerns about the negative impact of synthetic nematicides on the environment and on general public health led to a re-evaluation of these products. For example, high use of the soil fumigant methyl bromide and resulting contamination of ground, surface and drinking water in The Netherlands led to a ban on its use in the 1980's (Mus & Huygen, 1992). Later, methyl bromide was listed as an ozone-depleting compound at the 4th meeting of the Montreal Protocol in Copenhagen, 1992, and in accordance with the US Clean Air Act its use as a fumigant is now banned in several nations.

Methyl bromide was previously used as a pre-plant broad-spectrum soil fumigant to control soil-borne diseases, nematodes, insects and weeds in high value crops such as tomato, strawberry, cucurbits, nursery crops, flowers. It was also used to avoid re-plant problems in vineyards and orchards (Rodríguez-Kábana, 1997).

With the disappearance of nematicides or restrictions on their allowed use, the interest in the development of safe, sustainable, and economically viable nematode management strategies has increased. One such a strategy that potentially fulfills

A. Ciancio & K. G. Mukerji (eds.), Integrated Management and Biocontrol of Vegetable and Grain Crops Nematodes, 239–248.

these requirements is biofumigation. This method was included as a non-chemical alternative to methyl bromide by the "Methyl Bromide Technical Options Committee" (MBTOC, 1997).

2. BRASSICA BIOFUMIGATION MECHANISM

Kirkegaard, Angus, Gardner, and Cresswell, (1993a) described a process of 'biological fumigation' that was later called 'biofumigation' (Kirkegaard, Gardner, Desmarchelier, & Angus, 1993b; Matthiessen & Kirkegaard, 1993). The first article on 'biofumigation' in a refereed international journal dealt with the inhibition of the fungus *Gaeumannomyces graminis* by root tissue of *Brassica* species (Angus, Gardner, Kirkegaard, & Desmarchelier, 1994). In their paper, biofumigation is referred to as the release of volatile breakdown products, mainly isothiocyanates, from Brassica roots (Angus et al., 1994). Initially, the term biofumigation was limited to

> the suppression of soil-borne pests and pathogens by biocidal compounds released ... when glucosinolates in Brassica ... crops are hydrolized (Kirkegaard & Sarwar, 1998).

The mechanism responsible for the biocidal effect of decomposing Brassica crops is thought to be based on a chain of chemical reactions ultimately resulting in the formation of biologically active products (Underhill, 1980). Brassica crops contain glucosinolates located in the cell vacuoles. Glucosinolates are sulphur containing stable and non-toxic compounds, but upon tissue disruption they come in contact with myrosinase (= thioglucosidase), an enzyme endogenously present in Brassica tissues, but stored in the cell walls or the cytoplasm, away from the glucosinolates (Poulton & Moller, 1993). The enzymatic hydrolysis of glucosinolates produces volatile isothiocyanates (ITCs), nitriles, and thiocyanates (Cole, 1976; Fenwick, Heaney, & Mullin, 1983). The ITCs, in particular, have general biocidal properties (Kirkegaard & Sarwar, 1998). Isothiocyanates also form the active ingredient of some synthetic nematicides (methyl isothiocyanate releasers).

There are over 100 different glucosinolates (Manici et al., 2000; Underhill, 1980). A single *Brassica* species can contain several different types of glucosinulates (Sang, Minchinton, Johnstone, & Truscott, 1984), and the types and quantities of glucosinolates are highly variable between species and even varieties of Brassicas (Rosa, Heaney, Fenwich, & Portas, 1997). As a result, the quantities and types of biocidal ITCs resulting from the breakdown of glucosinolates are higly variable. Furthermore, the concentration of ITCs produced is also influenced by soil texture, moisture, temperature, microbial community and pH (Bending & Lincoln, 1999; Morra & Kirkegaard, 2002; Price, 1999).

To increase the efficiency of biofumigation using *Brassica* species, initial research focused on the selection and characterization of varieties with high contents of glucosinolates (Potter, Davies, & Rathjen, 1998; Potter, Vanstone, Davies, & Rathjen, 2000; Kirkegaard & Sarwar, 1998, 1999). However, results from a number of studies indicated that the content of glucosinolates in the biofumigant material did not predict the biocidal activity (Angus et al., 1994; Bending & Lincoln, 1999; Charron & Sams, 1999; Harvey, Hannahan, & Sams, 2002; Lazzeri & Manici, 2001;

McLeod & Steel, 1999; Morra & Kirkegaard, 2002). One reason for this apparent discrepancy could be that the conversion of glucosinolates into the biocidal ITCs is low. Laboratory experiments indicated that the efficacy of the conversion to ITCs was only 5% of the potential when using tissue disruption methods (cutting and chopping) similar to those frequently used under field conditions (Gardiner, Morra, Eberlein, Brown, & Borek, 1999; Morra & Kirkegaard, 2002). As a result, ITC concentrations in soil after biofumigation were generally much lower than those after application of a synthetic ITC-releasing fumigant (Brown, Morra, McCaffrey, Auld, & Williams, 1991). Morra and Kirkegaard (2002) and recently Matthiessen, Warton, and Shackleton, (2004) recognized the importance of increasing the efficacy of the glucosinolate to ITC conversion, and focused on methods to improve the disruption of the biofumigant tissue. Using a tractor-drawn tissue pulverizing implement, soil ITC levels increased 20-fold (100 nmol per g soil) compared to when using a cutting and chopping implement. In addition, they showed that adding excess water to the pulverized tissue was necessary for maximum ITC release (Matthiessen et al., 2004). Another possible reason for lack of a correlation between tissue glucosinlate content and pest suppressive activity is that compounds other than ITCs, such as alkenals or alkenols (Potter et al., 1998) or sulphur-containing compounds such as dimethyl-disulphide (Bending & Lincoln, 1999) may also play an important role.

3. STUDIES INVOLVING NEMATODES

3.1. Root-knot Nematodes

The negative impact of Brassica tissues on soil-borne pathogens and parasites has repeatedly been demonstrated and was reviewed by Brown and Morra (1997). Mojtahedi, Santo, Hang, and Wilson (1991); Mojtahedi, Santo, Wilson, and Hang (1993) were among the first to study in detail the effects of amending soil with *Brassica* tissue on a plant-parasitic nematode (the root-knot nematode *Meloidogyne chitwoodi*). They reported that incorporation of *B. napus* shoots into *M. chitwoodi*-infested soil reduced nematodes to very low levels and that amending with *B. napus* was more effective than amending with wheat or corn. The effect of the amendment, however, did not extend into the soil layers below the zone of incorporation, and the amendment was more effective against second-stage juveniles than against egg masses. The amendment protected host plant roots growing in the zone of *B. napus* incorporation from nematode infestation for up to six weeks. Soil incorporation rates of 4% (w/w) killed nearly all second-stage juveniles, whereas rates of 6% were required to prevent hatching of juveniles from egg masses (Mojtahedi et al., 1991, 1993).

A limited fumigant action was also reported by Roubtsova, López-Pérez, Edwards, and Ploeg (2007) after incorporating broccoli tissue into *M. incognita*-infested soil. They found in studies using soil columns that reductions in the number of *M. incognita* were much greater in soil layers containing the broccoli tissue than in layers immediately adjacent to the tissue-amended layers, and concluded that a thorough mixing of the biofumigant tissue with the soil containing the target nematodes is essential.

Stapleton and Duncan (1998) used biofumigation with Brassica tissue to control *M. incognita* and found that the efficacy was much higher when soils were heated to sub-lethal temperatures, and suggested to combine biofumigation with soil-solarization. Similar results were obtained by Ploeg and Stapleton (2001) who reported that biofumigation with broccoli failed to control *M. incognita* or *M. javanica* at 20°C, but resulted in near complete control at temperatures of 30 and 35°C. In addition, they found that the time necessary to achieve control was shorter as soil temperatures increased, and concluded that biofumigation to control *M. incognita* or *M. javanica* should be done when soil temperatures of about 25°C can be achieved. Tsror, Lebiush, Meshulam, Matan, and Lazzeri (2006) also reported much improved control of *Meloidogyne* spp. on tomato by combining solarization with biofumigation compared to solarization alone. Bello, López-Pérez, García-Álvarez, Sanz, and Lacasa (2004) also recommended to use biofumigation when soil temperatures are above 20°C. In contrast, Mojtahedi et al. (1993) reported that *M. chitwoodi* was controlled by Brassica amendments at average soil temperatures of 19°C, and speculated that the slower decomposition of the plant tissues under cool conditions resulted in an extended period of nematode control. It is unknown if differences in temperature requirements of the different nematode species affect their susceptibility to biofumigation. However, Mojtahedi et al. (1993) showed that second-stage juveniles of *M. chitwoodi* were more susceptible to biofumigation than egg masses, suggesting that biofumigation would be most effective when the nematodes are active.

McLeod and Steel (1999) significantly reduced *M. javanica* soil levels by incorporation of plant tissue from a range of *Brassica* species at rates of 1–2% (w/w), and produced evidence from *in vitro* experiments that glucosinolate-derived volatiles played an important role. However, they also concluded that in a soil environment other mechanisms, possibly the stimulation of nematode antagonistic organisms, played an important role in the observed nematicidal effects (McLeod & Steel, 1999).

In a 3-year field study on the effect of biofumigation with five different winter-grown crops on *M. incognita* infestation of a summer-grown tomato crop in Southern California, López-Pérez, Roubtsova, de Cara-García, and Ploeg (2007) reported that broccoli reduced tomato root-galling and increased yields compared to the other non-Brassica treatments, but generally did not lower *M. incognita* soil population levels. They suggested that biofumigation with broccoli tissue prevented an immediate attack of the tomato transplants, allowing plants to become established and rendering them more tolerant to the nematodes (López-Pérez et al., 2007).

Thus, the cultivation of Brassicas as green manure crops and their subsequent soil incorporation as a biofumigant appears to be an attractive option to control root-knot nematode species. However, a major drawback is that most Brassicas are hosts to root-knot nematodes (McLeod & Steel, 1999; McLeod, Kirkegaard, & Steel, 2001; McSorley & Frederick, 1995), and consequently there is a danger of nematode increase rather than decrease (McLeod & Steel, 1999; McLeod & Warren, 1993; Stirling & Stirling, 2003). To avoid nematode build-up on Brassicas one possible strategy would be to grow crops during the cool season when nematode activity is limited or they are inactive and/or multiplication is slow.

In a field study with *M. incognita* a winter crop of broccoli did not result in nematode build-up in Southern-California, suggesting that this strategy is effective (López-Pérez et al., 2007). However, ideally Brassica varieties should be used that combine a high glucosinolate content with resistance to locally occurring root-knot nematode populations (Kirkegaard & Sarwar, 1998; McLeod & Steel, 1999; Stirling & Stirling, 2003). Pattison, Versteeg, Akiew, and Kirkegaard (2006) recently screened the host status of 43 Brassica varieties for *M. arenaria* and *M. javanica* and reported large differences between varieties. Although most varieties were only moderately good hosts for the *Meloidogyne* populations used in their study, some varieties were as good a host as tomato for the *M. javanica* population used. However, some fodder radish varieties (*Raphanus sativus*) that combined a high level of resistance to *Meloidogyne* were identified, showing good biofumigant activity. They therefore could be grown as biofumigant crops without risking any *Meloidogyne* buildup (Pattison et al., 2006).

The Brassica crop arugula (*Eruca sativa*) var. Nemat also received interest as a potentially useful biofumigant crop to manage root-knot nematodes, as it appears to act as a trap crop for *M. hapla, M. chitwoodi* and *M. incognita* (Curto Lazzeri, Dallavalle, Santi, & Malaguti, 2006; Melakeberhan, Xu, Kravchenko, Mennan, & Riga, 2006; Riga & Wilson, 2006; Riga, Pierce, & Collins, 2006). Positive effects on yields of nematode-susceptible crops grown after arugula cultivation and biofumigation were reported for tomato, carrot and potato (Curto et al., 2006; Riga et al., 2006).

Few studies have analyzed the cost of biofumigation as a nematode management strategy. Riga et al. (2006) reported a strategy in which an arugula cover crop was combined with lower rates of a synthetic nematicide to manage *M. chitwoodi* in potato, reducing pest management costs by 50%.

3.2. Other Nematode Groups

Reports on the management of other plant-parasitic nematodes using soil incorporation of Brassica tissue include those by Halbrendt (1996) who reported lowering *Xiphinema americanum* population levels after incorporation of a rapeseed green manure crop into infested orchard soil, and several reports by Potter et al. (1998); Potter, Vanstone, Davies, Kirkegaard, and Rathjen (1999); Potter, Vanstone, Davies, and Rathjen (2000) on the control of *Pratylenchus neglectus* using canola (*B. napus*) and other Brassica species.

Management of sugarbeet cyst nematodes (*Heterodera schachtii*) with Brassica crops is a common strategy in North-western Europe and in some parts of the US (Hafez and Sundararaj, 2004; Muller, 1999). However, rather than using the crops primarily as biofumigants, they are grown as trap crops, attracting the nematodes to their roots and allowing root invasion by the nematodes, but preventing their multiplication. For applications against this nematode applied in Northern Italy, see Chapter 11 of this volume.

4. NON-BRASSICA BIOFUMIGATION

The management of soil-borne pathogens and pests, including plant-parasitic nematodes, by amending soil with organic material is a well-known and long-practiced strategy and was reviewed by Bridge (1996), Stirling (1991), Hoitink (1988) and Lazarovits et al. (2001). The initial definition of "biofumigation" as a process referring to the breakdown of Brassica tissue, was expanded by Halbrendt (1996) and Bello, López-Pérez, and Díaz-Viruliche (2000a); Bello, López-Pérez, Sanz, Escuer, and Herrero (2000b); Bello et al. (2004) to describe the process of biological decomposition of plant or animal byproducts, leading to the production of volatile compounds with disease and pest suppressive properties.

As biofumigation relies on the production of volatile substances during the decomposition process, organic material used for biofumigation should not be (fully) decomposed prior to use, and should preferably have a C/N ratio between 8 and 20. In addition, after incorporation of organic matter, soils should be watered to field capacity and sealed with plastic to increase temperature and trap developing gases (Bello et al., 2004; Rodríguez-Kábana, 1997).

A general recommended dose for incorporation is 50 t/ha, although the efficacy of materials can vary depending on the biochemical and biological properties, and the method of application (Bello et al., 2004). For example, Bello et al. (2004) tested the biofumigant effect of a range of agro-industrial byproducts and livestock manures, in different doses and combinations, on the levels of *M. incognita* control. These authors concluded that the majority of materials could effectively be used. In commercial greenhouse trials in Spain an integrated management system was developed, including biofumigation with sheep manure and mushroom residue and the cultivation of short-cycle vegetables acting as trap crops. Using this strategy, initial very high levels of *M. incognita* were reduced to near zero in the main susceptible cucumber and tomato crops (Bello, 1998). Similarly, biofumigation combined with soil solarization in a greenhouse in Spain provided levels of *M. incognita* control in bell pepper, that were similar to levels achieved using methyl bromide (Bello et al., 2004).

In potato field trials in Idaho and Washington, cropping sequences including the cultivation and incorporation of sudangrass, a crop also known to release nematicidal compounds, dramatically reduced *M. chitwoodi* infestation levels in potato (Riga, Mojtahedi, Ingham, & McGuire, 2004). Results from a vineyard with a high incidence of grapevine-fanleaf virus and its vector nematode species *Xiphinema index* (Bello et al., 2004), suggested that biofumigation may also be useful to reduce the fallow period necessary to eliminate vector nematodes in uprooted vineyards. Bello et al. (2004) also obtained promising levels of control of *M. incognita* and *M. javanica* in banana plantations in the Canary Islands, of *M. incognita* in peach orchards, and of the citrus nematode *Tylenchulus semipenetrans* in an uprooted orange orchard in Spain with biofumigation using urban waste and urea, or manure and banana residues.

5. CONCLUSIONS AND OUTLOOK

In conclusion, biofumigation using *Brassica* tissue or other sources of organic material, appears as a promising strategy for the management of soil-borne diseases pests and weeds. The general benefits of amending soil with organic matter are well known and include improvements in the soil nutrient status and water-holding capacity, and an increase in the presence and activity of beneficial soil organisms including those that are antagonistic to plant-parasitic nematodes. In addition, it may provide a use for agro-industrial and some kind of municipal "waste" products (Bridge, 1996; Lazarovits, Tenuta, & Conn, 2001; Stirling, 1991).

The mechanisms of pest and pathogen control by biofumigation are still largely unknown, and although the production of biocidal gases is undoubtedly important, several researchers have indicated that other mechanisms are also likely to play an important role (Bending & Lincoln, 1999; Potter et al., 1998). In fact, few studies have compared the efficacy of biofumigation under plastic to trap gases, and biofumigation without plastic. In one study, biofumigation with manure under plastic to control *M. incognita* in tomato only gave a slight reduction in root-galling indices compared to biofumigation without plastic, and *M. incognita* soil populations were controlled to very similar levels by both methods (Bello, 1998).

Researches on optimizing the methods to apply biofumigant materials under different soil types and climates and on developing or identifying crop varieties with high biofumigant activity that are resistant, non-hosts, or trap crops for the locally occurring target nematodes are likely to enhance the potential of this strategy. It is unlikely that biofumigation alone will provide sufficient levels of nematode control over multiple seasons, but advantages include that this method is also useful to manage other soil-borne problems, and that it can easily be combined with other strategies such as soil solarization, and the use of resistant varieties.

REFERENCES

Angus, J. F., Gardner, P. A., Kirkegaard, J. A., & Desmarchelier, J. M. (1994). Biofumigation: Isothiocyanates released from *Brassica* roots inhibit growth of the take-all fungus. *Plant and Soil*, 162, 107–112.

Bello, A. (1998). Biofumigation and integrated crop management. In A. Bello, J. A. González, M. Arias, & R. Rodríguez-Kábana (Eds.), *Alternatives to methyl bromide for the Southern European countries* (pp. 99–126). Valencia, Spain: Phytoma-España, DG XI EU, CSIC.

Bello, A., López-Pérez, J. A., & Díaz-Viruliche, L. (2000a). Biofumigación y solarización como alternativas al bromuro de metilo. In J. Z. Castellanos & F. Guerra O'Hart (Eds.), *Memorias del Simposium Internacional de la Fresa* (pp. 24–50), Zamora, INCAPA, Celaya, Guanajuato, México.

Bello, A., López-Pérez, J. A., Sanz, R., Escuer, M., & Herrero, J. (2000b). Biofumigation and organic amendments. *Regional workshop on methyl bromide alternatives for North Africa and Southern European countries* (pp. 113–141). Paris, France: United Nations Environment Program (UNEP).

Bello, A., López-Pérez, J. A., García-Álvarez, A., Sanz, R., & Lacasa, A. (2004). Biofumigation and nematode control in the Mediterranean region. In R. C. Cook & D. J. Hunt (Eds.), *Proceedings of the fourth international congress of nematology*, 8–13 June, 2002, Tenerife, Spain. Nematology monographs and perspectives Vol. 2 pp. 133–149) Leiden and Boston: Brill.

Bending, G. D., & Lincoln, S. D. (1999). Characterisation of volatile sulphur-containing compounds produced during decomposition of *Brassica juncea* tissues in soil. *Soil Biology and Biochemistry*, 31, 695–703.

Bridge, J. (1996). Nematode management in sustainable and subsistence agriculture. *Annual Review of Phytopathology*, 34, 201–225.

Brown, P. D., & Morra, M. J. (1997). Control of soil-borne plant pests using glucosinolate-containing plants. *Advances in Agronomy*, 61, 167–231.

Brown, P. D., Morra, M. J., McCaffrey, J. P., Auld, D. L., & Williams, L. W. (1991). Allelochemicals produced during glucosinolate degradation in soil. *Journal of Chemical Ecology*, 17, 2021–2034.

Charron, C. S., & Sams, C. E. (1999). Inhibition of *Pythium ultimum* and *Rhizoctonia solani* by shredded leaves of *Brassica* species. *Journal of the American Society of Horticultural Sciences*, 124, 462–467.

Cole, R. A. (1976). Isothiocyanates, nitriles and thiocyanates as products of autolysis of glucosinolates in Cruciferae. *Phytochemistry*, 15, 759–762.

Curto, G., Lazzeri, L., Dallavalle, E., Santi, R., & Malaguti, L. (2006). Effectiveness of crop rotation with Brassicaceae species for the management of the southern root-knot nematode *Meloidogyne incognita*. Abstracts of the second international biofumigation symposium (p. 51). June 25–29, Moscow, Idaho.

Fenwick, G. R., Heaney, R. K., & Mullin, W. J. (1983). Glucosinolates and their breakdown products in food and food plants. In T. E. Furia (Ed.), *Critical reviews in food science and nutrition* (pp. 123–201). Boca Raton: CRC Press.

Gardiner, J., Morra, M. J., Eberlein, C. V., Brown, P. D., & Borek, V. (1999). Allelochemicals released in soil following incorporation of rapeseed (*Brassica napus*) green manures. *Journal of Agriculture and Food Chemistry*, 47, 3837–3842.

Hafez, S. L., & Sundararaj, P. (2004). Biological and chemical management strategies in the sugarbeet cyst nematode management. Proceedings of the Winter Commodity Schools-2004, University of Idaho, USA, pp. 243–248.

Halbrendt, J. M. (1996). Allelopathy in the management of plant-parasitic nematodes. *Journal of Nematology*, 28, 8–14.

Harvey, S. G., Hannahan, H., & Sams, C. E. (2002). Indian mustard and allyl isothiocyanate inhibit *Sclerotium rolfsii*. *Journal of the American Society of Horticultural Sciences*, 127, 27–31.

Hoitink, H. A. (1988). Basis for the control of soilborne plant pathogens with composts. *Annual Review of Phytopathology*, 24, 93–114.

Kirkegaard, J. A., & Sarwar, M. (1998). Biofumigation potential of brassicas. *Plant and Soil*, 201, 71–89.

Kirkegaard, J. A., & Sarwar, M. (1999). Glucosinolate profiles of Australian canola (*Brassica napus annua* L.) and Indian mustard (*Brassica juncea* L.) cultivars: implications for biofumigation. *Australian Journal of Agricultural Research*, 50, 315–324.

Kirkegaard, J. A., Angus, J. F., Gardner, P. A., & Cresswell, H. P. (1993a). Benefits of Brassica break crops in the Southeast wheat belt. Proceedings of the 7th Australian agronomy conference, Adelaide, Australia (pp. 282–285).

Kirkegaard, J. A., Gardner, P. A., Desmarchelier, J. M., & Angus, J. F. (1993b). Biofumigation – using Brassica species to control pests and diseases in horticulture and agriculture. In N. Wratten & R. Mailer (Eds.), *Proceedings of the 9th Australian assembly on brassicas*. Wagga Wagga, 5–7 October, (pp. 77–82).

Lazarovits, G., Tenuta, M., & Conn, K. L. (2001). Organic amendments as a disease control strategy for soilborne diseases of high-value agricultural crops. *Australasian Plant Pathology*, 30, 111–117.

Lazzeri, L., & Manici, L. M. (2001). Allelopathic effect of glucosinolate-containing plant green manure on *Pythium* sp. and total fungal population in soil. *HortScience*, 36, 1283–1289.

López-Pérez, J. A., Roubtsova, T., de Cara-García, M., & Ploeg, A. T. (2008). Rotation and biofumigation with five winter-growen crops and subsequent root-knot nematode infestation and yield of summer-grown tomato. *Agriculture, Ecosystems and Environment*, 123, (in press).

Manici, L. A., Lazzeri, L., Baruzzi, G., Leoni, O., Galletti, S., & Palmieri, S. (2000). Suppressive activity of some glucosinolate enzyme degradation products on *Phythium irregulare* and *Rhizoctonia solani* in sterile soil. *Pest Management Science*, 56, 921–926.

Matthiessen, J. N., & Kirkegaard, J. A. (1993). Biofumigation, a new concept for 'clean and green' pest and disease control. Western Australian Potato Grower, October issue, 14–15.

Matthiessen, J. N., Warton, B., & Shackleton, M. A. (2004). The importance of plant maceration and water addition in achieving high *Brassica*-derived isothiocyanate levels in soil. *Agroindustria*, 3, 277–280.

McLeod, R. W., & Steel, C. C. (1999). Effects of brassica-leaf green manures and crops on activity and reproduction of *Meloidogyne javanica*. *Nematology*, 1, 613–624.

McLeod, R. W., Kirkegaard, J. A., & Steel, C. C. (2001). Invasion, development, growth and egg laying by *Meloidogyne javanica* in Brassicaceae crops. *Nematology*, 3, 463–472.

Mcleod, R., & Warren, M. (1993). Effects of covercrops on inter-row nematode infestation in vineyards. 1. Relative increase of root knot nematode *Meloidogyne incognita* and *M. javanica* on legume, cereal and brassica crops. *The Australian Grapegrower and Winemaker*, 357, 28–30.

McSorley, R., & Frederick, J. J. (1995). Responses of some common Cruciferae to root-knot nematodes. *Journal of Nematology*, 27, 550–554.

Melakeberhan, H., Xu, A., Kravchenko, A., Mennan, S., & Riga, E. (2006). Potential use of arugula (*Eruca sativa* L.) as a trap crop for *Meloidogyne hapla*. *Nematology*, 8, 793–799.

Mojtahedi, H., Santo, G. S., Hang, A. N., & Wilson, J. H. (1991). Suppression of root-knot nematode populations with selected rapeseed cultivars as green manure. *Journal of Nematology*, 23, 170–170.

Mojtahedi, H., Santo, G. S., Wilson, J. H., & Hang, A. N. (1993). Managing *Meloidogyne chitwoodi* on potato with rapeseed as green manure. *Plant Disease*, 77, 42–46.

Morra, M. J., & Kirkegaard, J. A. (2002). Isothiocyanate release from soil-incorporated Brassica tissues. *Soil Biology and Biochemistry*, 34, 1683–1690.

MBTOC (1997). Report of the Methyl Bromide Technical Options Committee. Nairobi, Kenya, UNEP, 221pp.

Muller, J. (1999). The economic importance of *Heterodera schachtii* in Europe. *Helminthologia*, 36, 205–213.

Mus, A., & Huygen, C. (1992). Methyl Bromide. The Dutch Environmental Situation and Policy. TNO. Institute of Environmental Sciences. order No. 50554. 13pp.

Pattison, A. B., Versteeg, C., Akiew, S., & Kirkegaard, J. (2006). Resistance of Brassicaceae plants to root-knot nematode (*Meloidogyne* spp.) in northern Australia. *International Journal of Pest Management*, 52, 53–62.

Ploeg, A. T., & Stapleton, J. J. (2001). Glasshouse studies on the effects of time, temperature and amendment of soil with broccoli plant residues on the infestation of melon plants by *Meloidogyne incognita* and *M. javanica*. *Nematology*, 3, 855–861.

Potter, M. J., Davies, K., & Rathjen, A. J. (1998). Suppressive impact of glucosinolates in Brassica vegetative tissues on root lesion nematode *Pratylenchus neglectus*. *Journal of Chemical Ecology*, 24, 67–80.

Potter, M. J., Vanstone, V., Davies, K., Kirkegaard, J., & Rathjen, A. (1999). Reduced susceptibility of *Brassica napus* to *Pratylenchus neglectus* in plants with elevated root concentrations of 2-phenylethyl glucosinolate. *Journal of Nematology*, 31, 291–298.

Potter, M. J., Vanstone, V. A., Davies, K. A., & Rathjen, A. J. (2000). Breeding to increase the concentration of 2-phenylethyl glucosinolate in the roots of *Brassica napus*. *Journal of Chemical Ecology*, 26, 1811–1820.

Poulton, J. E., & Moller, B. L. (1993). Glucosinolates. In P. J. Lea (Ed.), *Methods in plant biochemistry* (Vol. 9, pp. 209–237), London: Academic Press.

Price, A. (1999). Quantification of volatile compounds produced during simulated biofumigation utilizing Indian mustard degrading in soil under different environmental conditions, MS Thesis. University of Tennessee, Knoxville.

Riga, E., & Wilson, J. (2006). New nematode management issues and options in Washington state. *Journal of Nematology*, 38, 289.

Riga, E., Mojtahedi, H., Ingham, R., & McGuire, A. (2004). Green manure amendments and management of root-knot nematodes on potato in the Pacific North West of USA. In R. C. Cook & D. J. Hunt (Eds.), *Proceedings of the Fourth International Congress of Nematology*, June 8–13, 2002, Tenerife, Spain. *Nematology Monographs and Perspectives* (Vol. 2, pp. 151–158). Leiden and Boston: Brill.

Riga, E., Pierce, F., & Collins, H. (2006). The use of arugula on its own and in combination with synthetic nematicides against plant parasitic nematodes of potatoes. Abstracts of the Second International Biofumigation Symposium, June 25–29, Moscow, Idaho, US, 42.

Rodríguez-Kábana, R. (1997). Alternatives to methyl bromide (MB) soil fumigation. In A. Bello, J. A. González, M. Arias, & R. Rodríguez-Kábana (Eds.), *Alternatives to methyl bromide for the Southern European countries* (pp. 17–33). Valencia, Spain,.

Rosa, E. A. S., Heaney, R. K., Fenwich, G. R., & Portas, C. A. M. (1997). Glucosinolates in crop plants. *Horticultural Reviews*, 19, 99–215.

Roubtsova, T., López-Pérez, J. A., Edwards, S., & Ploeg, A. T. (2007). Effect of broccoli (*Brassica oleracea*) tissue, incorporated at different depths in a soil column, on *Meloidogyne incognita. Journal of Nematology*, 39, (in press).

Sang, J. P., Minchinton, I. R., Johnstone, P. K., & Truscott, R. J. W. (1984). Glucosinolate profiles in the seed, root and leaf tissue of cabbage, mustard, rapeseed, radish and swede. *Canadian Journal of Plant Science*, 64, 77–93.

Sasser, J. N., & Freckman, D. W. (1987). A world perspective of nematology: the role of the Society. In Veech, J. A. & Dickson, D. W. (Eds.), *Vistas on Nematology* (pp. 7–14). Hyattsville, MD: Society of Nematology.

Stapleton, J. J., & Duncan, R. A. (1998). Soil disinfestation with cruciferous amendments and sublethal heating: effects on *Meloidogyne incognita, Sclerotium rolfsii* and *Pythium ultimum. Plant Pathology*, 47, 737–742.

Stirling, G. R. (1991). *Biological control of plant-parasitic nematodes: progress, problems and prospects.* Wallingford, UK: CAB International.

Stirling, G. R., & Stirling, A. M. (2003). The potential of Brassica green manure crops for controlling root-knot nematode (*Meloidogyne javanica*) on horticultural crops in a subtropical environment. *Australian Journal of Experimental Agriculture*, 43, 623–630.

Tsror, L., Lebiush, S., Meshulam, M., Matan, E., & Lazzeri, L. (2006). Biofumigation for controlling soilborne diseases of tomato, potato and olive. Abstracts of the second international biofumigation symposium (p. 46). June 25–29, Moscow, Idaho.

Underhill, E. W. (1980). Glucosinolates. In E. A. Bell & B. V. Charlwood (Eds.), *Secondary plant products. Encyclopedia of Plant Physiology*, New Series (Vol. 8, pp. 493–511). Berlin: Springer-Verlag.

Section 4

Data Analysis and Knowledge-based Applications

13

JULIE M. NICOL[1*] AND ROGER RIVOAL[2]

GLOBAL KNOWLEDGE AND ITS APPLICATION FOR THE INTEGRATED CONTROL AND MANAGEMENT OF NEMATODES ON WHEAT

[1]International Wheat and Maize Improvement Center (CIMMYT), Wheat Program, PO Box 39, Emek, 06511, Ankara, Turkey
[2]UMR INRA/ENSAR, Biologie des Organismes et des Populations Appliquée à la Protection des Plantes (BiO3P), BP 35327, 35653 Le Rheu, France

Abstract. Importance of cereals and wheat nematodes in the world is revised. Distribution of cereal nematodes, species and pathotypes includes root lesion, cereal cyst nematodes and other cereal parasitic species. Life cycle, symptoms of damage and yield losses are also revised for root knot, stem and seed gall nematodes. Integrated control of cereal nematodes and some chemical, biological and cultural practices, including grass free rotations and fallowing with cultivation, are discussed. The effects of time of sowing, crop rotations and cultivation of resistant/tolerant varieties are also revised.

1. INTRODUCTION

1.1. Importance of Cereals and Wheat in the World

Cereals constitute the world's most important source of food. Amongst cereals, wheat, maize and rice occupy the most eminent position in terms of production, acreage and source of nutrition, particularly in developing countries. It has been estimated that about 70% of the land cultivated for food crops is devoted to cereal crops. By 2030, world population is expected to increase to 8 billion and world wheat (*Triticum aestivum*) production to increase from 584 million tonnes (1995–1999 average) to 860 million tonnes (Marathee & Gomez-MacPherson, 2001). The world wheat deficit during these three decades is expected to rise by 2.5 times, particularly in the developing world, where 84% of the population increase is expected and where wheat is a staple. To compensate for the additional demand for wheat, methods must be employed to minimise yield production constraints.

Plant parasitic nematodes are recognised as one such constraint, with at least seventeen important species in three major genera (*Heterodera*, *Pratylenchus* and

Corresponding author e-mail: j.nicol@cgiar.org

A. Ciancio & K. G. Mukerji (eds.), Integrated Management and Biocontrol of Vegetable and Grain Crops Nematodes, 251–294.

Meloidogyne). Although the introduction of new cultivars of wheat has boosted agricultural output, the yield potential of the new cultivars has not been fully expressed and is often far below theoretical maximum yields. This disparity between actual and theoretical yield expression can be attributed to "production constraints". Attention has therefore been focused on minimizing these constraints to increase production. Although insect pests and diseases have long been recognized as important constraints affecting crop production, extensive research on the "weak linkages" in the plant-pest system are lacking.

As most nematodes live in the soil, they represent one of the most difficult pest problems to identify, demonstrate and control. Farmers, agronomists and pest management consultants commonly underestimate their effects but it has been estimated that some 10% of the world crop production is lost as a result of plant nematode damage (Whitehead, 1998). It is also pertinent to consider in many of the cereal systems discussed in this chapter the interaction of nematodes with other plant pathogens, particularly soil borne fungi, and in many cases the synergism which results in more damage than either pathogen alone.

Management of nematodes may be approached by using a complement of methods in an integrated pest management system or may involve only one of these methods. Some of the most commonly practised methods will be discussed, including crop rotation, use of resistant and tolerant cultivars or varieties, cultural practices and chemicals. It is important to stress that the most appropriate control method will be determined by the nematode involved and the economic feasibility of implementing a possible management practice.

The purpose of this chapter is to provide an insight into the economically important nematodes on cereals. Information is presented here on their currently known distribution, damage potential, economic importance and management options that exist for their control. This review will focus on the primary nematodes of global economic importance on wheat and particularly Cereal Cyst Nematode (CCN, *Heterodera*) and Root Lesion Nematode (*Pratylenchus*). Other important genera including Root Knot (*Meloidogyne*), Stem (*Ditylenchus*) and Seed Gall (*Anguina*) will be mentioned, but in much less detail. Efforts have been made in this chapter to capture information from scientists from West Asia, North Africa, India and China which often is not internationally published, however it is of significant importance to the wheat productivity in especially the rainfed or marginal wheat growing regions of this countries. For further references and illustration of many of these nematodes, refer to the reviews of Kort (1972), Griffin (1984), Sikora (1988), Swarup and Sosa Moss (1990), Rivoal and Cook (1993), De Waele and Mc Donald (2000), Kollo (2002), Nicol (2002) and McDonald and Nicol (2005).

2. DISTRIBUTION OF CEREAL NEMATODES, SPECIES AND PATHOTYPES

2.1. Cereal Cyst Nematode

Although the cereal cyst nematode complex is represented by a group of twelve valid and several undescribed species, three main species are documented to be the most economically important: *Heterodera avenae*, *H. filipjevi* and *H. latipons* (Rivoal & Cook, 1993; McDonald & Nicol, 2005). Their common name is due to the fact that grasses (within family Graminearum) are hosts and the nematode adult female structure is a cyst.

The identification of cyst nematodes is complex and has traditionally been based on comparative morphology, through several diagnostic keys (Mulvey, 1972; Wouts, Schoemaker, Sturhan, & Burrows, 1995). However, more recently, techniques based on protein (Rumpenhorst & Sturhan, 1996) or DNA differences have been implemented, with most recently the use of DNA polymorphisms (Bekal, Gauthier, & Rivoal, 1997; Subbotin, Waeyenberge, Molokanova, & Moens, 1999; Subbotin, Waeyenberge, & Moens, 2000) allowing the identification to species level. One of the major obstacles to controlling CCN is the fact that a number of pathotypes occur, and this is further complicated by the presence of ecotypes. The major method to identify pathotype variation is the use of a Host Differential set, using specific barley, oat and wheat varieties, developed by Andersen & Andersen (1982). This was effective at time differentiating pathotypes of the known *H. avenae*, however since then many new pathotypes and additional species have been reported.

Within *H. avenae*, three groups of pathotypes have been distinguished using host reactions of the barley cultivars Drost4, Siri and Morocco with the resistance genes *Rha1*, *Rha2* and *Rha3*, respectively. Pathotypes belonging to groups 1 and 2 are the most numerous and widely distributed in Europe, North Africa and Asia (Andersen & Andersen, 1982; Al-Hazmi, Cook, & Ibrahim, 2001; Mokabli, Valette, Gauthier, & Rivoal, 2002). Pathotypes of group 3 (from Australia and Europe) are virulent to both the *Rha1* and *Rha2* genes (Andersen & Andersen, 1982). However, there appears to be mis-identification with some of these *H. avenae* pathotypes, particularly from Spain and Sweden, where populations previously known as the "Gotland" strain (Bekal et al., 1997) are actually *H. filipjevi*. A new group of pathotypes in *H. avenae* virulent to the *Rha3* gene have been shown to occur in North Africa (Mokabli et al., 2002).

Heterodera avenae is the most widely distributed and damaging species on cereals cultivated on more or less temperate regions. It has been detected in many countries, including Australia, Canada, Israel, South Africa, Japan and most European countries, as well as India (Sharma & Swarup, 1984; Handa, Mathur, Mathur , & Yadav, 1985b; Sikora, 1988), China (Peng et al., 2007) and several countries within North Africa and Western Asia, including Morocco, Tunisia, Libya and Pakistan (Sikora, 1988), Iran (Tanha Maafi, Subbotin, & Moens, 2003), Turkey (Nicol et al., 2002; Abidou et al., 2005), Algeria (Mokabli, Valette, & Rivoal, 2001), Saudi Arabia (Ibrahim, Al-Hazmi, Al-Yahya, & Alderfasi, 1999) and Israel (Mor, Cohn, & Spiegel, 1992).

Heterodera latipons is essentially only Mediterranean in distribution, being found in Syria (Sikora & Oostendorp, 1986; Scholz, 2001), Israel (Kort, 1972; Mor et al., 1992), Cyprus (Sikora, 1988), Turkey (Rumpenhorst, Elekçioglu, Sturhan, Öztürk, & Enneli, 1996), Italy and Libya (Kort, 1972). However, it is also known to occur in northern Europe (Sabova, Valocka, Liskova, & Vargova, 1988) and also in Bulgaria (Stoyanov, 1982). In Iran *H. latipons* is found in Mazandaran, East and West Azarbayejan, Ardabil, Hamadan, Lorestan and Kermanshah provinces (Tanha Maafi,

Sturhan, Kheiri, & Geraert, 2007; Talatchian, Akhiani, Grayeli, Shah-Mohammadi, & Teimouri, 1976; Noori, Talatchian, & Teimoori, 1980; Sturhan, unpubl.).

Another species with an increasingly wide distribution is *H. filipjevi*, formerly know as Gotland strain of *H. avenae* (Ferris et al., 1999; Bekal et al., 1997), which appears to be found in more continental climates such as Russia (Balakhnina, 1989; Subbotin, Rumpenhorst, & Sturhan, 1996), Tadzhikistan (Madzhidov, 1981; Subbotin et al., 1996), Sweden (Cook & Noel, 2002; Holgado, Rowe, Andersson, & Magnusson, 2004), Norway (Holgado et al., 2004), Turkey (Rumpenhorst et al., 1996; Nicol et al., 2002), and Greece (Mandani, Vovlas, Castillo, Subbotin, & Moens, 2004). In Iran *H. filipjevi* is widespread, being found in Ardabil, East and West Azarbayejan, Mazandaran, Golestan, Zanjan, Lorestan, Kermanshah, Kordestan, Hamadan, Esfahan, Kerman, Yazd, Fars, Systan and Blouchestan provinces (Tanha Maafi et al., 2007). A relatively new report also finds this species from Himachal Pradesh in India (S. P. Bishnoi, pers. com.)

Other *Heterodera* species known to be of importance to cereals include *H. hordecalis* in Sweden, Germany and Britain (Andersson, 1974; Sturhan, 1982; Cook & York, 1982a) and from the Ardabil province in Iran (Tanha Maafi et al., 2007), *H. zeae*, which is found in India, Pakistan (Sharma & Swarup, 1984; Maqbool, 1988) and Iraq (Stephan, 1988) and various others including *H. mani*, *H. bifenestra* and *H. pakistanensis*, as well as an unrelated species of cyst nematode, *Punctodera punctata* (Sikora, 1988).

Considering China and India which are the two largest wheat producers in the world, *H. avenae* appears to be widespread and damaging in both countries in the bread basket of their wheat production regions. In India *H. avenae* was first reported from Sikar district of Rajasthan in 1958 by Vasudeva, however now it has been reported from north rainfed wheat production region of Rajasthan (Koshy & Swarup, 1971; Mathur, 1969); Haryana (Bhatti, Dahiya, Gupta, & Malhan, 1980); Punjab (Koshy & Swarup, 1971; Chhabra, 1973; Singh, Sharma, & Sakhuja, 1977) and Himachal Pradesh (Koshy & Swarup, 1971). It is speculated that this nematode is continuing its spread slowly and gradually towards the Indo-Gangetic plains of Uttar Pradesh. Bekal, Jahier, and Rivoal (1998) attributed Nazafgarh, Delhi population to Ha 71 pathotype. More recently Bishnoi and Bajaj (2004) concluded on the basis of international host differential, biochemical and morphometric studies of eight geographical populations that the isolates from Jaipur, Udaipur, Narnaul, Sirsa and Delhi belong to pathotype Ha21, whilst Punjab (Ludhiana) and Ambala (Haryana) populations belong to pathotype Ha 41 and the Himachal Pradesh population belongs to *H. filipjevi.*

In China *H. avenae* was first reported from Hubei province in the centre of China in 1987, and now it has been reported in at least eight provinces in high frequencies including Henan, Hebei, Beijing suburb, Inner Mongolia, Shanxi, Qinghai, Anhui and Shandong (Peng et al., 2007). This wheat production area represents about 20 million ha which is around two thirds of China's total wheat production (120Mt). Survey data of more than 500 samples indicate population densities of CCN much higher than reported in other countries where economic damage is reported. Morphological and molecular characterization of the selected populations revealed a close relatedness to species within the *H. avenae* group. Restriction Fragment Length Polymorphism of the ITS regions within the ribosomal DNA classified these populations as "type B" *H avenae*. Using the host differential pathotypes test developed by Andersen and Andersen (1982), it appears there are at least three pathotypes (CH1, CH2, CH3) which are different from other known pathotypes. In neighbouring Iran molecular studies of specimens already have been reported and supported the presence of *H. avenae* (type B) in Iran (Tanha Maafi et al., 2003).

2.2. Root Lesion Nematodes

The genus *Pratylenchus* contains 63 valid species (Handoo & Golden, 1989), with at least eight species infesting small grains (Rivoal & Cook, 1993). Of these, *P. thornei*, *P. neglectus*, *P. penetrans* and *P. crenatus* are polyphagous and have a worldwide distribution. On cereals, *P. thornei* is the most studied species, being found in Syria, Yugoslavia, Mexico, Australia, Canada, Israel, Morocco, Turkey, Pakistan, India, Algeria, Italy (Nicol, 2002) and the USA (Smiley, Whittaker, Gourlie, Easley, & Ingham, 2005). *P. neglectus* has been reported in Australia (Taylor, Hollaway, & Hunt, 2000; Vanstone, Rathjen, Ware, & Wheeler, 1998), North America (Townshend, Potter, & Willis, 1978; Timper & Brodie 1997), Europe (Lasserre, Rivoal, & Cook, 1994; Hogger, 1990) and Turkey (Nicol et al., 2002). Both *P. neglectus* and *P. thornei* have also been identified in wheat fields in Gilan province of Iran (Tanha Maafi, 1998). *Pratylenchus penetrans* is largely associated with horticultural crops but has been recorded on wheat in Canada (Kimpinski, Anderson, Johnston, & Martin, 1989). *Pratylenchus pratensis* has been identified to be pathogenic on winter wheat in Azerbaijan (Kasimova & Atakishieva, 1981).

As with CCN, the identification of lesion nematodes considers traditional keys relating to morphology (Corbett, 1974; Loof, 1978; Handoo & Golden, 1989) as well as the new DNA based tools (Orui & Mizukubo, 1999). As reviewed by De Waele & Elsen (2002), biological diversity among populations of the same species has been reported in *P. brachyurus*, *P. goodeyi*, *P. loosi*, *P. neglectus*, *P. penetrans* and *P. vulnus*. Unlike CCN, in which many pathotypes exist, to date there is no formal report or evidence to indicate pathotypes in either *P. thornei* or *P. neglectus*. Furthermore, screening of identified resistant accessions in Australia, Mexico and Turkey with local populations reveals the resistance to pertain under greenhouse and field conditions. However, caution should be taken to examine the reproductive fitness between root lesion nematode populations from the field and also in greenhouse studies to be sure about the availability of plant resistance reactions, as nematodes in culture collections for an extended period of time can lose their pathogencity (De Waele & Elsen, 2002).

2.3. Other Cereal Nematodes – Root Knot, Stem and Seed Gall

Root-knot (RK), are the most economically important group of plant parasitic nematodes worldwide, attacking nearly every crop (Sasser & Freckman, 1987). Several species attack *Poaceae* in cool climates, including *Meloidogyne artiellia*,

M. chitwoodi, *M. naasi*, *M. microtyla* and *M. ottersoni* (Sikora, 1988). In warmer climates, *M. graminicola*, *M. graminis*, *M. kikuyensis* and *M. spartinae* are important (Taylor & Sasser, 1978). In tropical and subtropical areas, *M. incognita*, *M. javanica* and *M. arenaria* are all known to attack cereal crops (Swarup & Sosa Moss, 1990). To date, only *M. naasi* and *M. artiellia* have been shown to cause significant damage to wheat and barley in the winter growing season (Sikora, 1988).

Meloidogyne naasi is reported from Britain, Belgium, the Netherlands, France, Germany, Yugoslavia, Iran, U.S.A. and former U.S.S.R, occurring mostly in temperate climates (Kort, 1972). However, it has also been found in Mediterranean areas, on barley in the Maltese islands (Inserra, Lamberti, Volvas, & Dandria, 1975) and in New Zealand and Chile on small grains (Jepson, 1987). It is probably the most important root-knot nematode affecting grain in most European countries (Kort, 1972). It does not appear to be widespread in temperate, semi-arid regions such as Western Asia and Northern Africa (Sikora, 1988). *Meloidogyne naasi* is a polyphagous nematode, reproducing on at least 100 species of plants (Gooris & D'Herde, 1977) including barley, wheat, rye, sugar beet, onion and several broadleaf and monocot weeds (Kort, 1972). Generally *Poaceae* are considered to be better hosts (Gooris, 1968). In Europe, oat is a poor host compared with other cereals, whereas in the USA oat is an excellent host of *M. naasi* (Kort, 1972). Host races of *M. naasi* have been identified in the USA by using differential hosts (Michel, Malek, Taylor, & Edwards, 1973), which makes control of this nematode more difficult.

Other species of root knot nematodes attacking cereals include *M. artiellia,* which has a wide host range including crucifers, cereals and legumes, especially chickpea (Ritter, 1972; Di Vito, Greco, and Zaccheo, 1985). It is known to reproduce well on cereals and severely damages legumes (Kyrou, 1969; Sikora, 1988). This nematode is chiefly known from Mediterranean Europe in Italy, France, Greece and Spain (Di Vito & Zaccheo, 1987), but also west Asia (Sikora, 1988), Syria (Mamluk, Augustin, & Bellar, 1983) and Israel (Mor & Cohn, 1989).

Meloidogyne chitwoodi is a pest on cereals in the Pacific North West of the USA and is also found in Mexico, South Africa and Australia (Eisenback & Triantaphyllou, 1991). Many cereals, including wheat, oat, barley and maize and a number of dicotiledons are known to be hosts (Santo & O'Bannon, 1981). The three species, *M. incognita, M. javanica* and *M. arenaria* were found to be good hosts on a range of cereal cultivars including wheat, oat, rye and barley under greenhouse conditions (Johnson & Motsinger, 1989). *Meloidogyne graminis* is not known to be widely distributed, being limited to the southern United States, where it is associated with cereals and more often turfgrasses (Eriksson, 1972).

Stem nematodes (SN), belonging to the genus *Ditylenchus* comprise many species which are prevalent in a wide range of climatic conditions from temperate, subtropical to tropical, where moisture regimes enable nematode infection, multiplication and dispersal (Plowright, Caubel, & Mizen, 2002). *Ditylenchus dipsaci* is by far the most common and important species of stem nematode on cereals, particularly on oat, maize and rye and is widespread throughout western and central Europe, USA, Canada, Australia, Brazil, Argentina and North and South Africa (Plowright et al., 2002).

Another species, *D. radicicola* is distributed throughout the Scandinavian countries, Britain, the Netherlands, Germany, Poland, former USSR, USA and Canada. This nematode also occurs on many grasses of economic importance but is not considered important in subtropical or tropical environments (Plowright et al., 2002). S'Jacob (1962) suggested that biological races of this species occur.

The seed gall (SG) nematode (*Anguina tritici*), is of historical importance since it is the first plant parasitic nematode recorded in the literature. It is commonly known as "ear cockle" in many countries, but in India several names are used, including seed gall, Gegla, Mamni, Sehun and Dhanak. It is frequently found on small grain cereals, and is a problem where farm saved seed is sown without the use of modern cleaning systems. Cereals are infected throughout Western Asia and North Africa (Sikora, 1988; Elmali, 2002) including Iraq (Stephan, 1988), Turkey (Yüksel, Güncan, & Döken, 1980), Pakistan (Maqbool, 1988) and also on winter wheat in Azerbaijan (Kasimova & Atakishieva, 1981).

Iran wheat gall nematode (*Anguina tritici*) was observed in wheat fields of Isfahan and Kerman provinces for the first time in 1949 (Davachi, 1949). Recent surveyed regions of Isfahan province indicated 21.7% of fields were infested with *Anguina tritici*. In addition a closely related species, *A. agrostis*, was found in barley fields causing heavy infection of gum disease in Fars province of Iran for the first time in 2003 (Pakniat & Sahandpour, 2004). In the Indian sub-continent *A. tritici* is widespread in Bihar, Uttar Pradesh, and at a few places in Rajasthan, Haryana and Punjab (Bishnoi, pers. com.). It is also reported from China, parts of Eastern Europe (Tesic, 1969; Swarup, 1986; Urek & Sirca, 2003), Russia, Australia, New Zealand, Egypt, Brazil and several areas in the United States, as reviewed by Swarup and Sosa Moss (1990).

It is important to mention the bacterial related interaction which occurs with "ear cockle" nematode. The disease was first recorded from India by Hutchinson (1917), where the nematode is associated with a bacterium *Corynebacterium michiganense* pv. *tritici*. It has only much later been detected on barley in northern Iraq, where infestations reached 90% (Al-Talib, Al-Taae, Neiner, Stephan, & Al-Baldawi, 1986; Stephan, 1988). The bacterium is frequently present along with juveniles in galls and is responsible for expression of the disease. The bacterium is only capable of producing yellow streaks on leaves on its own, that run parallel to the veins. The nematode carries the bacterium to the growing point as an external body contaminant (Gupta & Swarup, 1972). The bacterium multiplies very quickly under favourable environmental conditions, increasing its concentration in a plant and forming a thick, viscous fluid in which nematode juveniles are not able to survive. Under such conditions, emerging ears are totally sterile and are covered with yellow slime. Economic losses associated with this combination are increased because of the lower price for infected grain (Rivoal & Cook, 1993).

2.4. Other Nematodes

There are other plant parasitic nematodes such as *Longidorus elongatus*, *Merlinius brevidens* and species of *Tylenchorhynchus* and *Paratrichodorus*, which have been found or are implicated to potentially cause yield losses on cereals, although their global distribution and economic importance to date have not been clearly defined. *Tylenchorhynchus nudus*, *T. vulgaris* and *M. brevidens* are responsible for poor growth in limited areas of USA and India (Smolik, 1972; Upadhyaya & Swarup, 1981). *Paratrichodorus anemones* and *P. minor* are two species reported to cause damage to cereal crops in USA, with wheat seeded early in autumn in sandy soils being highly susceptible to *P. minor*. Elekcioglu and Gozel (1997) clearly demonstrated field population dynamics in relation to wheat growth for the nematode complex *P. thornei*, *Paratrophurus acristylus* and *Paratylenhchus* species in the southeast of Turkey, concluding the importance of the two latter genera requires further investigation.

3. LIFE CYCLE, SYMPTOMS OF DAMAGE AND YIELD LOSS

Damage caused by nematodes may be affected by a number of biotic and abiotic factors. In general both cyst and lesion nematodes have a greater damage potential where plant growth is stressed, i.e., with poor soil nutrition or structure, temperature or water stress (Barker & Noe, 1987; Nicol & Ortiz-Monasterio, 2004), or where other pathogen pressure occurs (Taheri, Hollamby, & Vanstone, 1994). Damage caused by nematodes may also be greater where limited rotation or cultivar options exist. The damage threshold of cereal nematodes varies with plant cultivar, soil type, nematodes pathotype and ecotype and climatic conditions within a geographical area (Rivoal & Cook, 1993).

Many abiotic factors, for example fertility, pH, soil type and organic matter content influence nematode population development and damage severity (Duggan, 1961). Moderate nematode population levels, under favourable environmental conditions for plant growth, may not cause as much damage as when plant growth is restricted by moisture stress or low fertility levels (Kornobis, Wolny, & Wilski, 1980). Increased nitrogen application is known to reduce the intensity of nematode damage to the crop, but at high nematode population levels this may no longer hold true (Germershauzen, Kastner, & Schmidt, 1976).

The damage threshold (i.e. the given population of a pathogen to cause a given yield loss) must be determined under many environmental and genotypic factors, such as water and nutrient availability and tolerance and/or resistance reaction of a given cultivar or variety. Furthermore, interpretation of the damage threshold between specific nematological studies should be done with extreme caution, as very few studies are truly comparable, with inherent differences in sampling protocol, extraction procedure and nematodes counting (Duggan, 1961; Stone, 1968; Dixon, 1969; Gill & Swarup, 1971; Meagher & Brown, 1974; Simon & Rovira, 1982; Handa et al., 1985b; Dhawan & Nagesh, 1987, Rivoal & Sarr, 1987; Fisher & Hancock, 1991; Zancada & Althöfer, 1994; Al-Hazmi, Al-Yahya, & Abdul-Razig, 1999; Ibrahim et al., 1999).

3.1. Cereal Cyst Nematode

The life cycle of *H. avenae* involves only one generation during a cropping season, irrespective of geographical region and the host range of this nematode is restricted to gramineaceous plants. There is sexual dimorphism, with males remaining worm-like, whereas females become lemon-shaped and spend their life inside or attached to a root. An adult, white female is clearly visible on roots with a swollen body, about 1 mm across, protruding from the root surface (Fig. 1.). Eggs are retained within the female's body and after the female has died the body wall hardens to a resistant brown cyst, which protects the eggs and juveniles. The eggs within a cyst remain viable for several years (Kort, 1972).

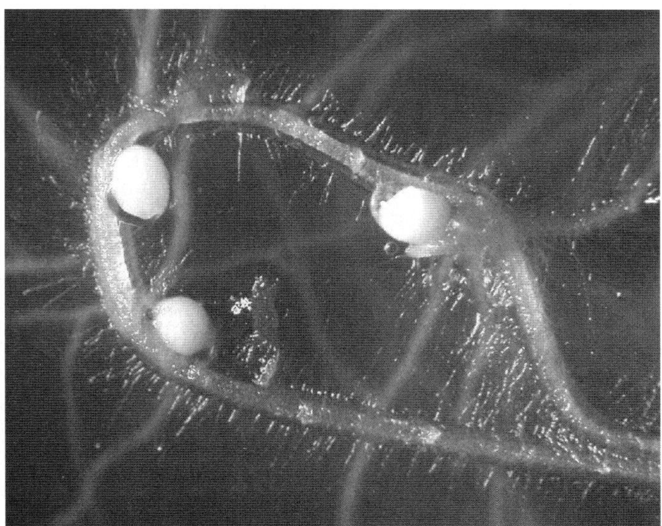

Figure 1. White Heterodera avenae *females clearly visible on roots with a swollen body (1mm) protruding across root surface (photo: R. Rivoal).*

Comparative studies on populations of *H. avenae* from different origins have revealed the existence of ecotypes differing in their hatching cycles, a result of the induction or suppression of dormancy (diapause) by different temperature conditions. Hatch of *H. avenae* in Mediterranean climates is characterized by juvenile emergence from autumn to the beginning of spring, whereas in more or less temperate climates (cooler, usually with snow), the majority of juveniles emerge in spring as the soil temperatures rise (Rivoal, 1982, 1986). The hatching requirements of other species are less understood but are essential to the understand of biology and control of those species.

The above ground symptoms caused by CCN occur early in the season as pale green patches with the lower leaves of the plant being yellow and generally plants with few tillers (Fig. 2). These patches of infestation may vary in size from 1 to 100 m^2 or more. The symptoms can easily be confused with nitrogen deficiency and poor soils and the root damage exacerbates the effect of any other stress, e.g. water and nutrient stress. The below ground symptoms may be slightly different depending on the type of grass host. Wheat attacked by *H. avenae* shows increased root production such that roots have a "bushy-knotted" appearance usually with several females visible at each root (Fig. 3). Oat roots are shortened and thickened, while barley roots appear less affected. The cysts are glistening white-grey initially and dark brown when mature. Attached loosely at their necks, many cysts are dislodged when roots are harvested for examination. Root symptoms are recognisable within one to two months after sowing in Mediterranean environments and often later in more or less temperate climates.

Figure 2. Patches of poor growth caused by Heterodera avenae *on winter wheat in Pacific Northwest of USA (photo: R. Smiley).*

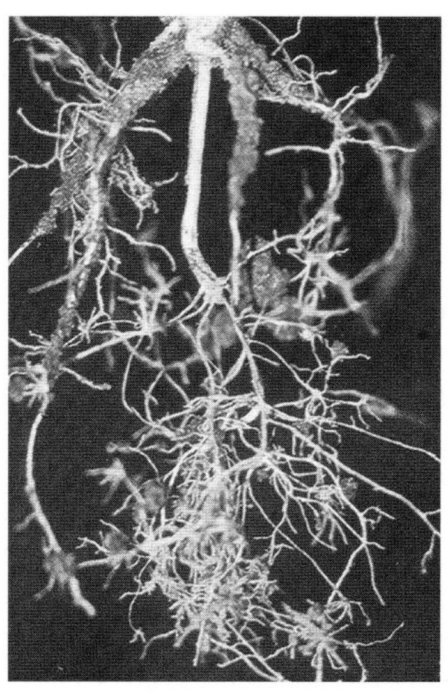

Figure 3. 'Bushy-knotted' roots attacked by Heterodera avenae, *with white female visible (photo: R. Rivoal).*

Heterodera avenae in the northwestern part of India and in southern Australia is considered a major limiting factor of wheat and barley. Figures in India suggest that for every 10 eggs/g soil, there is a loss of 188 kg/ha in wheat and 75 kg/ha in barley (Duggan, 1961; Dixon, 1969). Mathur, Handa, and Swarup (1986) reported avoidable loss in wheat ranging from 32.4 to 66.5% with inoculum varying from 4.6 to 10.6 eggs/ml soil. In China, recent yield loss studies conducted in three provinces including Anhui, Henan and Hebei using aldicarb to provide CCN control, indicated losses of the order 10–40% (Peng et al., 2007).

Yield losses due to this nematode are 15–20% on wheat in Pakistan (Maqbool, 1988), 40–92% on wheat and 17–77% on barley in Saudi Arabia (Ibrahim et al., 1999) and 20% on barley and 23–50% on wheat in Australia (Meagher, 1972). Recent studies in Oregon in Pacific North West have indicated losses on spring wheats of 24% (Smiley et al., 2005). In Tunisia *H. avenae* suppressed grain yields of initial population densities (Pi) on the yield of wheat cultivar Karim by 26–96% and 19–86% on barley cultivar Rihan (Namouchi-Kachouri, B'Chir, & Hajji, 2006). Staggering annual yield losses of 3 million pounds sterling in Europe and 72 million Australian dollars in Australia have been calculated as being caused by *H. avenae* (Wallace, 1965; Brown, 1981). The losses in Australia are now greatly reduced due to their control with resistant and tolerant cultivars.

Little is known about the economic importance of the species *H. latipons* even though it was first described in 1969 (Sikora, 1988). Recent studies by Scholz (2001) implicate yield loss with both barley and durum wheat with *H. latipons*. Field studies in Cyprus indicated a 50% yield loss on barley (Philis, 1988). Because the cysts are similar in size and shape it is possible that previous findings of this recently described nematode species have erroneously been attributed to the economically important *H. avenae* (Kort, 1972). In West Asia and North Africa *H. latipons* has been found on wheat and barley in four countries (Sikora, 1988). It has also recently been confirmed in Turkey (Rumpenhorst, 1996; Nicol et al., 2002) and from several Mediterranean countries, associated with poor growth of wheat (Kort, 1972). Unfortunately this nematode has not been studied in detail and information on its host range, biology and pathogenicity is scarce, but it is suspected to be an important constraint on barley and durum wheat production in temperate, semi-arid regions (Sikora, 1988; Scholz, 2001; Scholz & Sikora, 2004; Ismail, Sikora, & Schuster, 2001).

Similarly *H. filipjevi* is most likely an economically important nematode on cereals due to its widespread distribution and previous misidentifications as *H. avenae* in the former USSR and also Sweden. In Turkey significant yield losses (average 42%) in several rainfed winter wheat locations have been reported (Oztürk, Yildirim, & Kepenekci, 2000; Nicol et al., 2005; Nicol, unpubl.). Natural field trials conducted over several seasons have clearly indicated greater losses under drought conditions (Nicol, unpubl.). Given the increased recognition and incidence, these species are now being identified as a constraint to cereal production (Philis, 1988; Oztürk et al., 2000; Scholz, 2001).

As mentioned water stress is one of the key environmental conditions that can exacerbate damage caused by *H. avenae* and has been demonstrated by the use of radiothermometry technique to detect nematode attacks (Nicolas, Rivoal, Duchesne, & Lili, 1991). At milky dough stage, plant height, total chlorophyll content and light interception by leaves were suppressed but the temperature of plant canopy increased compared to the non-infected controls (Al-Yahya, Alderfasi, Al-Hazmi, Ibrahim, & Abdul-Razig, 1998). Pot experiments in controlled environments revealed a dramatic, negative effect of various populations of CCN on wheat root growth, associated with decreased shoot growth and decreased rates of transpiration (Amir & Sinclair, 1996).

3.2. Root Lesion Nematodes

Pratylenchus species are polycyclic, polyphagous, migratory root endoparasites, which are not confined to fixed places for their development and reproduction. Eggs are laid in the soil or inside plant roots. The nematode invades the tissues of the plant root, migrating and feeding inside a root. Secondary attack by fungi frequently occurs at these lesions. The life cycle is variable between species and environment and ranges from 45 to 65 days (Agrios, 1988).

Pratylenchus feeds on and destroy roots, resulting in characteristic dark brown or black lesions on the root surface, hence their name "lesion" nematodes (Fig. 4). Aboveground symptoms of *Pratylenchus* on cereals, like other cereal root nematodes are non-specific, with infected plants appearing stunted and unthrifty, sometimes with reduced numbers of tillers and yellowed lower leaves (Fig. 5).

The lesion nematode *P. thornei,* causes yield losses in wheat from 38–85% in Australia (Thompson & Clewett, 1986; Doyle, McLeod, Wong, Hetherington, & Southwell, 1987; Nicol, 1996; Nicol, Davies, Hancock, & Fisher, 1999; Taylor et al., 1999), 12–37% in Mexico (Nicol, 2002; Nicol & Ortiz-Monasterio, 2004), 70% in Israel (Orion, Amir, & Krikun, 1984) and most recently in Pacific Northwest USA (Smiley et al., 2005). While *P. thornei* has mainly been reported from regions with a Mediterranean climate, it is possible similar losses may also occur in other countries. *P. neglectus* and *P. penetrans* appear to be less widespread and damaging on cereals compared with *P. thornei.* In southern Australia, losses in wheat caused by *P. neglectus* ranged from 16–23% (Taylor et al., 1999). Vanstone et al. (1998) showed yield loss in wheat of 56–74% in some sites infested with both *P. thornei* and *P. neglectus*. In North America and Germany, *P. neglectus* has been shown to be a weak pathogen to cereals (Heide 1975; Mojtahedi & Santo, 1992). *Pratylenchus penetrans* has been reported to cause losses of 10–19% in wheat in Canada (Kimpinski et al., 1989) indicating that this nematode may be a problem in small grain cereals. Sikora (1988) identified *P. neglectus* and *P. penetrans* in addition to *P. thornei* on wheat and barley in Northern Africa, and all these plus *P. zeae* in western Asia. Further work is necessary to determine the significance of these species in these regions.

Although *Pratylenchus* is capable of multiplying for several generations during a single season, they spread only from plant to plant due to their relative immobility. The impact of plant parasitic nematodes on plant health and crop yield varies with biogeographic location, cropping sequence and intensity, cultivar selection, soil characteristics and nematode community structure (McKenry & Ferris, 1983). As mentioned previously, the economic threshold for plant damage will depend on many such factors and interpretation of the damage threshold between specific nematological studies should be done with extreme caution, as very few studies are truly comparable. There are inherent differences in sampling protocol, extraction procedure and nematode renumerification. It is for this reason the studies conducted are listed, however the reader should interpret these accordingly (Van Gundy, Perez, Stolzy, & Thomason, 1974; Orion et al., 1984; Doyle et al., 1987; Lasserre et al., 1994; Nicol et al., 1999; Taylor et al., 1999; Nicol & Ortiz-Monasterio, 2004).

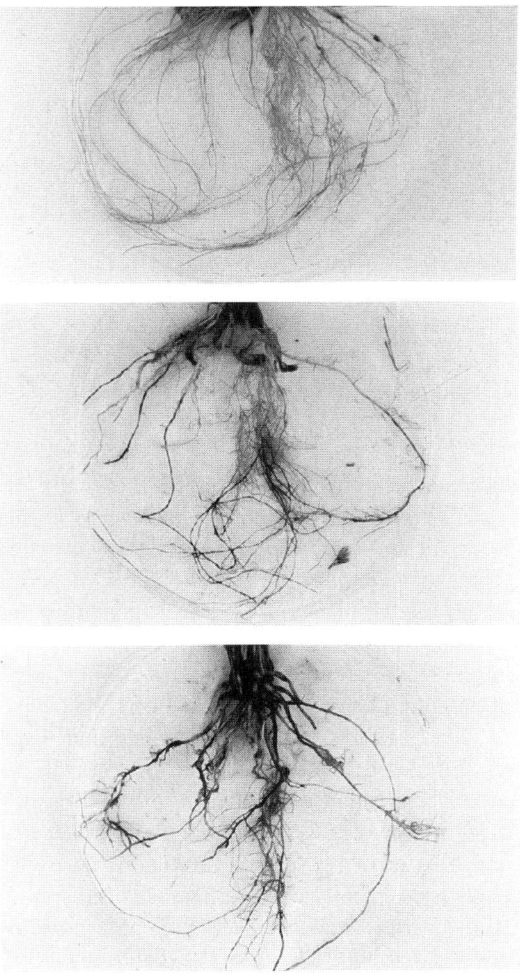

Figure 4. Symptoms of root lesion nematode, Pratylenchus thornei, *on susceptible wheat, showing extensive lesions, cortical degradation and reduction in both seminal and lateral root systems with increasing nematode density from top to bottom under natural field infestation (photo: J. M. Nicol, CIMMYT).*

Figure 5. Winter wheat attacked by root lesion nematode, Pratylenchus neglectus, *showing patchy distribution, reduced tillering and emergence of infected plants (photo: R Rivoal & R. Cook).*

3.3. Other Cereal Nematodes – Root Knot, Stem and Seed Gall

3.3.1. Root Knot Nematodes

Root knot nematodes cause typical small sized root galls on roots. Egg masses attached to the posterior end of protruding females are normally transparent, but darken on exposure to air and can resemble cysts of *H. avenae*. Young juveniles of *M. naasi* invade roots of cereals within 30–45 days of germination, after which small galls on root tips can be observed. *M. naasi* generally has one generation per season (Rivoal & Cook, 1993). Egg masses in galls survive in the soil. Eggs have a diapause, broken by increasing temperature after a cool period (Antoniou, 1989). In warmer regions on perennial or volunteer grass hosts more than one generation per season is possible (Kort, 1972). Juveniles develop and females become almost spherical in shape. Females deposit eggs in an egg sac and usually appear 8–10 weeks after sowing and are found embedded in the gall tissue (Kort, 1972). Large galls may contain 100 or more egg-laying females (Rivoal & Cook, 1993).

Towards the end of a growing season galling of the roots, especially the root tips, is common. Galls are typically curved, horseshoe or spiral shaped (Kort, 1972; Fig. 6). Symptoms of *M. naasi* attack closely resemble those caused by *H. avenae*, with patches of poorly growing, yellowing plants that may vary in size from a few square metres to larger areas. Other root knot nematodes attacking cereals are suspected to produce similar symptoms, but most are much less studied than *M. naasi*.

Information on the economic importance of root knot nematodes on cereals is limited to a few studied species. *M. naasi* can seriously affect wheat yield in Chile (Kilpatrick, Gilchrist, & Golden, 1976) and Europe (Person-Dedryver, 1986). On barley it has been known to cause up to 75% yield loss in California, USA (Allen, Hart, & Baghott, 1970). It is also associated with yield loss in barley in France (Caubel, Ritter, & Rivoal, 1972), Belgium (Gooris & D'Herde, 1977) and Great Britain (York, 1980). Severe losses can occur, with entire crops of spring barley lost in the Netherlands and France (Schneider, 1967). *M. naasi* damage is not known to be widespread in temperate semi-arid regions (Sikora, 1988).

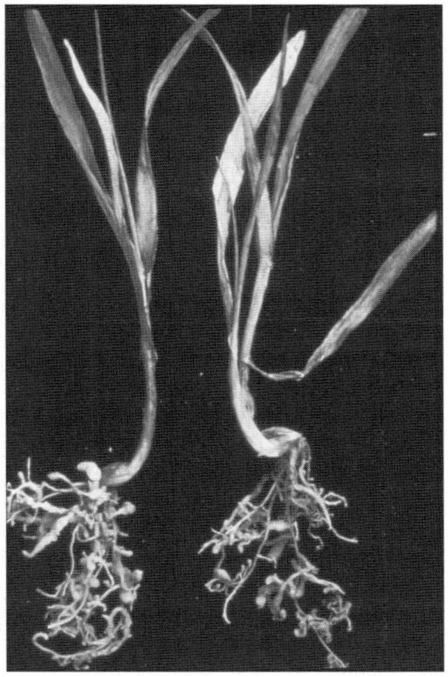

Figure 6. Typical galling of barley roots caused by Meloidogyne naasi *(photo: R. Rivoal).*

Damage to wheat by *M. artiellia* is known from Greece, southern Israel and Italy (Kyrou, 1969; Mor & Cohn, 1989). In Italy 90% yield losses on wheat have been recorded (Di Vito & Greco, 1988). *M. chitwoodi*, an important pathogen of potato also damages cereals in Utah, USA (Inserra, Vovlas, O'Bannon, & Griffin, 1985) and Mexico (Cuevas & Sosa Moss, 1990). In controlled laboratory studies, *M. incognita* and *M. javanica* have been shown to reduce plant growth of wheat (Abdel Hamid, Ramadan, Salem, & Osman, 1981; Roberts, Van Gundy, & McKinney, 1981; Sharma, 1981) and similarly *M. chitwoodi* (Nyczepir, Inserra, O'Bannon, & Santo, 1984). *M. incognita* is a known field problem on wheat in northwestern India (Swarup & Sosa Moss, 1990).

3.3.2. Stem Nematode

Ditylenchus dipsaci is a migratory endoparasite and invades foliage at the base of stems of cereal plants, where it migrates through tissues and feeds on adjacent cells. Reproduction continues inside a plant almost all year round but is minimal at low temperatures. When an infected plant dies, nematodes return to the soil from where they infect neighbouring plants. Typical symptoms of stem nematode attack include basal swellings, dwarfing and twisting of stalks and leaves, shortening of internodes and many axillary buds, producing an abnormal number of tillers to give a plant a bushy appearance (Fig. 7). Heavily infected plants may die in the seedling stage, resulting in bare patches in a field, while other attacked plants fail to produce flower spikes (Kort, 1972).

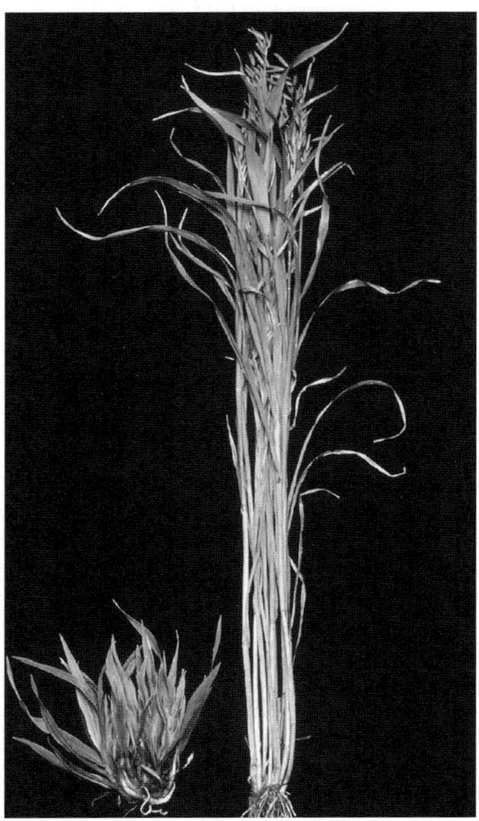

Figure 7. Close up of stem nematode, Ditylenchus dipsaci, *damage on susceptible oats indicating severe dwarfing, twisting of leaves, an abnormal number of tillers giving the plant a bushy stunted appearance (photo: S. Taylor).*

The nematodes are highly motile in soil and can cover a distance of 10 cm within two hours (Kort, 1972), hence their ability to spread from one plant to another is rapid. There are a number of biological races or strains of *D. dipsaci*, which are morphologically indistinguishable but differ in host range. Kort (1972) stated that the rye strain is more common in Europe and the oat strain is more common in Britain. Rye strains attack rye and oats as well as several other crops, including bean, maize, onion, tobacco, clover and also a number of weed species commonly associated with the growth of cereals in many countries (Kort, 1972). The oat strain attacks oats, onion, pea, bean and several weed species but not rye (Kort, 1972). Wheat is also attacked by *D. dipsaci* in central and eastern Europe (Rivoal & Cook, 1993), and central Asia in Azerbaijan (Kasimova & Atakishieva, 1981). The giant race of *D. dipsaci* is widely distributed throughout North Africa and the Near East on many crops and needs to be monitored for effects on cereals.

Economic damage by *D. dipsaci* depends on a combination of factors such as host plant susceptibility, infection level of soil, soil type and weather conditions. This is further complicated by the extensive intraspecific variation which is known in this species (Janssen, 1994). Furthermore, environmental conditions such as extended soil moisture content in the surface layer of soil provide optimum nematode activity, hence increasing the chance of a heavy attack. It is a problem with cereal crops growing on heavy soils in high rainfall areas (Griffin, 1984). The nematode is economically important on rye and oat but not on wheat and barley (Sikora, 1988). Although few studies have looked at the economic importance of this nematode, work on oats in England attributed a 37% yield loss to *D. dipsaci* (Whitehead, Tite, & Fraser, 1983) and in Italy was considered an important factor in poor wheat yields, where damage caused by *D. dipsaci* was associated with the presence of *Fusarium* (Belloni, 1954). In the seventies of the last Century, *D. dipsaci* had severely affected the maize crop in northen Europe when this culture has replaced oat production (Caubel, Person, & Rivoal, 1980).

3.3.3. Gall Nematode

This nematode disease is generally associated with situations where agricultural practices are not advanced. Within the infected cereal heads (florets), the nematode galls replace the grains. These galls are brown or black in colour and contain large numbers of second stage juveniles whose population ranges between 3000 and 12000, with an average of approximately 6000 juveniles per gall. These galls and their contents (second stage juveniles) are resistant to dry weather (anhydrobiosis) and it has been reported that they do not lose viability even up to 30 years. On getting favourable weather, like soil temperature (15°C ± 2), soil depth (2 cms), 20% soil moisture and 51% soil pore spaces, these galls rupture and discharge juveniles which in turn search the host and attack the plants.

Nematode-infected seed galls, which may be present already in the soil or sown into the soil at planting with contaminated seed, become moist and soft, with soil moisture facilitating the release of juveniles. Approximately one week after seed galls infected with nematode are placed in the soil, juveniles can be traced in the growing point of a germinating plant. These juveniles move upward passively on the growing point as the plant grows. They do not exhibit any morphological change until approximately two months. Nematode morphological changes take place only when the juveniles penetrate a flower primordial after two to three months and then turn into adults. As a result, ovules and other flowering parts of a plant are transmuted into galls or 'cockles' (Fig. 8). Nematodes mature inside galls and females lay thousands of eggs from which juveniles hatch and remain dormant in seed. The total life cycle is completed in around four months (Swarup & Sosa Moss, 1990). Temperature, humidity, planting depth and the source of galls are the major determinants in symptom expression. The nematode favours wet and cool weather (Kort, 1972). These environmental conditions and the source of galls are particularly important for development of yellow ear rot. This nematode-vectored bacterial disease, vernacularly known as "tundu" or "tannan" in India, is also commonly found associated with the ear-cockle nematode problem. The disease was first recorded from India by Hutchinson (1917), where the nematode is associated with *Corynebacterium michiganense* pv. *tritici*. This bacterium is frequently present along with juveniles in galls and is responsible for expression of the disease. The bacterium is only capable of producing yellow streaks on leaves on its own that run parallel to the veins. The nematode carries the bacterium to the growing point as an external body contaminant (Gupta & Swarup, 1972). Atmospheric temperatures between 5–10°C and a relative humidity of 95–100% favour multiplication of the bacterium in plants.

Figure 8. A healthy wheat ear (left), moderate infestation of gall nematode Anguina tritici, *and severe infestation of galls into 'cockles' (right) (photo: M. Ritter).*

Symptoms of *A. tritici* attack may be indicated by small and dying plants with leaves generally twisted due to nematode infection (Swarup & Sosa Moss, 1990). Infected ears are easily recognized by their smaller size and darkened colour compared with normal seeds, but infected seeds may be easily confused with bunt (*Tilletia tritici*). Under dry conditions juveniles may survive for decades (Kort, 1972).

In both ear-cockle and yellow ear-rot, the first observable symptom is an enlargement of the basal stem portion near the soil base, visible in three week old wheat seedlings. The emerging leaves are twisted and crinkled. Frequently, some leaves remain folded with their tips held near the growing point. These leaves, after about 30–45 days straighten out and many appear normal, with faint ridges on the surface. In comparison to healthy seedlings, the affected plants are dwarfed, with a spreading habit. These symptoms are more clearly discernible on young seedlings and decrease with plant age. Under very low infestation levels plants may not exhibit any visible symptoms, even though a few seed galls are produced in the ears, whereas severely infested plants may die without heading. Infested seedlings produce more tillers and grow faster than normal plants but not necessarily with an increase in the number of ears (Swarup & Sosa Moss, 1990).

Furthermore, ears emerge roughly a month earlier in diseased plants. Such ears are short and broad, with very small or no awns on the glumes. Nematode galls replace either all or some of the grains. In the yellow ear-rot disease, the characteristic feature is the production of a bright yellow slime- or gum-like substance on the abortive ears as well as leaves, which remains in contact with such ears while still in the boot leaf stage. Under humid conditions the bacterial slime trickles down tissues (Swarup & Sosa Moss, 1990) and upon drying it appears brown in colour. An infected spike is narrow and short, with wheat grains partially or completely replaced by slime. In the latter event an emerging spike remains sterile. The stalk of an infected spike is always distorted.

Worldwide, wheat, barley and rye are commonly attacked, but barley is less attacked in India (Paruthi & Gupta, 1987). Severely affected areas in India may suffer crop loss up to 80% (Bishnoi, pers. comm.), particularly in some regions and years such as in 1992 and 1997 several districts in Bihar and similar 1999 in Pawai Tehsil in Panna district of M. P. Significant losses of 20% in Ardestan wheat fields have also been reported (Behdad, 1982). Further studies of *A. tritici* on Roshan cultivar was studied under field conditions with different galled treatment of 0, 1, 2 and 4% infested with galls leading to damage of 0, 11, 21 and 35% respectively (Ahmadi & Akhiyani, 2001).

In Iraq, ear cockle is an important pest on wheat, with infection ranging from 0.03 to 22.9% and causing yield losses up to 30% (Stephan, 1988). Barley is also attacked in Iraq and Turkey (Yüksel et al., 1980; Al-Talib et al., 1986). In Pakistan, ear cockle is a known pest on wheat and barley and is found in nearly all parts of the country, causing losses of 2–3%. However, in association with the yellow ear-rot bacterium it produces serious yield losses on wheat (Maqbool, 1988). In China, Chu (1945) found yield losses between 10 and 30% on wheat.

4. INTEGRATED CONTROL OF CEREAL NEMATODES

In many of the countries where these nematodes occur wheat is often one of the major food staple, and the control of the nematode is of considerable importance to improve the production and livelihood of the farming communities. Furthermore much of West Asia and North Africa is characterised by wheat monoculture systems, where rainfall or irrigation is limited and options for crops rotation are not used or restricted. Such cropping systems frequently suffer moisture or drought stress and in these environments the effects of the nematode damage can be increased, and hence control of nematodes in these cropping systems is of paramount importance (Yadav, Bishnoi, & Chand, 2002)

Many different control options such as chemical, cultural, genetic (resistance/tolerance) and biological control are available and their need effect should be aimed at decreasing and maintain population densities under damage thresholds, so as to maintain or reach the attainable yield. However in order to this a clear understanding of nematode threshold densities that result in yield loss and the interaction of these thresholds with biotic and abiotic factors is required (Rivoal & Sarr, 1987).

Cultural practices represent efficient methods based on rotational combinations of non-hosts crops or cultivars and clean fallows. Frequencies of such combinations should be calculated upon data inferring from specific studies of population dynamics according to the targeted inputs. Application of fertilizers and soil amendments may compensate the reducing effect of nematodes on wheat yields but their use is frequently limited by financial constraints. Adjustment of sowing dates to escape synchrony of peak emergence with the more sensitive stage of the crop could maximise the final yield. Trap cropping could constitute efficient measures to decrease nematode densities. Allelopathy techniques based on toxic plant root exudates and microbial secretions offer also some alternative controlling measures. Control of stem and foliages nematodes could be effective by sanitation based on grain sieving or other discarding process.

Even if in the past low rates of nematicides applied to both soil and seed provided effective and economical control (e.g. in Australia, India and Israel), however, the present day cost and environmental concerns associated with these chemicals do not make them a viable economic alternative for almost all farmers. However, their use in scientific experiments to understand the importance of these nematodes will remain vital. For this reason we will not provide them as an option in this section and again refer to previous reviews cited at the start of this section which mention this work.

The use of resistant/tolerant varieties which ensure both reduction/inhibition of nematode multiplication within the plants and stable crop production offer the best control capabilities. In addition it requires no additional equipment or cost. However, the use of resistant cultivars requires a sound knowledge of the virulence spectrum of the targeted species and pathotypes. Engineering of transformed plants with inhibitors to the development of nematodes may be part of the future options for some countries.

The prospects for using biological antagonists within an IPM strategy for wheat nematodes is still considered promising with the development of natural populations of enemies (e.g. *Pochonia, Nematophtora*) or application of exogenic pathogens i.e. *Trichoderma viride* (Indra-Rajvanshi, 2003), however their ultimate use relies greatly on the agroecology of the cropping systems for persistence and effectiveness.

This section will focus particularly on CCN and RLN, but also consider the other three nematodes and what limited published information is available about their control and the options to combine these in an integrated manner. However, as found with many nematodes there are only a few well published studies which allow a good understanding of population dynamics to establish the most effective integrated management combinations to control a given nematode (Caubel et al., 1980; Dowe & Decker, 1977). More targeted research is needed to consider the holistic system of nematode control in balance with the cropping system options, the agro-ecological conditions and especially the use of resistance, which still offers one of the best cost effective means of control.

4.1. Cereal Cyst Nematode

4.1.1. Chemical

The present day cost and environmental concerns associated with these chemicals do not make them a viable economic alternative for almost all farmers. However, their use in scientific experiments to understand the importance of these nematodes will remain vital. For this reason we will not provide them as an option in this section and again refer to previous reviews mentioned at the start of this section.

4.1.2. Cultural Practices

4.1.2.1. Grass Free Rotations and Fallowing with Cultivation

One of the most efficient methods of controlling *H. avenae* is with the use of grass-free rotations using non-host crops. In long term experiments, non-host or resistant cereal frequencies of 50% (80% in lighter soils) keep populations below damaging thresholds (Rivoal & Besse, 1982; Fisher & Hancock, 1991). Similarly, in India, it was found that nematode population decreased by 70% with continued rotation of non host crops like mustard, carrot, fenugreek and gram or by fallowing, and this resulted in a corresponding 56% increase in barley yields with two year rotation of non host crops (Handa, Mathur, & Mathur, 1975a). Using natural *H. avenae* field infested soil in Hubei province in China, small grained cereals (wheat, barley, oats and grass weeds) were susceptible, whilst maize was infected but the life cycle not completed, and pastures (*Trifolium and Medicago*) were non-hosts (MingZhu, Zhi Feng, & YanNong, 1996). In Spain under natural field conditions the use of vetch in rotation and use of fallow with cereals was effective (Nombela, Navas, & Bello, 1998). Monitoring a 30 year rotation trial over several seasons under rainfed wheat cropping systems in Turkey clearly demonstrated the use of legumes (vetch, lentil), sunflower or safflower in wheat rotation system provided a significant reduction in cyst population, whereas fallowing had little effect and cereal rotation increased significantly cyst populations (Elekçioğlu et al., 2004). In Europe a four-year rotation can be practiced for nematode control, but economic factors do not permit such long rotations in most subtropical and tropical countries.

Clean fallow can reduce population densities of the nematode and one to five deep ploughings during hot summer months can cause reductions in nematode populations between 9.3 and 42.4%, with a corresponding yield increase of 4.4–97.5% (Mathur, Handa, & Swarup, 1987), but are not always economically and environmentally sound. In arid climates, the decrease in population is attributable to killing of cyst contents by intense solar heat and to desiccation of eggs and juveniles by hot winds. In contrast, reducing effect of fallow on population densities could be increased by maintaining humidity of soil which favours emergence of juveniles during the hatching period. Soil sanitation could be achieved by a straw mulch management which was demonstrated to decrease soil evaporation and this resulted in higher levels of soil water and decreased nematode inhibition of rooting (Amir & Sinclair, 1996; Sinclair & Amir, 1996). Studies in rainfed wheat system in Australia under natural CCN populations found no significant differences in the number of cysts produced with normal cultivation versus direct drill, or the timing and number of cultivations with rotary hoe (Boer, Kollmorgen, Macauley, & Franz, 1991), however similar studies by Roget and Rovira (1985) indicated early damage in wheat was reduced with direct drill than normal cultivation inferring these agronomic studies are to some degree site and location specific.

Recent research in India has focused on the identification of new chemicals from botanicals (Kanwar & Walia, 2004). The compositae *Chrysanthemum coronarium* has demonstrated efficient nematostatic activity to *H. avenae* (Bar-Eyal, Sharon, & Spiegel, 2006).

4.1.2.2. Irrigation

Mathur, Arya, Handa, and Mathur (1981) reported higher multiplication of nematodes in well irrigated fields in wheat and barley as compared to soil with low moisture. They found that sandy loam soil resulted in more yield of barley with reducing the irrigation gap i.e. 20 days with maximum post harvest population build up of nematode in question.

4.1.2.3. Time of Sowing

Mathur (1969) tried sowing wheat and barley from 18th October to 26th December at weekly intervals in pots and concluded that change in the date of sowing did not influence the incidence of CCN and their multiplication. Conflicting studies however demonstrated delay in sowing time could escape synchrony between peak emergence of juveniles and the more sensitive stages of the hosting crop, which permitted to maximize the production of wheat (Brigbhan & Kanwar, 2003; Singh & Singh, 2005).

4.1.2.4. Trap and Mixed Cropping

Natural trap cropping was observed when maize replaced oat production in northern Europe. Hypersensitive to *H. avenae* attacks, maize was nevertheless a poor host and provoked sound decreases of soil densities of this nematode (Caubel et al., 1980; Rivoal & Sarr, 1987). It has been also demonstrated that winter maize is also a poor host in India and can be similarly exploited as a trap crop of *H. avenae* and *H. filipjevi* (Bajaj & Kanwar, 2005). Resistant Italian ryegrass has been bred to be introduced in areas where *H. avenae* has a high intrinsic capacity to develop and when the crop season corresponds to the hatching period of the nematode (winter in southern France). As a forage crop and catch crop for nematodes and nitrate excesses, this ryegrass will contribute to control the nematode thus protecting subsequent cereal crops as bread and durum wheats (Rivoal & Bourdon, 2005).

Mixed cropping of wheat and barley as “Gojra” is common practice in the northern region of Rajasthan. Handa, Mathur, Mathur, Sharma, and Yadav (1985a) reported the beneficial effect of resistant variety of barley (Rajkiran) with susceptible variety of wheat (Kalyansona) for increase in grain yield and decrease in nematode population as compared to susceptible crop of wheat/barley. They further indicated the possibility of use a combination of different crops with varying nematode susceptibility to decrease the population and obtaining optimum yield. Similar studies in Rajasthan intercropping wheat and barley with Indian mustard indicated maximum grain yield in addition to highest reduction in cyst populations (Rajvanshi, Mathur, & Sharma, 2002).

4.1.2.5. Organic Amendments and Inorganic Fertilizers

Mathur (1969) in India reported that oil cakes, farm yard manure, compost and saw dust applications improved plant growth and subdued multiplication of CCN. Nitrogenous fertilizer resulted in better plant growth and more nematode multiplication, however no change was found with phosphorus and potash (Mathur, 1969).

4.1.3. Resistance (and Tolerance)

Plant resistance is defined as a reduction/inhibition of nematode multiplication within plants (Trudgill, Kerry, & Phillips, 1992), and is one of the best control methods for CCN due to its wide application as it usually requires no additional equipment or cost. Ideally the resistance should be combined with tolerance (plants which have the ability to yield despite the attack of the nematode). The effectiveness of CCN resistance however will depend on the effectiveness and durability of the resistance source and on correct identification of the nematode species and/or pathotype(s). In addition, an understanding of nematode threshold densities that result in yield loss and the interaction of these thresholds with biotic and abiotic factors is required (Rivoal & Sarr, 1987; Rivoal, Person-Dedryver, Doussinault, & Morlet, 1986).

As mentioned above in order to classify the pathotype variation for *H. avenae*, an International Test Assortment of barley, oat and wheat was developed by Andersen and Andersen (1982). *H. avenae* pathotypes have usually been characterised by virulence on barley genotypes, but geographically different populations can also be differentiated by virulence on wheat (Bekal et al., 1998; Cook & Rivoal, 1998; Rivoal et al., 2001). However, as mentioned this test is more than thirty years old and does not cater for the wider variation of species and pathotypes which are presently reported. Very few studies have been achieved on the two other species *H. filipjevi* and *H. latipons* but preliminary researches indicated heterogeneous responses between populations to different resistant germplasm. It was also demonstrated that populations of *H. avenae* differed in the capacity of juveniles to produce females (part of the fitness component) which was important for designing virulence/resistance investigations and for the management of nematode densities (Rivoal et al., 2001).

A summary of the CCN cereal resistance sources and their genetic control of cyst and lesion nematodes is provided in Table 1. The progress in understanding and locating resistance sources in cereals is more advanced for cyst (*H. avenae*) than lesion (*Pratylenchus* spp.) nematodes, in part due to the specific host-parasite relationship that cyst nematodes form with their hosts (Cook & Evans, 1987), whereby all published sources are controlled by a single gene. In contrast, the relationship of migratory lesion nematodes with their hosts is less specialized and therefore less likely to follow a gene for gene model. The identified sources of resistance to *H. avenae* have been found predominantly in wild relatives of wheat in the *Aegilops* genus (Dosba & Rivoal, 1982; Eastwood, Lagudah, Appels, Hannah, & Kollmorgen, 1991; Dhaliwal, Singh, Gill, & Randhawa, 1993; Delibes et al., 1993; Rivoal & Cook, 1993; Bekal et al., 1998; Jahier et al., 1998, 2001; Romero et al., 1998; Ogbonnaya et al., 2001a; Zaharieva et al., 2001; Barloy et al., 2007). Six out of the seven named *Cre* genes for *H. avenae* resistance in wheat as well as *Rkn2* for resistance to both *M. naasi* and *H. avenae* came from four *Aegilops* species (Table 1) and have already been introgressed into hexaploid wheat backgrounds for breeding purposes. The effectiveness of these designated *Cre* genes is depending on both the species of CCN and pathotype. It has been clearly demonstrated in Australia that *Cre3* has the greatest impact on reducing the Ha13 population followed by *Cre1* and *Cre8* (Safari et al., 2005). In order to understand the effectiveness of resistance to a given population such tests are necessary.

Molecular technologies have been applied to identify markers for various CCN plant resistance genes using techniques such as RAPD and RFLP, in both barley (Kretschemer et al., 1997; Barr et al., 1998) and wheat (Eagles et al., 2001; Ogbonnaya et al., 2001a, 2001b). McIntosh, Devos, Dubcovsky, and Rogers (2001) presented information about introgression, substitution and molecular characterisation of these resistance sources in cereals. In some Australian cereal breeding programmes, markers for both wheat and barley are being implemented using marker assisted selection (MAS) to pyramid resistance genes against *H. avenae*, pathotype Ha13 (Eagles et al., 2001; Ogbonnaya et al., 2001b). Identification and implementation of markers in this way requires sufficient understanding of the biology of the pathogen and genetic control of the resistance. In the future, it may be possible to transform wheat using resistance genes as a method to produce nematode resistant wheat cultivars (Lagudah et al., 1998).

4.1.4. Biological Control

Since a long time in several countries, it is known that populations of *H. avenae* could be naturally controlled by antagonistic fungi such as *Pochonia chlamydosporia* and *Nematophtora gynophila.* This control has been observed in long term experiments with monocultures of host cereals, however there have been marked contrasts in the results between areas which developed suppressive soils and the dryland areas (Kerry, 1981). Unfortunately the biocontrol treatments by these antagonists have never been commercially feasible. In Syria and Germany cereal soils were found to have high levels of natural suppressiveness against *H. latipons* with the pathogenic fungi *Fusarium* and *Acremonium*, with the level being higher in the Syrian semiarid soils (Ismail et al., 2001). Similarly in rainfed wheat soils of cereal crops in Turkey species of *Fusarium* have been isolated from *H. filipjevi* eggs which appear to be colonized and may play some role in suppressiveness (Nicol, unpubl.).

The use of parasites as the nematophagous fungus *Paecilomyces lilacinus*, predators as the trapping fungus *Monacrosporium lysipagum*, and the nematode *Seinura paratenuicaudata* which act on living and mobile stages provided, in laboratory experiments, offer some promise to control *H. avenae* and other nematodes as *Anguina* and *Meloidogyne* (Vats, Kanwar, & Bajaj, 2004; Khan, Williams, & Nevalainen, 2006).

Table 1. Principal sources of genes[a] used for wheat breeding resistance to Cereal Cyst Nematode (Heterodera avenae) *and Root Lesion Nematode* (Pratylenchus thornei *and* P. neglectus).

Species	Cultivar or line	Genetic information	References
		Cereal Cyst Nematode	
T. aestivum	Loros, AUS10894	*Cre1*[a] (formerly *Ccn1*), on chromosome 2BL.	Slootmaker, Lange, Jochemsen, and Schepers, 1974; Bekal et al., 1998.
	Festiguay	*Cre8 (formerly CreF)*, on chromosome 7L. Recent analysis suggests 6B.	Paull, Chalmers, and Karakousis, 1998; Williams et al., (unpub).
	AUS4930=Iraq 48	Possible identical genetic location as *Cre1*. Resistance to Pt.	Bekal et al., 1998; Nicol, Davies, and Eastwood, 1998, 2001; Green (pers. comm); Lagudah (pers. comm).
T. durum	Psathias 7654, 7655, Sansome, Khapli		Rivoal et al., 1986.
Triticosecale	T701-4-6	*CreR,* on chromosome 6RL.	Dundas, Frappell, Crack, and Fisher, 2001; Asiedu, Fisher, and Driscol, 1990.
Secale cereale	R173 Family	*CreR,* on chromosome 6RL	Taylor, Shepherd, and Langridge, 1998.

(continued)

Table 1. (continued)

Species	Cultivar or line	Genetic information	References
Ae. tauschii	CPI 110813	*Cre4,* deduced to be on chromosome 2D.	Eastwood et al., 1991; Rivoal et al., 2001.
Ae. tauschii	AUS18913	*Cre3,* on chromosome 2DL	Eastwood et al., 1991; Rivoal et al., 2001.
Ae. peregrina (Ae. variabilis)	1	*Cre(3S) with (Rkn2)* on chromosome 3S; *CreX* not yet located, *CreY*	Barloy, Martin, Rivoal, and Jahier, 1996; Jahier et al., 1998; Rivoal et al., 2001; Barloy et al., 2007; Lagudah (pers. comm).
Ae. longissima	18		Bekal et al., 1998.
Ae. geniculata	79 MZ1, MZ61, MZ77, MZ124		Bekal et al., 1998; Zaharieva et al., 2001.
Ae. triuncialis	TR-353	*Cre7 (*formerly *CreAet).*	Romero et al., 1998.
Ae. ventricosa	VPM 1	*Cre5* (formerly *CreX*), on chromosome 2AS.	Jahier et al., 2001; Ogbonnaya et al., 2001b.
	11, AP-1, H-93-8	*Cre2* (formerly *CreX*) on genome N^v.	Delibes et al., 1993; Andrés, Romero, Montes, and Delibes, 2001; Rivoal et al., 2001.
	11, AP-1, H-93-8, H-93-35	*Cre6,* on chromosome $5N^v$.	Ogbonnaya et al., 2001b; Rivoal et al., 2001.
		Root Lesion Nematode	
T. aestivum	GS50a	Resistance to Pt .	Thompson and Clewett, 1986.
	AUS4930=Iraq 48	Resistance to Pt but also portrays resistance to CCN.	Nicol et al. 1998.
	Excalibur	Resistance to Pn (*Rlnn1*), on chromosome 7AL.	Williams et al., 2002.
	Croc_1/*Ae. tausch.* (224)//Opata	Resistance to Pt. Unknown where resistance is derived from.	Nicol et al., 2001.
Ae tauschii	CPI 110872	Resistance to Pt and Pn.	Thompson (pers. comm).
Ae. geniculata	MZ10, MZ61, MZ96, MZ144	Moderate resistance to Pt. Several also portray resistance to CCN	Zaharieva et al., 2001.

Pt: *Pratylenchus thornei*, Pn: *Pratylenchus neglectus*; [a]: characterized single gene; for marker implemented in commercial breeding program refer to Ogbonnaya et al., 2001b; *Aegilops* classification used according to Van Slageren (1994). Information for other cereal species can be found in Nicol (2002).

4.1.5. True IPM Investigations

As previously applied with the SIRONEM bioassay in Australia (Brown, 1987), investigations to validate the resulting damage model and the correlation between the forecast damage and field rating of CCN were relatively frequent, both in northern Europe (Rivoal & Besse, 1982) and more recently in West Asia (Bonfil, Dolgin, Mufradi, & Asido, 2004). Long term experiments were initiated on the effects of resistance on the targeted nematode densities, the community of other nematodes and biological antagonists, recolonization by susceptible varieties and based on a population genetics approach for the first time using CCN (Lasserre et al., 1994; Rivoal, Lasserre, & Cook, 1995; Lasserre et al., 1996). However, the true integration of different controlling measures as nematicide, farm-yard manure, biological antagonist and resistant cultivar are rare and began in Asia (Pankaj Mishra & Sharma, 2002).

In India studies on integrated management of CCN on wheat and barley have been undertaken by integrating several methods, for example summer ploughing + irrigation, summer ploughing + nitrogenous fertilizers + seed treatment or soil application of nematicides (Handa et al., 1975a; Handa, Mathur, & Mathur, 1975b; Mangat, Gupta, & Ram, 1988). Integration of some of these methods has given encouraging results for increasing the crop yield and reducing nematode population.

4.2. Root Lesion Nematode

4.2.1. Chemical

The present day cost and environmental concerns associated with these chemicals do not make them a viable economic alternative for almost all farmers. However, their use in scientific experiments to understand the importance of these nematodes will remain vital. For this reason we will not provide them as an option in this section and again refer to previous reviews mentioned.

4.2.2. Cultural Practices

Cultural methods offer some control options, but are of limited effectiveness. To be of major significance these need to be integrated with other control measures.

4.2.2.1. Crop Rotation and Cultivation

The use of crop rotation is a limited option for root lesion nematodes, due to their polyphagous nature. Little is understood about the potential role of crop rotation in controlling these nematodes, although some field and laboratory work has been undertaken to better understand the hosting ability of both *P. thornei* (Van Gundy et al., 1974; O'Brien, 1983; Clewett, Thompson, & Fiske, 1993; Hollaway, Taylor, Eastwood, & Hunt, 2000) and *P. neglectus* (Vanstone, Nicol, & Taylor, 1993; Lasserre et al., 1994; Taylor et al., 1999, 2000) to utilise cereals and leguminous crops as hosts. Results from these studies indicate hosting ability is both species and cultivar specific, both with legumes and cereals. Therefore it is essential that hosting-ability studies are conducted with local/regional cultivars. It is possible, depending on crop rotation patterns and the population dynamics of nematodes, that resistant cultivars of cereals alone may not be sufficient to maintain nematode populations below economic levels of damage.

In Australia, cultivation reduced populations of *P. thornei* (Thompson, Mackenzie, & McCulloch, 1983) and in Israel Orion et al. (1984) found that biannual fallowing reduced *P. mediterraneus* populations by 90% and increased grain yields by 40–90%. Nombela et al. (1998) also found fallowing to be effective. An eleven-year management trial conducted in Queensland revealed that the topsoil of zero tillage fallow systems had higher *P. thornei* populations than mechanically cultivated treatments (Thompson et al., 1983).

Monitoring a 30 year rotation trial over several seasons under rainfed wheat cropping systems in Turkey with natural *P. thornei* populations clearly demonstrated the use of legumes (vetch, lentil) should be avoided due to increased populations, whilst sunflower or safflower and fallowing provided the best reduction of *P. thornei* in the wheat rotation system (Elekçioğlu et al., 2004). As with cereal cyst nematode, some triticale varieties such as Abacus and Muir in Australia are known to host fewer nematodes than with bread or durum wheats and hence may offer some useful rotational options (Farsi, Vanstone, Fisher, & Rathjen, 1995).

4.2.2.2. Time of Sowing

Van Gundy et al. (1974) found that delaying sowing of irrigated wheat by one month in Mexico gave maximum yields.

4.2.2.3. Other Cultural Practices

Di Vito, Greco, and Saxena (1991) found that mulching fields with polyethylene film for 6–8 weeks suppressed *P. thornei* populations by 50%.

4.2.3. Resistance (and Tolerance)

Unlike cereal cyst nematode, no commercially available sources of cereal resistance are available to *P. thornei*, although sources of tolerance have been used by cereal farmers in northern Australia for several years (Thompson, Brennan, Clewett, & Sheedy, 1997). As illustrated in Table 1, Thompson and Clewett (1986), Nicol, 1996; Nicol et al. (1999), and Nicol (2002) identified wheat lines that have proven field resistance and work is continuing to breed this resistance into suitable backgrounds. Recent work by Thompson and Haak (1997) identified twenty-nine accessions from the D-genome donor to wheat, *Aegilops tauschii*, suggesting there is future potential for gene introgression. Some of this material also contained the *Cre 3* and other different, unidentified sources of cereal cyst nematode resistance gene conferring resistance to some cereal cyst nematode pathotypes.

As with the cereal cyst nematode, molecular biology is being used to determine the genetic control, location and the subsequent identification of markers for resistance to both *P. thornei* and *P. neglectus*. Table 1 indicates that the significant gains in knowledge have been made with several sources of resistance in bread wheat against RLN. However, unlike CCN the genetic of resistance is quantitative so that development of QTL markers is required to use these in a marker assisted selection approach, which however will only explain part of the variation of resistance.

As with CCN, marker assisted selection is being used routinely with PCR based markers for *P. neglectus* (*rln1*), both in Australia and with CIMMYT International. Commercial cultivars with resistance and tolerance to RLN are now commercially available in Australia and soon within the international breeding programs at CIMMYT.

4.2.4. Biological Control

Successful biological control of *Pratylenchus* species is likely to be difficult due to their migratory behaviour. *Pratylenchus* species spend much of their lives in roots and tend to be found only in soil when their host plants are stressed, senescing or diseased, or when their hosts have been ploughed out after harvest (Stirling, 1991).

Currently, several commercial biological control products are available for the control of nematodes but their use for controlling lesion nematode on cereals is not reported in literature. However, as mentioned previously their application and use is more common on higher value, more intensive agricultural crops such as tomato. Trudgill et al. (1992) reinforces that the greatest value of biocontrol agents will be in combination with other control options.

4.2.5. True IPM Investigations

Unfortunately with RLN very few studies have looked at combining options for control, however it is common practice now in Australia to use resistant and/or tolerant cultivars in combination with rotation crops which are poor or non-hosts of RLN.

4.3. Other Cereal Nematodes – Root Knot, Stem and Seed Gall

Within the individual sections the known control methods for each nematode will be reported.

4.3.1. Chemical

The present day cost and environmental concerns associated with these chemicals do not make them a viable economic alternative for almost all farmers. However, their use in scientific experiments to understand the importance of these nematodes will remain vital. For this reason we will not provide them as an option in this section and again refer to previous reviews mentioned at the start of this section which mention this work.

Sulphur dioxide (SO_2) has been demonstrated to provoke an antagonistic action on the *A. tritici* which resulted in a significant decreasing of number of cockles on different wheat cultivars (Kausar, Khan, & Raghav, 2005).

4.3.2. Cultural Practices

4.3.2.1. Grass Free Rotations and Fallowing with Cultivation

Use of poor or non-host crops is an option to control *M. naasi* (Cook, York, & Guile, 1986). Also the use of fallow during the hatching period (Allen et al., 1970; Gooris & D'Herde, 1972) has been found effective. Rotations also offer some options for *M. artiellia*. Di Vito et al. (1985) were able to demonstrate that, although most legumes and Graminanceae are hosts, cowpea, lupin, sainfoin and maize could be considered non-hosts.

For seed gall nematode oat, maize and sorghum are considered to be non-hosts (Limber, 1976; Paruthi & Gupta, 1987) and while they may offer some option for reducing populations by rotation, the diseases is not completely controlled.

Due to the polyphagous nature of stem nematode and the fact that *D. dipsaci* being a pest on lucerne (alfalfa), red and white clover, pea, bean and bulbous species of the Liliaceae, including garlic, onion, tulip and narcissus, the use of crop rotation in some cropping systems is limited. However, within lucerne, red and white clover, oat, garlic, strawberry and sweet potato resistant cultivars have been developed, as reviewed by Plowright et al. (2002). Rotational combinations of non-hosts including barley and wheat offer some control method for the rye and oat races of *D. dipsaci*. However, once susceptible oat crops have been damaged, rotations are largely ineffective (Rivoal & Cook, 1993).

4.3.2.2. Seed Hygiene

Since ear-cockles (seed galls) are the only source for perpetuation of seed gall therefore their removal from contaminated seed lots can completely eliminate this problem. *A. tritici* can most easily be controlled by seed hygiene. Clean, uninfected seed can be obtained either through use of certified seed or by cleaning infected seed by using modern seed cleaning techniques or by sieving and freshwater flotation (Singh & Agrawal, 1987). Although it has been eradicated from the Western Hemisphere through adoption of this approach, it remains a problem on the Indian sub-continent, in Western Asia and to some extent China (Swarup & Sosa Moss, 1990). Galls are lighter in weight than wheat seed and can be easily discarded through a winnowing process or by flotation of contaminated seeds in 20% brine solution. It is important, however, to wash wheat seed after brine treatment two or three times in water to remove adhering salt particles, otherwise seed germination is impaired.

To dispense with salt treatment, Byars (1920) suggested presoaking contaminated seeds in water, then soaking them at either 50°C for 30 min, 52°C for 20 min, 54°C for 10 min or at 56°C for 5 min. The principle is to reactivate quiescent juveniles before killing them with hot water. Leukel (1957) suggested presoaking galls for 4–6 hours in water and then expose them to hot water at 54°C for 10 min.

4.3.3. Resistance and Tolerance

Work with the most economically important RKN species *M. naasi* has reported partial resistance found in barley and also in *Triticum squarrosa* and *T. monococcum*, while full resistance was identified with *Hordeum chilense*, *H. jabatum*, *T. umbellulatum* and *T. variabile* (bread wheat) (Cook & York, 1982b; Roberts et al., 1982; Person-Dedryver & Jahier, 1985). Resistance has also been expressed in *H. chilense* (Person-Dedryver, Jahier, & Miller, 1990; Yu, Person-Dedryver, & Jahier, 1990).

For countries where hygiene practices are difficult to implement to seed gall nematode, host resistance and rotation offer some hope. The earliest record of a resistance source is the cultivar Kanred (Leukel, 1924) used in a breeding programme initiated by Shen, Tai, and Chia (1934). Crosses between Kanred and a highly susceptible wheat cultivar resulted in a few lines in the F_2 and F_3 free from nematode attack. Unfortunately, this work was not continued. However, since then, resistance to *A. tritici* has been identified in Iraq in both wheat and barley (Saleh & Fattah, 1990) and Pakistan (Shahina, Abid, & Maqbool, 1989) and was sought in India (Swarup & Sosa Moss, 1990). In Iraq, laboratory screening has identified sources of resistance in both wheat and barley (Stephan, 1988). In Iran the reaction of some bread and durum wheat and barley cultivars were evaluated to wheat gall nematode (*Anguina tritici*): among bread wheat cultivars Atrak was more resistant than Darab2 cultivar and among durum wheat cultivars Showa was infected less than Yavarus cultivar.

Occurrence of different biological races or strains of *D. dipsaci* makes it a difficult nematode to control and as a result the only economic and highly effective method is use of host resistance as reviewed by Rivoal et al. (1986). In Britain the most successful oat crop has resistance derived by the landrace cv. Grey Winter, which is controlled by a single dominant gene that is now bred into several commercial cultivars (Plowright et al., 2002). In other oat, resistance may be derived from Uruguayan land races. The wild oat, *Avena ludoviciana* has more than one gene for resistance (Plowright et al., 2002), whilst a number of other oat cultivars have been reported resistant (Whitehead, 1998) but many of these offer only partial resistance or tolerance.

4.3.4. Biological Control

Meloidogyne species have been demonstrated to be controlled by the bacterium *Pasteuria penetrans* although difficulties are perceived with its mass-production and specificity of populations (Gowen & Pembroke, 2004). Success of such control could be connected to the intraspecific variability in attachment of *P. penetrans* to juveniles as for *M. chitwoodi* (Wishart, Blok, Phillips, & Davies, 2004). Disturbancy of biological control could be caused by a distinct microbial community in the egg mass that may have a function in protecting the eggs from antagonists as *P. chlamydosporia* (Kok & Papert, 2001). To date however, there are no published report of its effectiveness on control RKN on wheat.

For both Seed Gall and Stem Nematode there are no published reports of successful control with biologicals on wheat.

4.3.5. True IPM Investigations

It would appear as these three nematodes are considered less important on wheat, that the overall global research in their control is much more limited, hence studies that apply the integration of different methods are not reported.

5. CONCLUSIONS

This chapter has clearly identified CCN and RLN are major biotic constraints to wheat production systems worldwide, especially where the plants suffer other biotic and abiotic stress, particularly drought. In particular the widespread global importance of complex of CCN species and pathotypes is of major economic constraint to wheat production, particularly throughout the region of West Asia, North Africa, China and India. The other three nematodes RK, SG and SN have local reports of economic importance.

Control of any of these cereal nematodes requires a very clear understanding of the biology and population dynamics of each nematode, and in the case of CCN and SN the pathotype and even ecotype. Without this very basic information it is hard to fully understand the value of components of control.

It is clear with CCN and RLN most of the global efforts of research have focussed on the use of non-hosts, and the identification of resistance within bread wheat and associated relatives of wheat. This is the most logical method as it is cost effective, environmentally sound, and particularly in developing countries does not require additional facilities. International breeding programs such as CIMMYT (International Wheat and Maize Improvement Center cimmyt.cgiar.org) and ICARDA (International Centre for Agricultural Research in Dryland Areas icarda.cgiar.org) have a key and integral role to play in providing the appropriate germplasm to National Program partners in developing countries, in addition to technical backup. In several countries with long standing research this is being combined with the use of crop rotation. Other integrated methods at this time do now seem be used, however it is clear that molecular tools both for nematode diagnostics and the identification of resistance are playing a catalytic role in fast tracking efforts in this area. Futuristically the use of plant transformation with genes of interest with resistance to target nematodes may offer tremendous potential where they are accepted. With respect to RK and SN most efforts have similarly focussed on host resistance and rotation. However, for SG seed hygiene is the major method of control, which can easily been implemented by farmers.

ACKNOWLEDGEMENTS

The authors sincerely thank colleagues in West Asia, North Africa, India and China for providing up to date local knowledge on these problems. Special acknowledgement is given to Prof Dr Deliang Peng (Head of nematology laboratory, CASS, Beijing China), Dr Zahra Tahna Maafi (Senior Nematologist, Plant Pest Disease Research Insitute, Iran), Dr S. P. Bishnoi (Nematologist, ARS Durgapura, Jaipur India), Dr A. K. Singh (Nematologist, DWR, Karnal India), Dr Anjum Munir (Senior Nematologist, National Agricultural Research Centre Islamabad, Pakistan), Prof Abdullah A. Aldoss (Head Cereal Plant Breeding Program, Saudi Arabia) and Prof. Ahmad Al-Hazmi (Nematologist, Saudi Arabia) and Dr Namouchi-Kachouri (INRAT, Tunis, Tunisia).

REFERENCES

Abdel Hamid, M. E., Ramadan, H. H., Salem, F. M., & Osman, G. Y. (1981). The susceptibility of different field crops to infestation by *Meloidogyne javanica* and *M. incognita acrita*. *Anzeiger für schädlingskunde, Pflanzenschutz, Umweltschutz, 54*, 81–82.

Abidou, H., El-Ahmed, A., Nicol, J. M., Bolat, N., Rivoal, R., & Yahyaoui, A. (2005). The occurrence and distribution of species of the *Heterodera avenae* group in Syria and Turkey. *Nematologia Mediterranea, 33*,197–203.

Agrios, G. N. (1988). *Plant Pathology* (3rd ed.). London, UK: Academic Press.

Ahmadi, A. R., & Akhiyani, A. (2001). Status of wheat seed gall nematode (*Anguina tritici*) in irrigated wheat fields of Esfahan province. *Applied Entomology and Phytopathology,* 69, 1–13.

Al-Hazmi, A. S., Al-Yahya, F. A., & Abdul-Razig, A. T. (1999). Damage and reproduction potentials of *Heterodera avenae* under outdoor conditions. *Journal of Nematology, 31*(Suppl.), 662–666.

Al-Hazmi, A. S., Cook, R., & Ibrahim, A. A. M. (2001). Pathotype characterisation of the cereal cyst nematode, *Heterodera avenae*, in Saudi Arabia. *Nematology, 3*, 379–382.

Allen, M. W., Hart, W. H., & Baghott, K.V. (1970). Crop rotation controls the barley root-knot nematode at Tulelake. *California Agriculture, 24*, 4–5.

Al-Talib, N. Y., Al-Taae, A. K. M., Neiner, S. M., Stephan, Z. A., & Al-Baldawi, A. S. (1986). New record of *Anguina tritici* on barley from Iraq. *International Nematology Network Newsletter, 3*, 25–27.

Al-Yahya, F. A, Alderfasi, A. A., Al-Hazmi, A. S., Ibrahim, A. A. M., & Abdul-Razig, A. T. (1998). Effects of cereal cyst nematode on growth and physiological aspects of wheat under field conditions. *Pakistan Journal of Nematology, 16*, 55–62.

Amir, J., & Sinclair, T. R. (1996). Cereal cyst nematode effects on wheat water use and on root and shoot growth. *Field Crop Research, 47*, 13–19.

Andersen, S., & Andersen, K. (1982). Suggestions for determination and terminologyof pathotypes and genes for resistance in cyst-forming nematodes, especially *Heterodera avenae*. *EPPO/OEPP Bulletin, 12*, 379–386.

Andersson, S. (1974). *Heterodera hordecalis* n.sp. (Nematoda: Heteroderidae) a cyst nematode of cereals and grasses in southern Sweden. *Nematologica, 20*, 445–454.

Andrés, M. F., Romero, M. D., Montes, M. J., & Delibes, A. (2001). Genetic relationships and isozyme variability in the *Heterodera avenae* complex determined by isoelectric focusing. *Plant Pathology, 50*, 270–279.

Antoniou, M. (1989). Arrested development in plant parasitic nematodes. *Helminthological Abstracts Series B, 58*, 1–9.

Asiedu, R., Fisher, J. M., & Driscol, C. J. (1990). Resistance to *Heterodera avenae* in the rye genome of triticale. *Theoretical and Applied Genetics, 79*, 331–336.

Bajaj, H. K., & Kanwar, R. S. (2005). Parasitization of maize by *Heterodera avenae* and *H. filipjevi*. *Nematologia Mediterranea, 33*, 203–207.

Balakhnina, V. P. (1989). Resistance of varieties of *Triticum durum* desf. and *Triticum aestivum* L. to the oat cyst nematode. In *Gel'mintologiya segodnya: Problemy i perspektivy*. Tezisy dokladov nauchnoi konferentsii, Moskva, 4–6 Aprelya 1989. Tom 2. Moscow, USSR, pp. 36–37.

Bar-Eyal, M., Sharon, E., & Spiegel, Y. (2006). Nematicidal activity of *Chrysanthemum coronarium*. *European Journal of Plant Pathology, 114*, 427–433.

Barker, K. R., & Noe, J. P. (1987). Establishing and using threshold populations levels. In J. A. Veech & S. W. Dickson (Eds.), *Vistas on Nematology* (pp. 75–81). Maryland, USA: Society of Nematologists.

Barloy, D., Lemoine, J., Abelard, P., Tanguy, A.M., Rivoal, R., & Jahier, J. (2007). Marker-assisted pyramiding of two cereal cyst nematode resistance genes from *Aegilops variabilis* in wheat. *Molecular Breeding, 20*, 31–40.

Barloy, D., Martin, J., Rivoal, R., & Jahier, J. (1996, July 7–12). Genetic and molecular characterization of lines of wheat resistant to the cereal cyst nematode *Heterodera avenae*. In *Proceeding 3rd International Nematology Congress*, Gosier, Guadeloupe, 164.

Barr, A. R., Chalmers, K. J., Karakousis, A., Manning, S., Lance, R. C. M., & Lewis, J. (1998). RFLP mapping of a new cereal cyst nematode resistance locus in barley. *Plant Breeding, 117*, 185–187.

Behdad, E. (1982). Disease of field crops in Iran (424pp.). Isfahan, Iran: Neshat publishing Co. (in Farsi).

Bekal, S., Gauthier, J. P., & Rivoal, R. (1997). Genetic diversity among a complex of cereal cyst nematodes inferred from RFLP analysis of the ribosomal internal transcribed spacer region. *Genome, 40*, 479–486.

Bekal, S., Jahier, J., & Rivoal, R. (1998). Host response of different triticeae to species of the cereal cyst nematode complex in relation to breeding resistant durum wheat. *Fundamental and Applied Nematology, 21*, 359–370.

Belloni, V. (1954). Comparsa in Italia di una anguillulosi del frumento e prove de lotta. *Notiziario sulle Malattie delle Piante, 27*, 3–6.

Bhatti, D. S., Dahiya, R. S., Gupta, D. C., & Malhan, I. (1980). Plant parasitic nematodes associated with various crops in Haryana. *Haryana Agricultural University Journal of Research, 10*, 413–414.

Bishnoi, S. P., & Bajaj, H. K. (2004). On the species and pathotypes of *Heterodera avenae* complex of India. *Indian Journal of Nematology, 34*, 147–152.

Boer, R. F. de, Kollmorgen, J. F., Macauley, B. J., & Franz, P. R. (1991). Effects of cultivation on *Rhizoctonia* root rot, cereal cyst nematode, common root rot and yield of wheat in the Victorian Mallee. *Australian Journal of Experimental Agriculture, 31*, 367–372.

Bonfil, D. J., Dolgin, B., Mufradi, I., & Asido, S. (2004). Bioassay to forecast cereal cyst nematode damage to wheat in fields. *Precision Agriculture, 5*, 329–344.

Brig, B. & Kanwar, R. S. (2003). Effect of sowing time on multiplication of *Heterodera avenae* and performance of wheat crop. *Indian Journal of Nematology, 33*, 172–174.

Brown, R. H. (1981). Nematode diseases. In *Economic Importance and Biology of Cereal Root Diseases in Australia.* Australia: Report to Plant Pathology Subcommittee of Standing Committee on Agriculture.

Brown, R. H. (1987). Control strategies of low value crops. In Brown, R. H. and Kerry, B. R. (Eds.), *Principles and Practice of Nematode Control in Crops* (pp. 351–387). Sydney, Australia: Hartcourt Brace Jovanovic.

Byars, L. P. (1920). *The Nematode Disease of Wheat Caused by Tylenchus tritici*. United States Department of Agriculture Bulletin 842 Washington, D.C.

Caubel, G., Person, F., & Rivoal, R. (1980). Les nématodes dans les rotations céréalières. *Perspectives Agricoles, 36*, 34–48.

Caubel, G., Ritter, M., & Rivoal, R. (1972). Observations relatives à des attaques du nématode *Meloidogyne naasi* Franklin sur céréales et graminées fourragères dans l'Ouest de la France en 1970. *Comptes Rendus des Séances de l'Académie d'Agriculture de France, 58*, 351–356.

Chhabra, H. K. (1973). Distribution of *Heterodera avenae* Wollenweber, cereal cyst nematode in the Punjab. *Current Science, 42*, 441.

Chu, V. M. (1945). The prevalence of the wheat nematode in China and its control. *Phytopathology, 35*, 288–295.

Clewett, T. G., Thompson, J. P., & Fiske, M. L. (1993). Crop rotation to control *Pratylenchus thornei*. In V. A. Vanstone, S. P. Taylor, & J. M. Nicol (Eds.), In *Proceedings of the Pratylenchus Workshop*. Hobart, Tasmania, Australia: Ninth Biennial Australasian Plant Pathology Conference.

Cook, R., & Evans, K. (1987). Resistance and tolerance. In R. H. Brown & B. R. Kerry (Eds.), *Principles and Practice of Nematode Control in Crops* (pp. 179–231). Orlando, FL: Academic Press.

Cook, R., & Noel, G. R. (2002). Cyst nematodes: *Globodera* and species. In J. L. Starr, R. Cook, & J. Bridge (Eds.), *Plant Resistance to Parasitic Nematodes* (pp. 71–106). Wallingford, UK: CABI Publishing.

Cook, R., & Rivoal, R. (1998). Genetics of resistance and parasitism. In S. B. Sharma (Ed.), *The Cyst Nematodes* (pp. 322–352). London, UK: Chapman & Hall.

Cook, R., & York, P. A. (1982a). Resistance of cereals to *Heterodera avenae*: Methods of investigation, sources and inheritance of resistance. *EPPO Bulletin, 12*, 423–434.

Cook, R., & York, P. A. (1982b). Genetics of resistance to *Heterodera avenae* and *Meloidogyne naasi*. In *Proceedings of the 4th International Barley Genetics Symposium*, Edinburgh, Scotland.

Cook, R., York, P. A., & Guile, C.T. (1986). Effects and control of cereal root-knot nematode in barley/grass rotations. In *Proceedings of the 1986 British Crop Protection Conference – Pests and Diseases,* 433–440 Brighton, England.

Corbett, D. C. M. (1974). *Pratylenchus pinguicaudatus* n. sp. (Pratylenchidae: Nematoda) with a key to the genus *Pratylenchus*. *Nematologica, 15*, 550–556.

Cuevas, O. Y. J., & Sosa Moss, C. (1990). Host plants of *Meloidogyne chitwoodi* in the states of Tlaxcala and Puebla, Mexico. *Current Nematology, 1*, 69–70.

Davachi, A. (1949). Important pests of agricultural crops and their control methods. Bureau of Agric Chem. Ministry of Agric. 205 pp. (in Farsi).

Delibes, A., Romero, D., Aguaded, S., Duce, A., Mena, M., Lopez-Brana, I., et al. (1993). Resistance to the cereal cyst nematode (*Heterodera avenae* Woll.) transferred from the wild grass *Aegilops ventricosa* to hexpaloid wheat by a stepping-stone procedure. *Theoretical and Applied Genetics, 87*, 402–408.

De Waele, D., & Elsen, A (2002). Migratory endoparasites: *Pratlylenchus* and *Radopholus* species. In J. L. Starr, R. Cook & J. Bridge (Eds.), *Plant Resistance to Parasitic Nematodes* (pp. 175–206). Wallingford, UK: CABI Publishing.

De Waele, D., & Mc Donald, A. H. (2000). Diseases caused by nematodes. In R. A. Frederiksen & G. N. Odvody (Eds.), *Compendium of Sorghum Diseases* (2nd ed., pp. 50–53). Minnesota, USA: American Phytopathological Society.

Dhaliwal, H. W., Singh, H., Gill, K. S., & Randhawa, H. S. (1993). Evaluation and cataloguing of wheat germplasm for disease resistance and quality. In A. B. Damania (Ed.), *Biodiversity and Wheat Improvement* (pp. 123–140). Chichester, UK: John Wiley and Sons.

Dhawan, S. C., & Nagesh, M. (1987). On the relationship between population densities of *Heterodera avenae*, growth of wheat and nematode multiplication. *Indian Journal of Nematology, 17*, 231–236.

Di Vito, M., & Greco, N. (1988). Investigation on the biology of *Meloidogyne artiella*. *Revue de Nématologie, 11*, 221–225.

Di Vito, M., Greco, N., & Saxena, M.C. (1991). Effectiveness of soil solarization for control of *Heterodera ciceri* and *Pratylenchus thornei* on chickpeas in Syria. *Nematologia Mediterranea, 19*, 109–111.

Di Vito, M., Greco, N., & Zaccheo, G. (1985). On the host range of *Meloidogyne artiellia*. *Nematologia Mediterranea, 13*, 207–212.

Di Vito, M., & Zaccheo, G. (1987). Responses of cultivars of wheat to *Meloidogyne artiellia*. *Nematologia Mediterranea, 15*, 405–408.

Dixon, G. M. (1969). The effect of cereal cyst eelworm on spring sown cereals. *Plant Pathology, 18*, 109–112.

Dosba, F., & Rivoal, R. (1982). Estimation des niveaux de resistance au development d'*Heterodera avenae* chez les triticinées. *EPPO Bulletin, 12*, 451–456.

Dowe, A., & Decker, H. (1977). Importance of plant parasitic nematodes in intensive cereal growing and ways of preventing losses due to nematodes. *Nachrichtenblatt fur den Pflanzenschutz in der DDR, 31*, 159–162.

Doyle, A. D., McLeod, R. W., Wong, P. T. W., Hetherington, S. E., & Southwell, R. J. (1987). Evidence for the involvement of the root lesion nematode *Pratylenchus thornei* in wheat yield decline in northern New South Wales. *Ausralian Journal Experimental Agriculture, 27*, 563–570.

Duggan, J. J. (1961). The effect of cereal root eelworm on its hosts. *Irish Journal of Agriculture Research, 1*, 7–16.

Dundas, I. S., Frappell, D. E., Crack, D. M., & Fisher, J. M. (2001). Deletion mapping of a nematode resistance gene on rye chromosome 6R in Wheat. *Crop Science, 41*, 1771–1778.

Eagles, H. A., Bariana, H. S., Ogbonnaya, F. C., Rebetzke, G. J., Hollamby, G. L., Henry, R. J. et al., (2001). Implementation of markers in Australian wheat breeding. *Australian Journal of Agricultural Research, 52*, 1349–1356.

Eastwood, R. F., Lagudah, E. S., Appels, R., Hannah, M., & Kollmorgen, J. F. (1991). *Triticum tauschii*: a novel source of resistance to cereal cyst nematode (*Heterodera avenae*). *Australian Journal of Agricultural Research, 42*, 69–77.

Eisenback, J. D., & Triantaphyllou, H. H. (1991). Root-knot nematode: *Meloidogyne* species and races. In W. R. Nickle (Ed.), *Manual of agricultural nematology* (pp. 191–274). New York, USA: Marcel Dekker Inc.

Elekcioglu, I. H, & Gozel, U. (1997). Effect of mixed population of *Paratrophurus acristylus*, *Pratylenchus thornei* and *Pratylenchus* sp. (Nematoda: Tylenchida) on yield parameters of wheat in Turkey. *Journal of Nematology, 7*, 217–220.

Elekçioğlu, İ. H., Avcı, M., Nicol, J., Meyveci, K., Bolat, N., Yorgancılar, A., et al. (2004, September 8–10). The use of crop rotation as a means to control the cyst and lesion nematode under rainfed wheat production systems. 1st National Pathology Congress, Samsun, Turkey.

Elmali, M. (2002). The distribution and damage of wheat gall nematode *Anguina tritici* (Steinbuch) (Tylenchida: Tylenchidae) in western part of Anatolia. *Turkyie Entomoloji Dergisi, 26*, 105–114.

Eriksson, K. B. (1972). Nematode diseases of pasture legumes and turf grasses. In J. M. Webster (Ed.), *Economic Nematology* (pp. 66–96). Academic Press, New York, USA,

Farsi, M, Vanstone, V. A., Fisher, J. M., & Rathjen, A. J. (1995). Genetic variation in resistance to *Pratylenchus neglectus* in wheat and triticales. *Australian Journal of Experimental Agriculture, 35*, 597–602.

Ferris, V. R., Subbotin, S. A., Ireholm, A., Spiegel, Y., Faghihi, J., & Ferris, J. M. (1999). Ribosomal DNA sequence analysis of *Heterodera filipjevi* and *H. latipons* isolates from Russia and comparisons with other nematode isolates. *Russian Journal of Nematology, 7*, 121–125.

Fisher, J. M., & Hancock, T. W. (1991). Population dynamics of *Heterodera avenae* Woll in South Australia. *Australian Journal of Agricultural Research, 42*, 53–68.

Germershauzen, K., Kastner, A., & Schmidt, W. (1976). Observations on the threshold of damage by *Heterodera avenae* on cereals under different rates of nitrogen fertilization. *Vortragstagung zu Aktuellen Problemen der Phytonematologie, 11*, 30.

Gill, J. S., & Swarup, G. (1971). On the host range of cereal cyst nematode, *Heterodera avenae*, the casual organism of 'molya' disease of wheat and barley in Rajasthan, India. *Indian Journal of Nematology, 1*, 63–67.

Gooris, J. (1968). Host plants and non-host plants of the Gramineae root-knot nematode *Meloidogyne naasi* Franklin. *Mededelingen van de Rijksfaculteit Landbouwwetenschappen te Gent, 33*, 85–100.

Gooris, J., & D'Herde, C. J. (1972). Mode d'hivernage de *Meloidogyne naasi* Franklin dans le sol et lutte par rotation culturale. *Revue de l'Agriculture* (Bruxelles), 25, 659–664.

Gooris, J., & D'Herde, C. J. (1977). Study on the biology of *Meloidogyne naasi* Franklin, 1965. Ministry of Agriculture, Agriculture Research Centre, Ghent, Belgium, 165p.

Gowen, S. R., & Pembroke, B. (2004). *Pasteuria penetrans* and the integrated control of root-knot nematodes. *Bulletin OILB/SROP, 27*, 75–77.

Griffin, G. D. (1984). Nematode parasites of alfalfa, cereals, and grasses. In W. E. Nickle (Ed.), *Plant and Insect Nematodes* (pp. 243–321). New York, USA: Marcel Dekker Inc.

Gupta, P., & Swarup, G. (1972). Ear-cockle and yellow rot disease of wheat. II. Nematode-bacterial association. *Nematologica, 18*, 320–324.

Handa, D. K., Mathur, R. L., & Mathur, B. N. (1975a, publ. 1976). The effect of crop rotation and nematicides on the cereal cyst nematode *Heterodera avenae* and yield of wheat and barley. *Indian Journal of Mycology and Plant Pathology, 5*, 20–21.

Handa, D. K., Mathur, R. L., & Mathur, B. N. (1975b, publ. 1976). Studies on the effect of deep summer ploughings on the cereal cyst nematode (*Heterodera avenae)* and yield of wheat and barley [Abstract]. *Indian Journal of Mycology and Plant Pathology, 5*, 18.

Handa, D. K., Mathur, R. L., Mathur, B. N., Sharma, G. L., & Yadav, B. D. (1985a). Host efficiency of tall and dwarf wheat and barley cops to cereal cyst nematode (*Heterodera avenae*). *Indian Journal of Nematology, 15*, 36–42.

Handa, D. K., Mathur, R. L., Mathur B. N., & Yadav, B. D. (1985b). Estimation of losses in barley due to cereal cyst nematode in sandy and sandy loamy soils. *Indian Journal of Nematology, 15*, 163–166.

Handoo, Z. A., & Golden, A. M. (1989). A key and diagnostic compendium to the species of the genus *Pratylenchus* Filpjev, 1936 (Lesion nematodes). *Journal of Nematology, 21*, 202–218.

Heide, A. (1975). Studies on the population dynamics of migratory root nematodes in cereal monocultures as well as in rotational cereal growing. *Archiv fur Phytopathologie und Pflanzenschutz, 11*, 225–232.

Hogger, C. H. (1990). Distribution of plant-parasitic nematodes on winter wheat in Switzerland *Landwirtschaft Schweiz, 9*, 477–483.

Holgado, R., Rowe, J., Andersson, S., & Magnusson, C. (2004). Electrophoresis and biotest studies on some populations of cereal cyst nematode, *Heterodera* spp. (Tylenchida: Heteroderidae) *Nematology, 6*, 857–865.

Hollaway, C. J., Taylor, S. P., Eastwood, R. F., & Hunt, C. H. (2000). Effect of field crops on density of *Pratylenchus* in southeastern Australia, Part 2: *P. thornei*. *Journal of Nematology, 32*, 600–608.

Hutchinson, C. M. (1917). A bacterial disease of wheat in Punjab. Memoirs of Department of Agriculture India, Bacteriological Series 1.

Ibrahim, A. A. M., Al-Hazmi, A. S., Al-Yahya, F. A., & Alderfasi, A. A. (1999). Damage potential and reproduction of *Heterodera avenae* on wheat and barley under Saudi field conditions. *Nematology, 1*, 625–630.

Indra-Rajvanshi (2003). Integrated management of cereal cyst nematode, *Heterodera avenae* in barley. *Current Nematology, 14*, 99–101.

Inserra, R. N., Lamberti, F., Volvas, N., & Dandria, D. (1975). *Meloidogyne naasi* nell'Italia meridionale e a Malta. *Nematologia Mediterranea, 3*, 163–166.

Inserra, R. N., Vovlas, N., O'Bannon, J. H., & Griffin, G. D. (1985). Development of *Meloidogyne chitwoodi* on wheat. *Journal of Nematology, 17*, 322–326.

Ismail, S., Sikora, R. A., & Schuster, R. P. (2001). Occurrence and biodiversity of egg pathogenic fungi of Mediterranean cereal cyst nematode *Heterodera latipons*. *Mededelingen-Faculteit Landbouwkundige en Toegepaste Biologische Wetenschappen, Universiteit Gent, 66*, 645–653.

Jahier, J., Abelard, P., Tanguy, A. M., Dedryver, F., Rivoal, R., Khatar, S., et al. (2001). The *Aegilops ventricosa* segment on chromsome 2AS of the wheat cultivar VPM1 carries the cereal cyst nematode resistance gene Cre5. *Plant Breeding, 120*, 125–128.

Jahier, J., Rivoal, R., Yu, M. Q., Abelard, P., Tanguy, A. M. & Barloy, D. (1998). Transfer of genes for resistance to cereal cyst nematode from *Aegilops variabilis* Eig to wheat. *Journal of Genetics and Breeding, 52*, 253–257.

Janssen, G. J. W. (1994). The relevance of races in *Ditylenchus dipsaci* (Kühn) Filipjev, the stem nematode. *Fundamental and Applied Nematology, 17*, 469–473.

Jepson, S. B. (1987). Identification of root-knot nematodes (*Meloidogyne* species). Wallingford, U.K.: C.A.B. International.

Johnson, A. W., & Motsinger, R. E. (1989). Suitability of small grains as hosts of *Meloidogyne* species. *Journal of Nematology, 21*(Suppl.), 650–653.

Kanwar, R. S., & Walia, R. K. (2004). Applied allelopathy techniques. In S. S. Narwal, R. Singh, & R. K. Walia (Eds.), *Research methods in plant sciences: allelopathy*. (Vol. 2, pp. 188–204). Plant Protection. Jodhpur, India: Scientific Publishers.

Kasimova, G. A, & Atakishieva, Y. Y. (1981). Nematode fauna and nematode dynamics on winter wheat in Apsheron. *Parazitologicheskie Issledovaniya v Azerbaidzhane*, 119–134.

Kausar, S., Khan, A. A., & Raghav, D. (2005). Interaction of sulphur and seed gall nematode, *Anguina tritici* on wheat. *Journal of Food, Agriculture and Environment, 3*, 130–132.

Kerry, B. R. (1981). Fungal parasites, a weapon against cyst nematodes. *Plant Disease, 65*, 390–393.

Khan, A., Williams, K. L., & Nevalainen, H. K. M. (2006). Control pf plant-parasitic nematodes by *Paecilomyces lilacinus* and *Monacrosporium lysipagum* in pot trials. *Biocontrol, 51*, 643–658.

Kilpatrick R. A., Gilchrist, L., & Golden A. M. (1976). Root knot on wheat in Chile. *Plant Disease Reporter, 60*, 135.

Kimpinski, J., Anderson, R. V., Johnston, H. W., & Martin, R. A. (1989). Nematodes and fungal diseases in barley and wheat on Prince Edward Island. *Crop Protection, 8*, 412–416.

Kok, C. J., & Papert, A. (2001). Microbial community of *Meloidogyne* egg masses. In R. J. Sikora (Ed.), *Proceedings of the IOBC/WPRS Study group "Integrated Control of Soil Pests". Tri-trophic interactions in the rhizosphere and root health nematode-fungal-bacterial interrelationships*, Bad Honnef, Germany, 3–5 November, (1999). *Bulletin OILB/SROP, 24*, 91–95.

Kollo, I. A. (2002). Plant-parasitic nematodes of sorghum and pearl millet: Emphasis on Africa. In J. F. Leslie (Ed.), *Sorghum and Millets Diseases* (pp. 259–266). Iowa, USA: Iowa State Press.

Kornobis, S., Wolny, S., & Wilski, A. (1980). The effect of *Heterodera avenae* population density in the soil on the yield and plant weight of oats. *Zeszyty Problemowe Postetow Nauk Rolniczych, 232*, 9–17.

Kort, J. (1972). Nematode diseases of cereals of temperate climates. In J. M. Webster (Ed.), *Economic Nematology* (pp. 97–126). London, UK: Academic Press.

Koshy, P. K., & Swarup, G. (1971). Distribution of *Heterodera avenae, H. zeae, H. cajani* and *Anguina tritici* in India. *Indian Journal of Nematology, 1*, 106–111.

Kretschemer, J. M., Chalmers, K. J., Manning, S., Karakousis, A., Barr, A. R., Islam, A. K. M. R., et al. (1997). RFLP mapping of the Ha2 cereal cyst nematode resistance gene in barley. *Theoretical and Applied Genetics, 94*, 1060–1064.

Kyrou, N. C. (1969). First record of occurrence of *Meloidogyne artiellia* on wheat in Greece. *Nematologica, 15*, 432–433.

Lagudah, E. S., Moullet, O., Ogbonnaya, F., Eastwood, R., Appels, R., Jahier, J., et al. (1998). Cyst nematode resistance genes in wheat. In *Proceedings from the 7th International Congress of Plant Pathology*, Edinburgh, Scotland.

Lasserre, F., Gigault, F., Gauthier, J. P., Henry, J. P. Sandmeier, M., & Rivoal, R. (1996). Genetic variation in natural populations of the cereal cyst nematode (*Heterodera avenae* Woll.) submitted to resistant and susceptible cultivars of cereals. *Theoretical and Applied Genetics, 93*, 1–8.

Lasserre, F., Rivoal, R., & Cook, R. (1994). Interactions between *Heterodera avenae* and *Pratylenchus neglectus* on wheat. *Journal of Nematology, 26*, 336–344.

Leukel, R.W. (1924). Investigations on the nematode disease of cereals caused by *Tylenchus tritici*. *Journal of Agricultural Research, 27*, 928–956.

Leukel, R. W. (1957). Nematode disease of wheat and rye. Farmers' Bulletin, USDA No. 1607.

Limber, D. (1976). Artificial infection of sweet corn seedlings with *Anguina tritici* Steinbuch (1799) Chitwood, 1935. *Proceedings of the Helminthological Society of Washington, 43*, 201–203.

Loof, P. A. A. (1978). The genus *Pratylenchus* Filipjev, 1936 (Nematoda: Pratylenchidae): a review of its anatomy, morphology, distribution, systematics and identification. Swedish University of Agricultural Sciences, Research Information Centre, Uppsala, Sweden, 50 pp.

Madzhidov, A. R. (1981). [*Bidera flipjevi* n.sp. (Heteroderina: Tylenchida) in Tadzhikistan]. *Izvestiya Akademii Nauk Tadzhikskoï SSR, Biologicheskie Nauki, 2*, 40–44.

Mamluk, O. F., Augustin, B., & Bellar, M. (1983). New record of cyst and root-knot nematodes on legume crops in the dry areas of Syria. *Phytopathologia Mediterranea, 22*, 80.

Mandani, M., Vovlas, N., Castillo, P., Subbotin, S. A., & Moens, M. (2004). Molecular characterization of cyst nematode species (*Heterodera* spp.) from the Mediterranean basin using RFLPs and sequences of ITS-rDNA. *Journal of Phytopathology, 152*, 229–234.

Mangat, B. P. S., Gupta, D. C., & Ram, K. (1988). Effect of deep summer ploughing and in combination with aldicarb and time of application of aldicarb on cyst population of *Heterodera avenae* and subsequent effect on wheat yield. *Indian Journal of Nematology, 18*, 345–346.

Maqbool, M. A. (1988). Present status of research on plant parasitic nematodes in cereals and food and forage legumes in Pakistan. In M. C. Saxena, R. A. Sikora, & J. P. Srivastava (Eds.), *Nematodes parasitic to cereals and legumes in temperate remi-arid regions* (pp. 173–180). Aleppo, Syria: ICARDA.

Marathee, J.-P., & Gomez-MacPherson, H. (2001). Future world supply and demand. In A. P. Bonjean & W. J. Angus (Eds.), *The world wheat book: A history of wheat breeding* (pp. 1107–1116). Paris: Lavoisier Publishing.

Mathur, B. N. (1969). Studies on cereal cyst nematode (*Heterodera avenae* Woll.) with special reference to 'molya' diseases of weath and barley in Rajesthan. Ph.D. thesis, University of Rajesthan, 233 pp.

Mathur, B. N., Arya, H. C., Handa, D. K., & Mathur, R. L. (1981). The biology of cereal-cyst nematode, *Heterodera avenae* in India. III. Factors affecting population level and damage to host crops. *Indian Journal of Mycology and Plant Pathology, 11*, 5–13.

Mathur, B. N., Handa, D. K., & Swarup, G. (1986). On the loss estimation and chemical control of 'Molya' disease of wheat caused by *Heterodera avenae* in India. *Indian Journal of Nematology, 16*, 152–159.

Mathur, B. N., Handa, D. K., & Swarup, G. (1987). Effect of deep summer ploughings on the cereal cyst nematode, *Heterodera avenae* and yield of wheat in Rajasthan, India. *Indian Journal of Nematology. 17*, 292–295.

McDonald, A., & Nicol, J. M. (2005). Nematode Parasites of Cereals. In J. Bridge, R. Sikora & M. Luc (Eds.), *Plant parasitic nematodes in tropical and subtropical agriculture* (pp. 131–191). CABI Publishing, Wallingford, Oxon, United Kingdom.

McIntosh, R. A., Devos, K. M., Dubcovsky, D., & Rogers, W. J. (2001). Catalogue of gene symbols for wheat. In A. E. Slinkard (Ed.), *Proceedings of the 9th Wheat Genetics Symposium.* Vol. 5, Saskatoon, Canada: University Extension Press, University of Saskatchewan.

McKenry, M. V., & Ferris, H. (1983). Nematodes. In: Challenging problems in plant health. T. Kommendahl (Ed.), (pp. 267–279). St Paul, USA: American Phytopathological Society Press.

Meagher, J. W. (1972). Cereal cyst nematode (*Heterodera avenae* Woll). Studies on ecology and content in Victoria. Technical Bulletin 24, Department of Agriculture, Victoria, Australia, 50p.

Meagher, J. W., & Brown, R. H. (1974). Microplot experiments on the effect of plant hosts on populations of cereal cyst nematode (*Heterodera aveane*) and its potential as a pathogen of wheat. *Nematologica, 20*, 337–346.

Michel, R. E., Malek, R. B., Taylor, D. P., & Edwards, D. I. (1973). Races of the barley root-knot nematode, *Meloidogyne naasi* I. Characterisation by host preference. *Journal of Nematology, 5*, 41–44.

MingZhu, W., Zhi Feng, L., & YanNong, X. (1996). Host range of wheat cereal cyst nematode in some plants. *Plant Protection, 22*, 3–5.

Mojtahedi, H., & Santo, G. S. (1992). *Pratylenchus neglectus* on dryland wheat in Washington. *Plant Disease, 76*, 323.

Mokabli, A., Valette, S., & Rivoal, R. (2001). Différenciation de quelques espèces de nématodes à kystes des céréales et des graminées par électrophorèse sur gel d'acétate de cellulose. *Nematologia Mediterranea, 29*, 103–108.

Mokabli, A., Valette, S., Gauthier, J. P., & Rivoal, R. (2002). Variation in virulence of cereal cyst nematode populations from North Africa and Asia. *Nematology, 4*, 521–525.

Mor, M., & Cohn, E. (1989). New nematode pathogens in Israel: *Meloidogyne* on wheat and *Hoplolaimus* on cotton. *Phytoparasitica, 17*, 221.

Mor, M., Cohn, E., & Spiegel, Y. (1992). Phenology, pathogenicity and pathotypes of cereal cyst nematodes, *Heterodera avenae* Woll. and *H. latipons* (Nematoda: Heteroderidae) in Israel. *Nematologica, 38*, 444–501.

Mulvey, R. H. (1972). Identification of *Heterodera* cysts by terminal and cone structures. *Canadian Journal of Zoology, 50*, 1277–1292.

Namouchi-Kachouri, N., B'Chir, M. M., & Hajji, A. (2006, 19–23 November). Damage potential and reproduction of *Heterodera avenae* on wheat and barley under Tunisian field conditions. 9th Arab Congress of Plant Protection Damascus- Syria, Abstract E, 114.

Nicol, J. M. (1996). The distribution, pathogenicity and population dynamics of *Pratylencus thornei* (Sher and Allen, 1954) on wheat in South Australia. Ph.D Thesis, The University of Adelaide, Adelaide, Australia, 236 pp.

Nicol, J. M. (2002). Important Nematode Pests of Cereals. In: B. C. Curtis (Ed.), *Wheat Production and Improvement* (pp. 345–366). Rome, Italy: FAO Plant Production and Protection Series, FAO.

Nicol, J. M., Bolat, N., Sahin, E., Tulek, A., Yildirim, A. F., Yorgancilar, A., et al. (2005). The Cereal Cyst Nematode is causing economic damage on rainfed wheat production systems of Turkey. Joint Meeting of the Annual Western Soil Fungus Conference and the American Phytopathology Society Pacific Division, Portland, United States of America, 28 June – 1st July. [Abstract].

Nicol, J. M., Davies, K. A., & Eastwood, R. (1998). AUS4930: A new source of resistance to *Pratylenchus thornei* in wheat. 24th International Symposium of the European Society of Nematologists, St. Andrews, Scotland, UK [Abstract].

Nicol, J. M., Davies, K. A., Hancock, T. W., & Fisher, J. M. (1999). Yield loss caused by *Pratylenchus thornei* on wheat in South Australia. *Journal of Nematology, 31*, 367–376.

Nicol, J. M., Rivoal, R., Trethowan, R. M., Van Ginkel, M., Mergoum, M., & Singh, R. P. (2001). CIMMYT's Approach to identity and use resistance to nematodes and soil-borne fungi, in developing superior wheat germplasm. In Z. Bedö & L. Langö (Eds.), *Wheat in a global environment* (pp. 381–389). The Netherlands: Kluwer Academic Publishers.

Nicol, J., & Ortiz-Monasterio, I. (2004). Effects of the root lesion nematode, *Pratylenchus thornei,* on wheat yields in Mexico. *Journal of Nematology, 6*, 485–493.

Nicol, J., Rivoal, R., Valette, S., Bolat, N., Aktas, H., Braun, H. J., et al. (2002). The frequency and diversity of the cyst and lesion nematode on wheat in the Turkish Central Anatolian Plateau. *Journal of Nematology, 4*, 272. [Abstract].

Nicolas, H., Rivoal, R., Duchesne, J., & Lili, Z. (1991). Detection of *Heterodera avenae* infestations on winter wheat by radiothermometry. *Revue de Nématologie, 14*, 285–290.

Nombela, G., Navas, A., & Bello, A. (1998). Effects of crop rotations of cereal with vetch and fallow on soil nematofauna in Central Spain. *Nematologica, 44*, 63–80.

Noori, P., Talatchian, P., & Teimoori, F. (1980). Survey on sugar beet fields' harmful nematodes in the west of Iran from 1975 to 1978. *Iranian Journal d'Entomologie et Phytopathologie Appliquées, 48*, 39–41.

Nyczepir, A. P., Inserra, R. N., O'Bannon, J. H., & Santo, G. S. (1984). Influence of *Meloidogyne chitwoodi* and *M. hapla* on wheat growth. *Journal of Nematology, 16*, 162–165.

O'Brien, P. C. (1983). A further study on host range of *Pratylenchus thornei*. *Australian Plant Pathology, 12*, 1–3.

Ogbonnaya, F. C., Seah, S., Delibes, A., Jahier, J., Lopez-Brana, I., Eastwood, R. F, et al. (2001a). Molecular-genetic characterisation of a new nematode resistance gene in wheat. *Theoretical and Applied Genetics, 102*, 623–629.

Ogbonnaya, F. C., Subrahmanyam, N. C., Moullet, O., de Majnik, J., Eagles, H. A., Brown, J. S., et al. (2001b). Diagnostic DNA markers for cereal cyst nematode resistance in bread wheat. *Australian Journal of Agricultural Research, 52*, 1367–1374.

Orion, D., Amir, J., & Krikun, J. (1984). Field observations on *Pratylenchus thornei* and its effects on wheat under arid conditions. *Revue de Nématologie, 7*, 341–345.

Orui, Y., & Mizukubo, T. (1999). Discrimination of seven *Pratylenchus* species (Nematoda: Pratylenchidae) in Japan by PCR-RFLP analysis. *Applied Entomology and Zoology, 34*, 205–211.

Oztürk, G., Yildirim, A. F., & Kepenekci, I. (2000). [Investigation on the effect of *Heterodera filipjevi* Madzhidov on cereal yields which is one of the important cyst nematodes in cereal planted areas in Konya province]. In H. Ekiz (Ed.), Orta Anadolu'da hububat tardotless imdotless indotless in sorunlardotless i ve cözüm yollardotless i Sempozyumu, Konya, Turkey, 8–11 Haziran, 477–482.

Pakniat, M., & Sahandpour, A. (2004). Occurrence of *Anguina agrostis* (Steinbuch, 1799) Filipjev, 1936 on Barley in Fars province. 16th Iranian Plant Protection Congress, Karaj, 77 [Abstract].

Pankaj Mishra, S. D., & Sharma, G. L. (2002, 11–13 November). Integrated management of *Heterodera avenae* infecting wheat. In R. V. Singh, S. C. Pankaj Dhawan, & H. S. Gaur (Eds.), *Proceedings of National Symposium on Biodiversity and Management of Nematodes in Cropping Systems for Sustainable Agriculture* (pp. 265–266). Jaipur, India.

Paruthi, I. J., & Gupta, D. C. (1987). Incidence of tundu in barley and kanki in wheat field infested with *Anguina tritici*. *Harayana Agricultural University Journal of Research, 17*, 78–79.

Paull, J. G., Chalmers, K. J., & Karakousis, A. (1998). Genetic diversity in Australian wheat varieties and breeding material based on RFLP data. *Theoretical and Applied Genetics, 96*, 435–446.

Peng, D., Nicol, J. M., Zhang, D., Chen, S., Waeyenberge, L., Moens, M., et al. (2007). Occurrence, distribution and research situation of cereal cyst nematode in China. International Plant Protection Conference, Scotland, Glasgow, Sept. 07 [Abstract].

Person-Dedryver, F. (1986). Incidence du nématode à galle *Meloidogyne naasi* en cultures céréalières intensives. Dix années d'études concertées INRA-ONIC-ITCF, 1973-1983. INRA, Paris, 175–187.

Person-Dedryver, F., & Jahier, J. (1985). Les céréales à paille hôtes de *Meloidogyne naasi* Franklin II. Variabilité du comportement multiplicateur ou résistant de variétés cultivées en France. *Agronomie, 5*, 573–578.

Person-Dedryver, F., Jahier, J., & Miller, T. E. (1990). Assessing the resistance to cereal root-knot nematode, *Meloidogyne naasi*, in a wheat line with the added chromosome arm 1HchS of *Hordeum chilense*. *Journal of Genetics and Breeding, 44*, 291–295.

Philis, I. (1988). Occurrence of *Heterodera latipons* on barley in Cyprus. *Nematologia Mediterranea, 16*, 223.

Plowright, R. A., Caubel, G., & Mizen, K. A. (2002). *Ditylenchus* Species. In J. L. Starr, R. Cook & J. Bridge (Eds.), *Plant Resistance to Parasitic Nematodes* (pp. 107–139). Wallingford, UK: CAB International.

Rajvanshi, I., Mathur, B. N., & Sharma, G. L. (2002). Effect of inter-cropping on indicidence of *Heterodera avenae* in wheat and barley crops. *Annals of Plant Protection Sciences, 10*, 406–407.

Ritter, M. (1972). Rôle économique et importance des *Meloidogyne* en Europe et dans le bassin Méditerranéen. *OEPP/EPPO Bulletin, 2*, 17–22.

Rivoal, R. (1982). Caractérisation de deux écotypes d'*Heterodera avenae* en France par leurs cycles et conditions thermiques d'éclosion. *Bulletin EPPO/OEPP, 12*, 353–359.

Rivoal, R. (1986). Biology of *Heterodera avenae* Wollenweber in France. IV. Comparative study of the hatching cycles of two ecotypes after their transfer to different climatic conditions. *Revue de Nématologie, 9*, 405–410.

Rivoal, R., Bekal, S., Valette, S., Gauthier, J. P., Bel Hadj Fradj, M., Mokabli, A., et al. (2001). Variation in reproductive capacity and virulence on different genotypes and resistance genes of Triticeae, in the cereal cyst nematode species complex. *Nematology, 3*, 581–592.

Rivoal, R., & Besse, T. (1982). Le nématode à kyste des céréales. *Perspectives Agricoles, 63*, 38–43.

Rivoal, R., & Bourdon P. (2005). Sélection du ray-grass d'Italie pour la résistance au nématode à kyste des céréales (*Heterodera avenae*). *Fourrages, 184*, 557–566.

Rivoal, R., & Cook, R. (1993). Nematode pests of cereals. In K. Evans, D. L. Trudgill & J. M. Webster (Eds.), *Plant parasitic nematodes in temperate agriculture* (pp. 259–303). U.K.: CAB International.

Rivoal, R., Lasserre, F., & Cook, R. (1995). Consequences of long-term cropping with resistant cultivars on the population dynamics of the endoparasitic nematodes *Heterodera aveanae* and *Pratylenchus neglectus* in a cereal production ecosystem. *Nematologica, 4*, 516–529.

Rivoal, R., Person-Dedryver, F., Doussinault, G., & Morlet, G. (1986). Significance and application of cereal resistance to nematodes. In F. Rapilly & G. Doussinault (Eds.), Les résistances génétiques dans le systèmes de protection des cultures céréalières contre les champignons, virus et nématodes. *Les Colloques de l'INRA*, 35, Paris, France, 57–64.

Rivoal, R., & Sarr, E. (1987). Field experiments on *Heterodera avenae* in France and implications for winter wheat performance. *Nematologica, 33*, 460–479.

Roberts, P. A., Van Gundy, S. D., & McKinney, H. E. (1981). Effects of soil temperature and planting date of wheat on *Meloidogyne incognita*, reproduction, soil populations and grain yield. *Journal of Nematology, 13*, 338–345.

Roberts, P. A., Van Gundy, S. D. & Waines, J. G. (1982). Reaction of wild and domesticated *Triticum* and *Aegilops* species to root-knot nematodes (*Meloidogyne*). *Nematologica*, 28, 182–191.

Roget, D. K, & Rovira, A. D. (1985). Effect of tillage on *Heterodera avenae* in wheat. In C. A. Parker, A. D. Rovira, K. L. Moore & P. T. W. Wong (Eds.), *Ecology and management of soilborne plant pathogens* (pp. 252–254). The American Phytopathological Society, St. Paul, MN.

Romero, M. D., Montes, M. J., Sin, E., Lopez-Braña, I., Duce, A., Martín-Sanchez, J. A. et al. (1998). A cereal cyst nematode (*Heterodera avenae* Woll.) resistance gene transferred from *Aegilops triuncalis* to hexaploid wheat. *Theoretical and Applied Genetics, 96*, 1135–1140.

Rumpenhorst, H. J., Elekçioglu, I. H., Sturhan, D., Öztürk, G., & Enneli, S. (1996). The cereal cyst nematode *Heterodera filipjevi* (Madzhidov) in Turkey. *Nematologia Mediterranea, 24*, 135–138.

Rumpenhorst, H. J., & Sturhan, D. (1996). Morphological and electrophoretic studies on populations of the *Heterodera avenae* complex from the former USSR. *Russian Journal of Nematology, 49*, 29–38.

Sabova, M., Valocka, B., Liskova, M., & Vargova, V. (1988). The first finding of *Heterodera latipons* Franklin, 1969 on grass stands in Czecholovakia. *Helminthologia, 25*, 201–206.

Safari, E., Gororo, N. N., Eastwood, R. F., Lewis, J., Eagles, H. A., & Ogbonnaya, F. C. (2005). Impact of *Cre1*, *Cre8* and *Cre3* genes on cereal cyst nematode resistance in wheat. *Theoretical and Applied Genetics, 110*, 567–572.

Saleh, H. M., & Fattah, F. A. (1990). Studies on the wheat seed gall nematode. *Nematologia Mediterranea, 18*, 59–62.

Santo, G. S., & O'Bannon, J. H. (1981). Pathogenicity of the Columbia root-knot nematode (*Meloidogyne chitwoodi*) on wheat, corn, oat and barley in the Pacific North West. *Journal of Nematology, 13*, 548–550.

Sasser, J. N., & Freckman, D. W. (1987). A world perspective on nematology: the role of the society. In J. A. Veech & D. W. Dickson (Eds.), *Vistas on Nematology* (pp. 7–14). Maryland: Society of Nematologists, Hyattsville.

Schneider, J. (1967). Un nouveau nématode du genre *Meloidogyne* parasite des céréales en France. *Phytoma, 185*, 21–25.

Scholz, U. (2001). Biology, pathogenicity and control of the cereal cyst nematode *Heterodera latipons* Franklin on wheat and barley under semiarid conditions, and interactions with common root rot *Bipolaris sorokinana* (Sacc.) Shoemaker [teleomorph: *Cochliobolus sativus* (Ito et Kurib.) Drechs. ex Dastur.]. Ph.D Thesis. University of Bonn, Germany, 159 pp.

Scholz, U., & Sikora, R. A. (2004). Hatching behavior of *Heterodera latipons* Franklin under Syrian agro-ecological conditions. *Nematology, 6*, 245–256.

Shahina, F., Abid, M., & Maqbool, M. A. (1989). Screening for resistance in corn cultivars against *Heterodera zeae*. *Pakistan Journal of Nematology, 7*, 75–79.

Sharma, R. D. (1981). Pathogenicity of *Meloidogyne javanica* to wheat. In Trabalhes Apresentados V Reunas Brasileira de Nematologia, 9 B. Publicao No. 5.

Sharma, S. B., & Swarup, G. (1984). Cyst forming nematodes of India. Cosmo Publication 1, New Delhi, India, 150 pp.

Shen, T. H., Tai, S. E., & Chia, W. L. (1934). A preliminary report on the inheritance of nematode resistance and length of beak in a certain wheat cross. *Bulletin College of Agriculture and Forestry*, No. 19, University of Nanking, China.

Sikora, R. A. (1988). Plant parasitic nematodes of wheat and barley in temperate and temperate semi-arid regions – a comparative analysis. In M. C. Saxena, R. A. Sikora & J. P. Srivastava (Eds.), *Nematodes Parasitic to Cereals and Legumes in Temperate Semi-arid Regions* (pp. 46–48). International Centre for Agricultural Research in the Dry Areas (ICARDA), Syria.

Sikora, R. A., & Oostendorp, M. (1986). Report: Occurrence of plant parasitic nematodes in ICARDA experimental fields. ICARDA, Aleppo, Syria, 4 pp.

Simon, A., & Rovira, A.D. (1982). The relation between wheat yield and early damage of root by cereal cyst nematode. *Australian Journal Agriculture and Animal Husbandry, 22*, 201–208.

Sinclair, T. R., & Amir, J. (1996). Model analysis of a mulch system for continuous wheat in an arid climate. *Field Crops Research, 47*, 33–41.

Singh, D., & Agrawal, K. (1987). Ear cockle disease (*Anguina tritici* (Steinbuch) Filipjev) of wheat in Rajasthan, India. *Seed Science and Technology, 15*, 777–784.

Singh, I., Sharma, N. K. & Sakhuja, P. K. (1977). Distribution of the cereal cyst nematode, *Heterodera avenae* Woll., on wheat in Ludhiana (Punjab). *Journal of Research, Punjab Agricultural University, 14*, 314–317.

Singh, D., & Singh, S. (2005). Effect of date of sowing on the pathogenicity of different populations of cereal cyst nematode (*Heterodera avenae*) on dry shoot weight, number of grains/earhead and on the yield of wheat. *Environment and Ecology, 23*, 180–183.

S'Jacob, J. J. (1962). Beobachtungen an *Ditylenchus radicicola* (Greeff). *Nematologica, 7*, 231–234.

Slootmaker, L. A. J., Lange, R., Jochemsen, G., & Schepers, J. (1974). Monosomic analysis in breadwheat of resistance to cereal root eelworm. *Euphytica, 23*, 497–503.

Smiley, R. W., Whittaker, R. G., Gourlie, J. A., Easley, S. A., & Ingham, R. E. (2005). Plant-parasitic nematodes associated with reduced wheat yield in Oregon: *Heterodera avenae*. *Journal of Nematology, 37*, 297–307.

Smolik, J. D. (1972). Reproduction of *Tylenchorhynchus nudus* and *Helicotylenchus leiocephalus* on spring wheat and effect of *T. nudus* on growth of spring wheat. *Proceedings South Dakota Academy of Science, 51*, 153–159.

Stephan, Z. A. (1988). Plant parasitic nematodes on cereals and legumes in Iraq. In M. C. Saxena, R. A. Sikora & J. P. Srivastava (Eds.), *Nematodes parasitic to cereals and legumes in temperate semi-arid regions* (pp. 155–159). ICARDA, Aleppo, Syria.

Stirling, G. R. (1991). Biological control of plant parasitic nematodes. Progress, problems and prospects. Wallingford, Oxon, UK: CAB International. 275 pp.

Stone, L. E. W. (1968). Cereal cyst nematode in spring barley damage assessments 1963–64. *Plant Pathology, 17*, 145–150.

Stoyanov, D. (1982). Cyst-forming nematodes on cereals in Bulgaria. *EPPO/OEPP Bulletin, 12*, 341–344.

Sturhan, D. (1982). Distribution of cereal and grass cyst nematodes in the Federal Republic of Germany. *EPPO/OEPP Bulletin, 12*, 321–324.

Subbotin, S. A., Rumpenhorst, H. J., & Sturhan, D. (1996). Morphological and electrophoretic studies on populations of the *Heterodera avenae* complex from the former USSR. *Russian Journal of Nematology, 49*, 29–38.

Subbotin, S. A., Waeyenberge, L., & Moens, M. (2000). Identification of cyst forming nematodes of the genus *Heterodera* (Nematoda: Heteroderidae) based on the ribosomal DNA-RFLP. *Nematology, 2*, 153–164.

Subbotin, S. A., Waeyenberge, L., Molokanova, I. A., & Moens, M. (1999). Identification of *Heterodera avenae* group species by morphometrics and rDNA-RFLPs. *Nematology, 1*, 195–207.

Swarup, G. (1986). Investigations on wheat nematodes. In J. P. Tandon & A. P. Sethi (Eds.), *Twenty-five years of coordinated wheat research 1961–86* (pp. 189–206). New Delhi: Wheat Project Directorate, Indian Council of Agricultural Research.

Swarup, G., & Sosa Moss, C. (1990). Nematode parasites of cereals. In M. Luc, R. A. Sikora & J. Bridge (Eds.), *Plant parasitic nematodes in subtropical and tropical agriculture* (pp. 109–136). Wallingford, UK: CAB International.

Tanha Maafi, Z. (1998). *Pratylenchus neglectus* and *P. thornei* endoparasitc nematodes associated with wheat in Gilan province. 13th Iranian Plant Protection Congress. p 66. [Abstract].

Tanha Maafi, Z., Sturhan, D., Kheiri, A., & Geraert, E. (2007). Species of the *Heterodera avenae* group (Nematoda: Heteroderidae) from Iran. *Russian Journal of Nematology*, 15, in press.

Tanha Maafi, Z., Subbotin, S. A., & Moens, M. (2003). Molecular identification of cyst-forming nematodes (Heteroderidae) from Iran and a phylogeny based on ITS-rDNA sequences. *Nematology, 5*, 99–111.

Taheri, A., Hollamby, G. J., & Vanstone, V. A. (1994). Interaction between root lesion nematode, *Pratylenchus neglectus* (Rensch 1924) Chitwood and Oteifa 1952, and root rotting fungi of wheat. *New Zealand Journal of Crop and Horticultural Science, 22*, 181–185.

Talatchian, P., Akhiani, A., Grayeli, Z., Shah-Mohammadi, M., & Teimouri, F. (1976). Survey on cyst forming nematodes in Iran in 1975 and their importance. *Iranian Journal of Plant Pathology, 12*, 42–43.

Taylor, S. P., Hollaway, G. J., & Hunt, C. H., (2000). Effect of field crops on population densities of *Pratylenchus neglectus* and *P. thornei* in southeastern Australia; Part 1: *P. neglectus*. *Journal of Nematology, 32*, 591–599.

Taylor, A. L., & Sasser, J. N. (1978). Biology, identification and control of root-knot nematodes (*Meloidogyne* species). North Carolina State University Department of Plant Pathology and USAID, Raleigh, U.S.A.

Taylor, C., Shepherd, K. W., & Langridge, P. (1998). A molecular genetic map on the long arm of chromosome 6R of rye incorporating the cereal cyst nematode resistance gene, *CreR*. *Theoretical and Applied Genetics, 97*, 1000–1012.

Taylor, S. P., Vanstone, V. A., Ware, A. H., McKay, A. C., Szot, D., & Russ, M. H. (1999). Measuring yield loss in cereals caused by root lesion nematodes (*Pratylenchus neglectus* and *P. thornei*) with and without nematicide. *Australian Journal of Agricultural Research, 50*, 617–622.

Tesic, T. (1969). [A study on the resistance of wheat varieties to wheat eelworm (*Anguina tritici* Stein.)]. *Savrenema Poljoprivreda, 17*, 541–543.

Thompson, J. P., Brennan, P. S., Clewett, T. G., & Sheedy, J. G. (1997). *Disease reactions. Root-lesion nematode*. Northern Region Wheat Variety Trials 1996. Brisbane, Queensland, Australia: Department of Primary Industries Publication.

Thompson, J. P., & Clewett, T. G. (1986). Research on root-lesion nematode. In Queensland Wheat Research Institute Biennial Report 1982–1984. (pp. 32–35). Queensland Dept. Primary Industries, Queensland. Government, Wheat Research Institute, Toowoomba, Queensland.

Thompson, J. P., & Haak, M. I. (1997). Resistance to root-lesion nematode (*Pratylenchus thornei*) in *Aegilops tauschii* Coss., the D-genome donor to wheat. *Australian Journal of Agricultural Research, 48*, 553–559.

Thompson, J. P., Mackenzie, J., & McCulloch, J. (1983). Root lesion nematode (*Pratylenchus thornei*) on Queensland wheat farms. Proceedings of the 46th International Congress of Plant Pathology, Melbourne, Australia, [Abstract].

Timper, P., & Brodie, B. B. (1997). First report of *Pratylenchus neglectus* in New York. *Plant Disease, 81*, 228.

Townshend, J. L., Potter, J. W., & Willis, C. B. (1978). Ranges in distribution of species of *Pratylenchus* in Northeastern North America. *Canadian Plant Disease Survey, 58*, 80–82.

Trudgill, D. L., Kerry, B. R., & Phillips, M. S. (1992). Seminar: Integrated control of nematodes (with particular reference to cyst and root knot nematodes). *Nematologica, 38*, 482–487.

Upadhyaya, K. D., & Swarup, G. (1981). Growth of wheat in the presence of *Merlinius brevidens* singly and in combination with *Tylenchorhynchus vulgaris*. *Indian Journal of Nematology, 11*, 42–46.

Urek, G., & Sirca, S. (2003). Plant parasitic nematode affecting the aboveground plant parts in Slovenia. Zbornik predavanj in referatov 6. Slovenskega Posvetovanje o Varstva Rastlin, Zrece, Slovenije, 4–6 marec, 486–488.

Van Gundy, S. D., Perez, B. J. G., Stolzy, L. H. A., & Thomason, I. J. (1974). A pest management approach to the control of *Pratylenchus thornei* on wheat in Mexico. *Journal of Nematology, 6*, 107–116.

Vanstone, V. A., Nicol, J. M., & Taylor, S. P. (1993). Multiplication of *Pratylenchus neglectus* and *P. thornei* on cereals and rotational crops. In V. A. Vanstone, S. P. Taylor & J. M. Nicol (Eds.), *Proceedings of the Pratylenchus Workshop, 9th Biennial Australasian Plant Pathology Conference*, Hobart, Tasmania, Australia.

Vanstone, V. A., Rathjen, A. J., Ware, A.H., & Wheeler, R. D. (1998). Relationship between root lesion nematodes (*Pratylenchus neglectus* and *P. thornei*) and performance of wheat varieties. *Australian Journal of Experimental Agriculture, 38*, 181–188.

Vats, R., Kanwar, R. S., & Bajaj, H. K. (2004). Biology of *Seinura paratenuicaudata* Geraert. *Nematologia Mediterranea, 32*, 117–121.

Wallace, H. R. (1965). The ecology and control of the cereal root nematode. *Journal of Australian Institute of Agricultural Science, 31*, 178–186.

Whitehead, A. G. (1998). Plant nematode control (384 pp.). Wallingford, U.K.: CAB International.

Whitehead, A. G., Tite, D. J., & Fraser, J. E. (1983). Control of stem nematode *Ditylenchus dipsaci* (oat race) by aldicarb and resistant crop plants. *Annals of Applied Biology, 103*, 291–299.

Williams, K. J., Taylor, S. P., Bogacki, P., Pallotta, M., Bariana, H. S., & Wallwork, H. (2002). Mapping of the root lesion nematode (*Pratylenchus neglectus*) resistance gene *Rlnn1* in wheat. *Theoretical and Applied Genetics, 104*, 874–879.

Wishart, J., Blok, V. C., Phillips, M. S., & Davies, K. G . (2004). *Pasteuria penetrans* and *P. nishizawae* attachment to *Meloidogyne chitwoodi, M. fallax* and *M. hapla*. *Nematology, 6*, 507–510.

Wouts, W. W., Schoemaker, A., Sturhan, D., & Burrows, P. R. (1995). *Heterodera spinicauda* sp. n. (Nematoda: Heteroderidae) from mud flats in the Netherlands, with a key to the species of the *H. avenae* group. *Journal of Nematology, 41*, 575–58.

Yadav B. D., Bishnoi, S. P., & Chand, R. (2002). Proceedings of the National Symposium on 'Biodiversity and Management of nematodes in Cropping Systems for Sustainable Agriculture – Nematology in Rajasthan at a glance'. Department of Nematology, Rajasthan Agricultural University, Agricultural Research Station, Durgapura, Jaipur, 51pp.

York, P. A. (1980). Relationship between cereal root-knot nematode *Meloidogyne naasi* and growth and yield of spring barley. *Nematologica, 26*, 220–229.

Yu, M. Q., Person-Dedryver, F., & Jahier, J. (1990). Resistance to root knot nematode, *Meloidogyne naasi* (Franklin) transferred from *Aegilops variabilis* Eig to bread wheat. *Agronomie, 6*, 451–456.

Yüksel, H., Güncan, A., & Döken, M. T. (1980). The distribution and damage of bunts (*Tilletia* spp.) and wheat gall nematode [*Anguina tritici* (Steinbuch) Chitwood] on wheat in the eastern part of Anatolia. *Journal of Turkish Phytopathology, 9*, 77–88.

Zancada, M. C., & Althöfer, M. V. (1994). Effect of *Heterodera avenae* on the yield of winter wheat. *Nematologica, 40*, 244–248.

Zaharieva, M., Monneveux, P., Henry, M., Rivoal, R., Valkoun, J. & Nachit, M. M. (2001). Evaluation of a collection of wild wheat relative *Aegilops geniculata* Roth. and identification of potential sources for useful traits. *Euphytica, 119*, 33–38.

14

C. ORNAT AND F. J. SORRIBAS

INTEGRATED MANAGEMENT OF ROOT-KNOT NEMATODES IN MEDITERRANEAN HORTICULTURAL CROPS

Departament d'Enginyeria Agroalimentària i Biotecnologia, Universitat Politecnica de Catalunya , ESAB-EUETAB 08860 Castelldefels, Barcelona, Spain

Abstract. Several vegetables are grown around the Mediterranean basin for fresh consumption as a basic component of the Mediterranean diet, as climate allows cropping thorough all year. A great socio-economic and cultural diversity makes of this area a mosaic, in which large and small-scale production systems are coexisting. *Meloidogyne* spp. are the main plant parasitic nematodes causing yield losses mainly in protected crops due to climate and intensive croppings. *M. javanica, M. incognita* and *M. arenaria* are the most frequent species found in almost all countries. The principles of control of root-knot nematodes are changing from the use of nematicides applied to eradicate them, towards integrated nematode management, accepting the pests presence at levels that do not cause economic yield losses, according to sustainable agricultural systems. Basic information concerning biology, plant-nematode interactions, potential yield losses and value, efficacy and costs of control methods, are necessary to elaborate prediction models to support and design integrated management strategies.

1. INTRODUCTION

The Mediterranean region comprises the Mediterranean Sea and its coastal area, including eighteen countries: northern basin countries (Albania, Former Yugoslavia, France, Greece, Italy, Monaco, Spain) and southern basin countries (Algeria, Cyprus, Egypt, Israel, Lebanon, Libya, Malta, Morocco, Syria, Tunisia, Turkey). The Mediterranean climate is characterized by mild temperatures, with a cold period in winter, annual rainfall between 250–800 mm distributed during spring and autumn and dry summers. The northern region is relatively more temperate and humid whereas the southern region is warmer and drier, with endemic water shortages due to the interaction of relatively low seasonal rainfall and high evapotranspiration rates.

The Mediterranean climate predominates in the countries surrounding the Mediterranean Sea, but it also occurs in others zones of the Earth: Cape Town in South Africa, central coast of Chile, central and southern coast of California and portions of southwestern Australia. These regions share climate conditions similar to the Mediterranean ones, thus, the crops and the problems for cropping, are also similar.

A. Ciancio & K. G. Mukerji (eds.), Integrated Management and Biocontrol of Vegetable and Grain Crops Nematodes, 295–319.

Table 1. Production of vegetable crops (tonnes) in the main producing countries of the Mediterranean basin in 2005 (FAO, 2006).

Crop	Egypt	Italy	Spain	Turkey
Lettuce and chicory	140.000	1.010.520	920.000	375.000
Melons (inc.cantaloupes)	565.000	611.501	1.176.900	1.700.000
Strawberries	100.000	147.049	308.000	160.000
Tomato	7.600.000	7.187.016	4.651.000	9.700.000
Watermelons	1.500.000	519.463	724.900	3.800.000
Eggplants (aubergines)	1.000.000	338.803	60.000	880.000
Asparagus		43.274	47.600	11
Cucumbers and gherkins	600.000	72.572	485.000	1.725.000
Chillies and peppers, green	460.000	362.994	953.200	1.745.000
Spinach	48.700	99.367	45.000	220.000
Artichokes	70.000	469.975	188.900	30.000
Pumpkins, squash and gourds	690.000	488.083	300.000	376.000

The countries of the Mediterranean area are important producers of vegetables for fresh consumption as a basic component of the Mediterranean diet. Production and harvested area increased by about 28% and 14% between 1995 and 2005, respectively. In 2005, the harvested area was about 2.5 million hectares and production was near 73 million Tm. Production and harvested area of Turkey, Egypt, Italy and Spain are more than 75% and 70%, respectively, of the Mediterranean basin (Table 1 and Table 2) (FAO, 2006).

Table 2. Area harvested (ha) in the main producing countries of the Mediterranean basin in 2005 (FAO, 2006).

Crop	*Egypt*	*Italy*	*Spain*	*Turkey*
Lettuce and chicory	6.000	50.008	39.000	19.700
Melons (inc.cantaloupes)	24.000	27.815	35.200	103.000
Strawberries	3.800	6.226	7.600	10.500
Tomato	195.000	138.756	70.400	260.000
Watermelons	62.000	14.193	16.100	137.000
Eggplants (aubergines)	43.000	12.164	1.500	35.000
Asparagus		6.365	12.000	3
Cucumbers and gherkins	28.000	1.989	7.200	60.000
Chillies and peppers, green	29.000	13.787	22.500	88.000
Spinach	2.500	7.367	3.000	22.500
Artichokes	3.500	50.127	18.600	2.500
Pumpkins, squash and gourds	39.200	16.732	7.000	22.000

The most important vegetables grown in plastic houses are: tomato, strawberry, pepper, squash, eggplant, that represent 89% of production (FAO, Grupo de cultivos hortícolas, 2002).

2. MELOIDOGYNE

Meloidogyne javanica, M. incognita, M. arenaria and *M. hapla* are the most frequent root-knot nematode species present in almost all mediterranean countries (Table 3). These species are worldwide distributed and have a wide host range that includes vegetable crops. In the Mediterranean countries they often represent the main soil pathogen problems for vegetable and flower crops, especially under protected conditions (Greco & Esmenjaud, 2004). *M. incognita* and *M. javanica* are commonly found in the tropics, whereas *M. arenaria* and *M. hapla* are more common in the subtropical and temperate climates, respectively.

Other *Meloidogyne* species present in the Mediterranean basin are *M. artiella*, pathogenic to legumes and cereals in southern Europe and the Near East, as well as *M. lusitanica*, *M. baetica,* and *M. hispanica* that are not parasites of vegetables. *M. baetica* has been reported in olive trees in Portugal (Abrantes, Vovlas, & Santos, 1991) and in Spain (Nico, Rapoport, Jiménez-Díaz, & Castillo, 2002). *M. baetica* has been detected on *Pistacia lentiscus,* and *Aristolochia baetica* in Spain (Castillo, Vovlas, Subbotin, & Troccoli, 2003), whereas *M. hispanica* has only been reported in peach orchards in south-eastern Spain (Hirschmann, 1986).

Finally, the other important root-knot nematode specie able to parasitise vegetables, *M. chitwoodi*, is not present in the Mediterranean basin although it has been detected in some temperate countries in Europe, and in South-Africa, but not in Asia (EPPO, 2006).

2.1. Symptoms

Galls in roots are the most characteristic symptom shown by plants infected by the most important *Meloidogyne* species. The size of galls is variable depending on quantity of inoculum and plant species. Low nematode density produces individual or scattered galls induced by one or few females and an egg mass, related to individual females, can be observed on the root surface. As density of nematodes increases, galls develop closer to each other and roots become deformed, their size increase considerably. In this case, only few egg masses can be seen on the root surface, as the majority of them are inside the root.

Plant species affects gall size, ranked from more to less discrete galls: *Alliaceae, Cruciferae, Asteraceae, Chenopodiaceae, Apiaceae, Fabaceae, Cucurbitaceae* and *Solanaceae*. However, plants from the same family do not have the same sensibility to root-knot nematodes. For example, galls on pepper are smaller than on aubergine or tomato, or galls on carrot are smaller than on celery. The severity of diseased roots for the same vegetable crop and for the same initial population density can also differ according to the time of the year. For instance, lettuce roots are more severely galled when the crop is grown in summer than in late autumn. Symptoms in the aboveground depend on disease severity, they can range from no symptoms, when initial population density is lower than the tolerance limit, to dead plants, at higher population densities. Damping-off can occur when seeds are planted in heavily infested soils. Nutrient and water absorption is affected in plants with severely galled roots. As a consequence, plant growth is retarded, leaves show nutrient deficiency, wilt, yellowing and necrosis, flowering can be reduced or flowers become dry, in addition, the number of fruits is reduced or fruit size does not attain marketable standards.

Table 3. Distribution of Meloidogyne javanica, M. incognita, M. arenaria *and* M. hapla *in countries from the Mediterranean basin.*

Country	*Meloidogyne* species			
	M. javanica	*M. incognita*	*M. arenaria*	*M.hapla*
Albania		Shepherd and Barker, 1990		
Algeria	Ibrahim, 1985	Scotto La Massese, 1961; Ibrahim, 1985	Sellami, Lonici, Eddoud, and Besenghir, 1999	Sellami et al., 1999
Cyprus	Philis, 1983; Ibrahim, 1985	Philis, 1983		
Egypt	Ibrahim, Ibrahim, and Rezk, 1972; Ibrahim and Rezk, 1988	Ibrahim et al., 1972; Ibrahim and Rezk, 1988	Taylor, Sasser, and Nelson, 1982	
Former Yugoslavia	Shepherd and Barker, 1990; Grujicic, 1974	Grujicic, 1974	Grujicic, 1975	Grujicic and Paunovic, 1971
France		Shepherd and Barker (1990)	Dalmasso, 1980	Berge, Dalmasso, and Ritter, 1972
Greece	Pyrowolakis, 1975	Kyrou, 1976	Koliopanos, 1982	Pyrowolakis, 1975
Israel	Tarjan, 1953	Tarjan, 1953; Orion, Nessim-Bistritsky, and Hochberg, 1982		Minz, 1956
Italy	Ibrahim, 1985	Ibrahim, 1985	Ibrahim, 1985	Ambrogioni, 1969
Jordan	Hashim, 1979	Hashim, 1979	Karajeh, Abu-Gharbieh, and Masoud, 2005b	
Lebanon	Saad and Tanveer, 1972	Saad and Tanveer, 1972	Macaron, Laterrot, Davet, Makkouk, and Revise, 1975	
Libya	Dabaj and Jenser, 1987	Khan and Dabaj, 1980; Dabaj and Jenser, 1987	Khan, 1982	Dabaj and Jenser, 1987

Malta	Ibrahim, 1985	Ibrahim, 1985		
Morocco	Ibrahim, 1985	Ibrahim, 1985	Taylor et al., 1982	
Spain	Jiménez-Millan, Bello, Arias, and López Pedregal, 1964	Jiménez-Millan et al., 1964	Jiménez-Millan et al., 1964	Jiménez-Millan et al., 1964
Syria	Ibrahim, 1985	Ibrahim, 1985	Tayar, 1982	
Tunisia	Waldmann, 1971	Moens, 1985		
Turkey	Ibrahim, 1985	Bora, 1970	Taylor et al., 1982	

The sequence of symptoms is faster in summer than in other seasons since plant requirements are greater. In vegetables cultivated for their tubers or roots *Meloidogyne* can cause deformations that result in severe losses in quality. Damage caused by *Meloidogyne* can be more severe when synergistic relationships with fungi or bacteria occur producing additional root-rot that accelerates the sequence of symptoms.

2.2. Biology and Ecology

Meloidogyne can survive without a host plant as juvenile of 1st (J1) or 2nd (J2) stage inside the egg, in the egg mass or as J2 in soil. When a crop is planted, J2 penetrate the root directly, just behind the root tip. Migration into the vascular cylinder is intercellular and non destructive. The J2 infects roots establishing a feeding site within the developing vascular cylinder if there is a compatible response, and if conditions are conducive to it. After infection, J2 moults three times before reaching maturity. In optimal conditions, juveniles of *Meloidogyne javanica*, *M. incognita*, or *M. arenaria* develop to females. Reproduction is by parthenogenesis, and each female produces an egg mass containing from 500 to 1500 eggs. Below optimal conditions, some juveniles develop to males, which do not feed on the plant, in order to regulate nematode population densities and avoid intraspecific competition.

Although *Meloidogyne* can occur wherever a plant can develop (Sasser & Carter, 1985), survival and development of root-knot nematodes are conditioned by the host plant and the environmental conditions in soil. Thus, a host plant allows build-up of nematode population densities, whereas a poor host hinders the build-up that, finally, doesn't take place if the plant is a non-host. Vegetables such as tomato, cucumber, lettuce, aubergine, or melon are hosts for root-knot nematodes, while vegetables such as cabbage and onion are poor or non-host (Netscher & Luc, 1974). However, there exists inter (Sasser, 1954) and intraspecific (Taylor & Sasser, 1978; Southards & Priest, 1973; Sasser, 1966; Riggs & Winstead, 1959) variability in the reproductive capacity of root-knot nematodes in selected hosts. This aspect suggested the existence of physiological races that can be recognized by the use of differential hosts (Hartman & Sasser, 1985). Nevertheless, more races can be differentiated when new hosts are included in the test (Noe, 1985; Southards &

Priest, 1973). Therefore, this test is not a practical tool to design plant rotations for managing nematode population densities.

In addition, the plant host influences the length of the life cycle of *Meloidogyne* (Godfrey & Oliveira, 1932) and length of embryogenesis (Bafokuzara, 1983).

Soil temperature is the most important environmental factor that regulates the life cycle of *Meloidogyne*. The length of the life cycle can be expressed as accumulated temperature over a minimal threshold temperature, this is the thermal time. The thermal time requirement of plant-parasitic nematodes and its ecological significance have been reviewed extensively (Trudgill, 1995a; Trudgill & Perry, 1994). *Meloidogyne* needs between 11,500 and 13,000 heat units to complete its life cycle. A heat unit is one degree centigrade over the minimal threshold temperature (10ºC) acting for an hour (Tyler, 1933). Ferris, Roberts, and Thomason, (1985) reported that the nematode needs between 600 and 700 degree days over 10ºC to complete one generation. Thermal time requirements for the complete life cycle differ between species and populations. The length of one generation of *M. hapla*, *M. javanica*, and *M. incognita* was 554, 343, and 400ºC-days over 8.25, 13.1, and 10.1ºC, respectively (Ploeg & Maris, 1999; Madulu & Trudgill, 1994; Lahtinen, Trudgill, & Tiilikkala, 1988).

Dao (1970) found that a population of *M. incognita* from the Netherlands was able to infect and reproduce at about 5ºC lower than a Venezuelan population, reason for which he suggested the existence of thermotypes defined as nematode population with a fixed difference in temperature requirements. Nematode survival in the absence of a host is also conditioned by temperature, in general the optimum temperatures for survival of eggs and juveniles ranged from 10 to 15ºC (Thomason, Van Gundy, & Kirkpatrick, 1964; Bergeson, 1959). Knowledge of thermal of time requirements of each species allows predictions the time necessary to reach certain events (Trudgill, 1995b) and to develop management strategies (Van Gundy, 1985).

The second-stage juvenile of *Meloidogyne* spp. lives free in the soil, hatching in this second-stage from eggs. These juveniles move in the moist soil searching for roots of a possible host plant, then penetrate through the growth zones and induce differentiation of plant cells into specialized feeding cells. Symptoms on roots are: root swellings called galls and general alteration of the root vascular system.

2.3. Yield Losses of Economic Importance

Meloidogyne spp. can cause yield losses of over 30% in various vegetable crops (Netscher & Sikora, 1990). In experimental conditions, yield reductions in aubergines caused by *M. incognita* were higher than 80% (Di Vito, Greco, & Carella, 1986). In northeastern Spain, an initial population density of 4,750 juveniles/250 cm^3 soil of *M. javanica* caused a 36% and a 61% reduction of yield in lettuce and tomato cropped in summer in plastic-houses, respectively (Verdejo-Lucas, Sorribas, & Puigdomènech, 1994). Cucumber yield loss caused by an initial population density of 1,100 juveniles/250 cm^3 soil of *M. javanica* was 60% (Ornat, Verdejo-Lucas, & Sorribas, 1997).

Maximum yield loss of tomato cropped in plastic-houses from March to July has been estimated at 36% in northern Spain (Sorribas, Ornat, Verdejo-Lucas,

Galeano, & Valero, 2005a), and 21% in the Balearic Islands (unpublished data). Differences in yield losses could be mainly explained by differences in initial population densities, environmental conditions, and crop management. Furthermore, yield losses could increase when *Meloidogyne* interacts with other plant pathogens such as *Fusarium oxysporum* or *Rhizoctonia solani,* producing a disease complex (Back, Haydock, & Jenkinson, 2002; Hussey & McGuire, 1987; Webster, 1985; Taylor, 1979) therefore increasing their severity of attack or overcoming plant resistance.

Economic importance of *Meloidogyne* on vegetables crops for a specific area can vary depending on the frequency of infestation and the population levels. For example, in northeastern Spain, 50% of 66 plastic-houses, and 27% of 59 open fields were infested by *Meloidogyne. M. javanica* was the most abundant species in plastic houses (41% of the sites) whereas *M. incognita* was the main species in open fields (50% of the sites). Nematode population densities at planting the spring crop ranged from 1 to 590 juveniles/250 cm^3 soil, and from 1 to 2,100 juveniles/250 cm^3 soil when planting the summer crop (Ornat, 1998; Sorribas, 1996). The tolerance limit, defined as the maximum density of inoculum that does not cause yield loss, for *M. javanica* on tomato, was 2 juveniles/250 cm^3 sandy soil in commercial plastic-houses (Sorribas, 1996). In pot experiments, tolerance threshold to *M. incognita* for eggplant, tomato, artichoke, pepper, and cabbage was 0.054, 0.55, 1.1, 0.3 and 0.5 eggs and juveniles/cm^3 soil, respectively (Di Vito, Cianciotta, & Zaccheo, 1992; Sasanelli, Di Vito, & Zaccheo, 1992; Di Vito et al., 1986). Tolerance limits have to be determined for each specific growing area since they are affected by plant nematode interaction, soil type, environmental conditions, and crop management.

3. ROOT-KNOT NEMATODES MANAGEMENT

Vegetable production systems in the Mediterranean basin are diverse and depending on the economy of the farm, similarly to other growing areas in tropical and subtropical climates (Sikora & Fernandez, 2005). Vegetables are grown in plastic houses and in open field in both, large and small-scale production systems. Large-scale production systems are managed by large enterprises or cooperatives for both export and national markets, are highly specialized, have access to technology and have permanent technical assistance. On the other hand, small- scale producers are family enterprises which produce and commercialize mainly for local markets, are moderate to scarcely specialized, have limited access to technology and occasionally they benefit from technical assistance.

Meloidogyne damage is greater in protected crops than in open field because of susceptibility of main crops, cropping intensity, and environmental factors. In protected crops, root knot nematode management is mainly based on fumigants and nematicides because implementation of variations in their predefined plans is difficult due to market requirements. However, restrictions or banning of some of the most effective nematicides have been led to the use of other techniques such as plant resistance, solarization and/or biofumigation, and cultural practices, mainly in ecological and integrated production systems. In open fields, the number of vegetables growing in rotation is higher than in plastic houses, environmental factors are less conducive to disease and in consequence the nematode can be managed more efficiently.

3.1. Plant Resistance

In Nematology, resistance is the ability of a plant to suppress development or reproduction of nematodes (Roberts, 2002). The use of resistant cultivars is an elegant, economical and environmentally safe method for controlling root knot nematodes (Netscher & Sikora,1990; Netscher & Mauboussin, 1973). In addition, plant resistance is particularly useful for organic farming or integrated production since these systems do not allow, or restrict, the use of chemical control, respectively. In commercial resistant cultivars yield is not significantly affected when cropped in nematode infested soils because of tolerance is coupled with resistance (Sorribas, et al., 2005a; Roberts, 2002; Rich & Olson, 1999; Ornat et al., 1997; Philis & Vakis, 1977). However, commercial resistant cultivars are only available for tomato and pepper, despite the fact that sources of resistance have been reported for other vegetables for example: complete resistance to *M. javanica* within *Solanum melongena* and *S. torvum* (Boiteux & Charchar, 1996), to *Meloidogyne hapla* in inbred lines of processing carrot (Wang & Goldman, 1996), and to *M. arenaria* and *M. javanica* in *Cucumis sativus* (Walters, Wehner, & Barker, 1996, 1999).

In Solanaceae, the expression of plant resistance to *Meloidogyne* spp. is characterized by a hypersensitive reaction, which consists in localized plant-cell necrosis around the nematode's head (Kaplan & Keen, 1980). Tomato is the most important vegetable cropped in the Mediterranean basin and commercial resistant cultivars and rootstocks are available. Resistance in tomato is conferred by the single dominant gene Mi, which was introgressed from the wild relative of tomato *Lycopersicon peruvianum* (Smith, 1944) and is present in all resistant commercial cultivars. The Mi-resistance gene confers resistance, but not immunity, to *Meloidogyne incognita*, *M. javanica* and *M. arenaria* (Roberts & Thomason, 1989). However, expression of resistance is affected by some factors such as soil temperature, species and populations of *Meloidogyne*, Mi-dosage, and tomato genetic background. The efficient use of resistance to manage root knot nematodes must take into consideration the following factors. First, soil temperatures higher than 28ºC suppress resistance expression (Dropkin, 1969). This limitation, due to temperature, suggests that, in the Mediterranean basin, the use of these varieties may have to be restricted to spring planting, when soil temperatures are lower.

Second, resistant tomatoes have a high level of resistance to populations of *M. incognita* and *M. arenaria*, but are less resistant to *M. javanica* (Ornat et al., 2001b; Sorribas & Verdejo-Lucas, 1999; Busquets, Sorribas, & Verdejo-Lucas, 1994). Considering that *M. javanica* is the most common species of root-knot nematode in the Mediterranean region (Verdejo-Lucas, Ornat, Sorribas, & Stchiegel, 2002; Ornat & Verdejo-Lucas, 1999; Sellami et al., 1999; Eddaoudi, Ammati, & Rammah, 1997; Tzortzakakis & Gowen, 1996; Sorribas & Verdejo-Lucas, 1994; Ibrahim, 1985; Philis, 1983; Lamberti, 1981) it will be necessary to combine resistance with other management techniques if resistance durability is to be assured. In addition, some *Meloidogyne* populations can overcome resistance.

Virulence, defined as the ability of nematodes to reproduce on a host plant that possesses one or more resistance genes, occurs naturally in *Meloidogyne* populations on tomato apparently without previous exposure to, or selection by, the Mi-resistance gene (Prot, 1984; Netscher, 1976). In Spain, over 30 root-knot nematode populations examined only one population of *M. javanica* was virulent to the Mi-resistance gene, occurring naturally without previous exposure to the resistance gene (Ornat et al., 2001). Virulent nematode populations may also be selected after repeated exposure to tomatoes with *Mi*-gene resistance (Roberts, 1995; Castagnone-Sereno, Bongiovanni, & Dalmasso, 1993; Netscher, 1976) or suddenly (Williamson 1998).

In the Mediterranean region, virulent populations to the *Mi* gene have been reported: *M. javanica* in Greece (Tzortzakakis & Gowen, 1996), Jordan (Karajeh, Abu-Gharbieh, & Masoud, 2005a), Morocco (Eddaoudi et al., 1997), and Spain (Ornat et al., 2001); and *M. incognita* in both Greece (Tzortzakakis, Adam, Blok, Paraskevopoulos, & Bourtzis, 2005), and Spain (Robertson et al., 2006).

Finally, Mi-gene dosage also influences resistance expression. Mi-gene can be in homozygosis (Mi Mi) or heterozygosis (Mi mi) in tomato cultivars. Tzortzakakis, Trudgill and Phillips (1998) reported that tomatoes carrying the Mi-gene in homozygosis were more resistant than in heterozygosis. However, in addition to gene dosage, genetic background especially in heterozygous condition could affect expression of resistance (Jacquet, Bongiovanni, Martinez, Verschave, & Wajnberg, 2005). Some experiments carried out in controlled conditions in Spain showed different expression of the resistance in heterozygous commercial tomato cultivars with reproduction indexes (Final population density/Initial population density) that ranged from 0 to 3.1 in the tomato cultivars Bandera and Carpy, respectively (Sorribas & Verdejo-Lucas, 1999).

In pepper (*Capsicum* spp.), a resistance to root-knot nematode is conferred by a dominant gene N (Thies & Fery, 2000; Di Vito & Saccardo, 1979; Hare, 1956) and a minimum of five dominant genes (Me1 to Me5) from accessions PM 127 and PM687 (Hendy, Pochard, & Dalmasso, 1985). Genes Me confer the same broad resistance spectrum as Mi as well as stability at high temperature (Djian-Caporalino et al., 1999). Peppers carrying the Me1 gene are resistant to both Mi-virulent and avirulent populations of *M. arenaria*, *M. incognita* and *M. javanica*, although Mi-virulent populations of *M. arenaria* and *M. incognita* can parasitize peppers containing the Me3 gene (Castagnone-Sereno, Bongiovanni, & Djian-Caporalino, 2001).

Although plant resistance is an economical, environmental safety and healthy control method, sensory characteristics not always are accepted by the market. When it occurs, grafting plants on resistant rootstocks could be an alternative. In addition, grafting could be a method to manage other important soil-born pathogens and a source conferring resistance in vegetables for which no commercial resistant cultivars are available. Its use in vegetable crops has increased in the last decade, mainly for tomato, cucurbits, pepper and eggplant (MBTOC, 2006). In 2003–2004 the use of grafted tomato in 2003–2004 was 45, 30, 28 and 12 millions plants in

Spain, Morocco, France, and Italy, respectively (Besri, 2005). Nevertheless grafted plants are more expensive than non-grafted ones, they give an extra production: i.e. grafted tomato in Morocco and Spain yielded 53 and 70 T per Ha more, respectively, than non-grafted ones (Besri, 2005; Verdejo-Lucas, Buñol, Sorribas, & Ornat, 2004). Grafting have to be used in an integrated manner to avoid the selection of virulent populations, i.e. the use of sweet pepper rootstocks resistant to *Phytophthora capsici* and *Meloidogyne incognita* selected virulent root-knot nematode populations, but not of *P. capsici* (Ros et al., 2005).

3.2. Heat

Thermal control is generally aimed at inducing internal injuries that will lead to death over a short period of time (Lagüe, Gill, & Péloquin, 2001). Steaming is the introduction of water vapour in soil to kill soilborn pests with the latent heat release when steam condenses into water (Bungay,1999). Treatments consist in increasing soil temperature to 70°C for at least half an hour (Runia, 2000). Death of *M. incognita* occurs when exposed for a few minutes to temperatures above 48°C (Noling, 1997). Negative pressure steaming or sheet steaming can be used for soil sterilization purposes. Negative pressure steaming allows treatment at more soil depth than sheet steaming, and uses almost half the fuel of sheet methods (Runia, 2000). Steaming requires high amounts of water, power or fuel (Crump, 2001) therefore this method is restricted to high value crops.

Soil solarization consists in increasing temperature of wet soil by covering the area with a transparent film that traps solar radiation. This tactic was developed by Katan and co-workers in Israel in the mid 70s (Katan, Greenberger, Laon, & Grinstein, 2007). Solarization can control many soil born pests, pathogens and weeds when extended over a period of 6–8 weeks under intense solar radiation. In the Mediterranean basin, the best conditions for pest control through soil solarization are given during the hot, dry summer, mainly in plastic houses which are not cultivated in most areas due to high temperatures.

The efficacy of solarization to control root-knot nematodes is variable e.g. Greco, Brandonisio, and Elia (1992) reported a 99% efficacy, despite the fact that it had not been effective in a previous experiment in Italy (Greco, Brandonisio, & Elia, 1985). The variability of results can be attributed to the complex mode of action and the influence of environmental conditions (Stapleton, 2000). To predict the effectiveness of this method, information is required on the survival of nematodes as affected by a range of temperatures and exposure time (Greco & Esmenjaud, 2004). Survival rate of *Meloidogyne arenaria, M. incognita* and *M. javanica* is inversely related to accumulated soil temperature. Degree days (°C) of 950, 1,200 and 1,900 were needed to reduce initial population densities of *M. javanica, M. incognita* and *M. arenaria*, respectively, to 90%. (Sorribas et al., 2005b). Solarization can provide excellent control under conducive conditions and proper use, but under marginal environmental conditions it would be more effective to use new technologies, and or combining it with other tactics in an integrated manner as organic amendments, nematicides, antagonists, cover crops, and plant resistance (MBTOC, 2006).

3.3. Soil-less Cultivation

Soil-less cultivation is a technique used for bypassing soil pathogens. However, this tactic has not resulted in the elimination of problems caused by plant-parasitic nematodes. The most probable sources of nematode infestation are: infested plant material, infested soil carried into soil-less system by wind, equipment, animals and humans and infested water (Hallmann, Hänisch, Braunsmann, & Klenner, 2005). All commonly used substrates are suitable for nematode infestation (Stapel & Amsing, 2004).

3.4. Crop Rotation

Crop rotation is the most important cultural method to control plant parasitic nematodes. The main aim of crop rotation for nematode management is the reduction of the initial population level of damaging nematode species to levels that allow the following crops to become established and complete early growth before being heavily attacked (Nusbaum & Ferris, 1973). However, some factors can restrict its use: On one hand, the occurrence of polyphagous nematodes or nematode communities can restrict the selection of suitable host plants. On the other hand, crop value also affects the use of rotation for nematode management because growers tend to produce the most economically important crops demanded by markets (e.g. tomato, pepper, cucumber, melon), and mainly fumigants and nematicides are used to control them. However, when the availability of nematicides is restricted or their use banned, rotation is often an important option (Duncan, 1991).

The aim of any rotation is to allow a sufficient interval of time between susceptible crops to reduce nematode population densities to a level that allows to grow and yield at an acceptable rate the next susceptible crop (Trivedi & Barker, 1986). Thus, vegetable species that are non-host, poor host, or resistant may be included in the rotation sequences. Crop rotation would be effective when a susceptible crop is planted once every four growing seasons and following a non-host or resistant crop when nematode densities in soil are low (Bridge, 1996).

Different sequences of vegetable rotation with susceptible host – poor host – poor host – susceptible host, that are normally conducted in commercial open fields in north-eastern Spain show that nematode population densities remain at similar levels at planting the next susceptible crop two years later. However, in plastic-houses, the choice of vegetables that can be cropped in the rotation sequence is reduced due to economic reasons.

In Crete, the option of vegetables in the rotation sequence for cultural management is reduced to resistant solanaceous crops in rotation with susceptible ones of the same family or cucurbits. In this context, growing a resistant tomato or pepper for one cropping cycle in a site infested by *M. javanica* and followed by a susceptible tomato resulted in final nematode population similar to those produced by fenamiphos applications to a sequence of two susceptible crops (Tzortzakakis et al., 2000). Cucumber cropped after a previous resistant tomato cultivar yielded 60% more than after a susceptible tomato in plastic-house infested by *M. javanica* in

Spain, (Ornat et al., 1997). Similar results have been reported for cantaloupe (Rich & Olson, 2004) or cucumber (Colyer, Kirkpatrick, Vernon, Barham, & Bateman, 1998; Hanna, Colyer, Kirkpatrick, Romaine, & Vernon, 1994) after resistant tomato, and cucumber and squash after resistant pepper (Thies et al., 2004).

3.5. Trap Crops

Trap cropping consists in planting a good host for a short period of time, enough to ensure high nematode penetration and initial development to a non-motile growth stage. After that, roots have to be removed or destroyed in order to kill nematodes before achieving reproduction (Sikora, Bridge, & Starr, 2005). This method is more attractive to growers if they can use a profitable short cycle vegetable included in their common crop rotation. For example, lettuce can act as a trap crop in the northeastern Spain when it is planted in October or November instead of September (Ornat et al., 2001). Lettuce and radish are used in organic peri-urban production in Cuba (Cuadra, Cruz, & Fajardo, 2000). Arugula (*Eruca sativa* L.) allows nematode infection but restricts its development and reproduction and can be used as a biofumigant when incorporated into the soil (Melakeberhan, Xu, Kravchenco, Mennan, & Riga, 2006).

3.6. Fallowing and Tillage

Meloidogyne is an obligate parasite that needs a host plant to complete its life cycle. Therefore, during the lack of a host plant the nematode has to consume their own reserves and could die by starvation. In the absence of a host plant, environmental conditions such as temperature and moisture are the main factors affecting survival rate of nematodes (Bergeson, 1959; Roberts, Van Gundy & McKinney, 1981; Towson & Apt, 1983; Goodell & Ferris, 1989). Soil tillage during the fallowing periods eliminates weeds and volunteer plants to prevent increases on nematode densities since *Meloidogyne* can reproduce on a wide range of weeds (Table 4).

Soil desiccation and direct heat from the sun may have an immediate impact on population decline. Decrease of population densities depends on the length of the fallow period related to the accumulated soil temperature during this period. In intensive agriculture the fallow period is limited to a few weeks. However when coupled with root destruction, even short-term fallowing has a significant impact on nematode populations (Verdejo-Lucas, 1999). For instance, fallowing during 8 weeks in summer, the hottest and driest season, reduced root knot nematode population about 50%, this reduction was about 80% when soil was tilled at the end of the crop just before fallowing (Ornat, Verdejo-Lucas, Sorribas & Tzortzakakis, 1999).

Table 4. Weed host species for Meloidogyne incognita *(Mi),* M. javanica *(Mj) and* M. arenaria *(Ma) associated with vegetable crops in northeast of Spain (Ornat, 1998; Barceló, Sorribas, Ornat, & Verdejo-Lucas, 1997).*

Botanic family	Species	
Amarantaceae	*Amaranthus albus* L.	Mi, Mj
	A. blitum	Mj
	A. graecizans L. *sylvestris* (Vill.)	Mi
	A. hybridus	Mi
	A. retroflexus L.	Mi, Ma
Caryophyllaceae	*Stellaria media* (L.) Vill.	Mi, Mj, Ma
Chenopodiaceae	*Atriplex patula* L.	Mi
	Chenopodium album L.	Mi, Mj
	Ch. murale L.	Mj
Compositae	*Cirsium arvense* (L.) Scop.	Mj
	Erigeron L. spp.	Mi, Mj
	Galisonga parviflora Cav.	Mi
	Senecio vulgaris L.	Mi,Mj
	Sonchus oleraceus L.	Mi,Mj
	S. tenerrimus L.	Mi, Mj, Ma
	Xanthium strumarium L.	Mi
Convolvulaceae	*Convolvulus arvensis* L.	Mi, Mj
Cruciferae	*Capsella bursa-pastoris* (L.) Medicus	Mi, Mj, Ma
	Coronopus didymus (L.) Sm.	Mi,Mj
	Lepidium draba L.	Mj
Cyperaceae	*Cyperus rotundus* L.	Mi
Euphorbiaceae	*Mercurialis annua* L.	Mj
Geraniaceae	*Geranium molle* L.	Mi
Gramineae	*Bromus wildenowii* Kunth	Mi
	Cynodon dactylon (L.) Pers.	Mi
	Digitaria sanguinalis (L.) Scop.	Mi, Mj, Ma
	Lolium perenne L.	Mi
	Poa annua L.	Mi, Ma
	Setaria verticillata (L.) Beauv.	Mi, Mj, Ma
	Sorghum halepense (L.) Pers.	Mi
Labiatae	*Lamium amplexicaule* L.	Mi
	Mentha L. spp.	Mj
Leguminosae	*Medicago arabica* (L.) Hudson	Mj, Ma
	Trifolium L. spp.	Mi, Mj.
	Vicia sativa L.	Mi
Malvaceae	*Malva sylvestris* (L.)	Not identified
Oxalidaceae	*Oxalys corniculata* (L.)	Mj
	O. corymbosa DC.	Ma
Polygonaceae	*Polygonum aviculare* L.	Mi
	Rumex crispus L.	Mj, Ma
Portulacaceae	*Portulaca oleracea* L.	Mi, Mj
Primulaceae	*Anagallis arvensis* L.	Ma
Rosaceae	*Potentilla reptans* L.	Mi
Scrophulariaceae	*Veronica hederifolia* L.	Mi
Solanaceae	*Solanum nigrum* L.	Mi, Mj
Urticaceae	*Urtica urens* L.	Mi

3.7. Biological

Several fungi antagonistic to nematodes have been detected in the Mediterranean area. *Pochonia chlamydosporia* var. *chlamydosporia*, *P. chlamydosporia* var. *catenulata*, *Fusarium oxysporum*, *F. solani, Fusarium* spp., *Acremonium strictum*, *Gliocladium roseum*, *Cylindrocarpon* spp., *Engiodontium album*, *Dactylella oviparasitica*, and other fungi non identified were isolated from *Meloidogyne* in two areas of vegetable production in Mediterranean coast of Spain (Verdejo-Lucas, et al., 2002).

Pochonia chlamydosporia was assessed as biological control agent of *Meloidogyne* in plastic house experiments in Greece and Spain. In Greece, *P. chlamydosporia* had a variable establishment in soil and did not control the nematode (Tzortzakakis & Petsas, 2003; Tzortzakakis, Phillips, & Trudgill, 2000). In Spain, the same fungal isolate used in Greece, consistently reduce root galling in tomato but reduction in eggs per gram root was only achieved when a native isolate was used in multiple fungal applications (Sorribas, Ornat, Galeano, & Verdejo-Lucas, 2003; Verdejo-Lucas, Sorribas, Ornat, & Galeano, 2003). However, in an open field experiment carried out in Italy, the same isolate of *P. chlamydosporia* showed encouraging results (Ciancio, Leonetti, & Alba, 2002).

The bacterial obligate parasite *Pasteuria penetrans,* has been studied extensively as a control agent. In Spain, *P. penetrans* was detected in 7% of the 93 sampled fields adhered to cuticle of *Meloidogyne*, as well as other phytoparasitic nematodes. In natural conditions, the percentage of *Meloidogyne* juveniles with spores fluctuated between 16 and a 50% (Verdejo-Lucas, Español, Ornat, & Sorribas, 1997). In plastic house experiments in Crete, *Pasteuria penetrans* applied at 20,000 or 25,000 spores/g of soil was able to parasitize 65%–75% of juveniles + females (Tzortzakakis, Verdejo-Lucas, Ornat, Sorribas, & Goumas, 1999; Tzortzakakis & Gowen, 1994).

Currently, biological control of nematodes is difficult because of the complexity of the soil biology and environment to promote or establish antagonists in soils that effectively suppress nematode populations (Starr, Bridge, & Cook, 2002). For more information see chapters 2, 3, 10 and 15 in this book and Lopez-Llorca, Jansson, Macià, and Salinas (2006), Sikora (1992) and Stirling (1991).

3.8. Biofumigation

The term biofumigation has been applied to the process where volatile toxic gases are released in the degradation of organic amendments, plant roots, and tissues and where such gases control diseases, nematodes, and weeds. Brassicaceae are commonly researched as biofumigant due to the production of glucosinolates and their isothiocyanate (ITC) derivates, which have herbicidal, fungicidal, insecticidal, and/or nematicidal properties (Bello, 1998; Kirkegaard & Sarwar, 1998; Gamliel & Stapleton, 1993). No suppression of nematodes or inconsistencies among studies are attributed to different concentrations of glucosinolates derivates (Zasada & Ferris, 2004; Kirkegaard & Sarwar, 1998). Biofumigation combined with solarization can improve its effectiveness and have been used successfully in the production of tomatoes, melons, peppers, and other vegetables (Bello, 1998; Sanz, Escuer & López-Pérez, 1998). For more information see Chapters 11 and 12 in this volume.

3.9. Chemical

Two major groups of nematicides are distinguished by the manner in which they spread through the soil. Soil fumigants are volatile chemicals that have to be applied before planting due to their phytotoxicity. Non-fumigant nematicides are liquids or solids that act by contact, or by plant systemic action, and can be applied at planting and after planting, depending on their degradation. Their distribution in soil depends on the soil water solution.

Methyl bromide was the most used fumigant by its effectiveness against soil born pathogens, pests and weeds. Nowadays, their use has been banned or restricted due to its effect on ozone depletion layer. The phase out for non Article 5 countries (developed countries) was in 2005, and will be in 2015 for Article 5 countries (developing countries). Alternatives to Methyl bromide are being assessed by the Methyl Bromide Technical Options Committee (MBTOC, 2006). Fumigants nematicides (1,3-dichloropropene, metham-sodium, chloropicrin) are more effective in the control of root-knot nematodes than non-fumigant nematicides (fenamiphos, cadusaphos, oxamyl). Fumigants could be adopted as an alternative to methyl bromide, but some of them could be banned in a short-term in some countries (see EC directive 91/414/CEE). Non-fumigants nematicides can lack their efficacy by microbial degradation due to repeated applications, i.e. ethoprophos and fenamiphos (Mojtahedi, Santo, & Pinkerton, 1991; Davis, Johnson, & Wauchope, 1993; Karpouzas, Giannakou, Walker, & Gowen, 1999; Karpouzas, Hatziapostolou, Papadopoulou-Mourkidou, Giannakou, & Geogiadou, 2004). Consequently, alternation of active compounds is required to prolong their efficacy (Sikora et al., 2005).

4. TOOLS FOR DECISION IN INTEGRATED PEST MANAGEMENT

All available management methods should be used in an integrated manner considering the biology of root-knot nematodes, nematode-plant interactions, agronomical practices, environmental characteristics, and socio-economic aspects of the specific growing area. The management of *Meloidogyne* with only one tactic may be partially effective (cultural practices) or may be no durable due to negative environmental effects (depletion of ozone layer, water contamination), lack of persistence (fast degradation in soil of some non-fumigant nematicides), occurrence of virulent populations or shift of nematological problems (plant resistance). Integrated nematode management (INM) is a strategy that uses all available tactics in a complementary and environmental safe manner to maintain nematode population levels below economic threshold.

Although guidelines for integrated pest management (i.e. IOBC/WPRS) and overviews for integrated nematode management (Sikora et al., 2005) have been published to help technicians and vegetable growers it is necessary to validate and adapt them for each specific growing area.

Economic threshold, defined as the nematode population density at which the value of the damage caused equals the cost of control (Ferris, 1978), is a basic requirement to develop integrated nematode management strategies (INM). Estimation of the economic threshold is based on Seinhorst nematode damage function (1965), that relates nematode densities in preplanting to relative yield, crop value, cost of control, and the control cost function (Ferris, 1978). To design INM strategies for specific cropping sequences it is needed to construct nematode damage functions for each *Meloidogyne* spp.-vegetable combinations, and to know the fluctuation of root-knot nematode population considering the particular agrosystem. Thus, sampling plans are required to estimate frequency and abundance of plant parasitic nematodes (McSorley & Parrado, 1982; McSorley & Dickson, 1991; Prot & Ferris, 1992). In addition, the use of accurate methods to extract plant parasitic nematodes is essential (see Hooper & Evans, 1993). Knowledge of all this information allows the development of predictive models to make decisions on nematode management (McSorley & Phillips, 1993; Ferris & Noling, 1987).

4.1. Management of Meloidogyne javanica *with Rotation in Plastic Houses in Northeastern Spain: an Example*

In the vegetable production area of northeastern Spain two to three crops are usually cropped at the same site from spring through winter. Tomato and lettuce are the most frequently cultivated vegetables for fresh market in plastic houses. Tomato is growing from March to July, and lettuce is growing from mid September, October or November to December, January or February. Between crops, there is a fallow period. Fluctuation of nematode population densities are shown in Fig. 1. Control of root-knot nematodes in the area was mainly based on fumigants or non-fumigants nematicides. However, a change has been produced when growers accepted and implemented sustainable production systems that restrict or ban the use of nematicides. In this context, more accurate information was needed to design strategies to manage nematode problems.

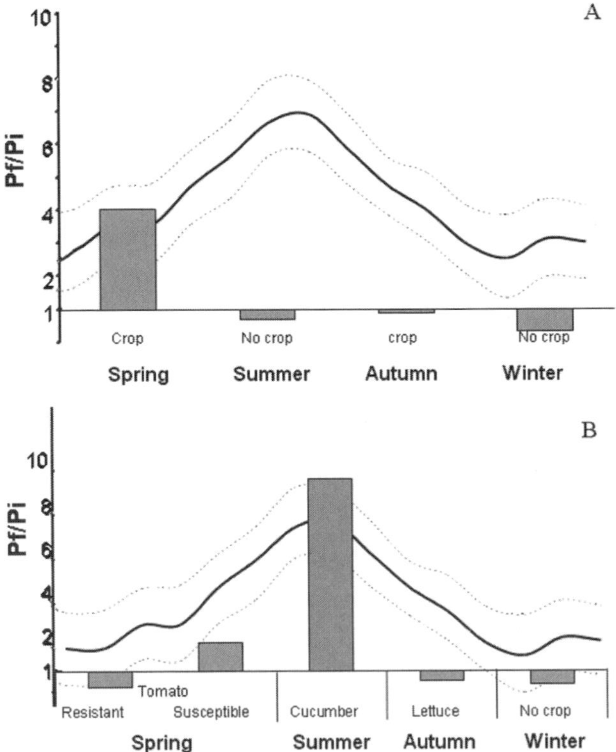

Figure 1. Average reproduction index of Meloidogyne *in a cropping season in plastic house, (A) two crops per season, (B) three crops per season.*

Considering two crops per season, a Seinhorst nematode damage function for *M. javanica*-tomato was obtained. Maximum yield losses and tolerance limit were estimated in 34% and 2 J2/250 cm^3 soil, respectively. Nematode achieves three generations during the tomato crop according to their thermal time model. At the end of the tomato crop, if plants are uprooted, nematode densities in soil decrease about 50% during the period between crops, and an additional 30% reduction occurs if soil is tilled. Conversely, nematode density does not decrease if the aboveground plants are cut and roots are left in soil. The following crop of lettuce can act as trap crop when it is planted in middle October or November reducing nematode densities between 50 and 20%, respectively, because the nematode can infect roots but does not accumulate enough degree days to reproduce at the end of the commercial crop.

Considering the usual range of nematode densities at the end of the susceptible tomato crops founds in this area (8 000 and 28 000 J2/250 cm^3 soil), the estimated increase in tomato yield using both agronomical methods, tillage and lettuce as trap crop, can range between 22 and 13%. The efficacy of these methods could be increased using other economically interesting methods to growers as resistant tomato cultivars, that prevent increases in nematode population densities, and/or increasing accumulated temperature during fallowing periods by solarization alone, because survival of root-knot without host is inversely related to accumulated soil temperature, or combined with biofumigation.

ACKNOWLEDGEMENT

Authors thank the support of Comision Interministerial de Ciencia y Tecnología (CICYT), project AGL2004-01207/AGR

REFERENCES

Abrantes, I. M. d. O., Vovlas, N., & Santos, M. S. N. D. (1991). *Meloidogyne lusitanica* sp. (Nematoda: *Meloidogynidae*), a root-knot nematode parasitizing olive tree (*Olea europaea* L.). *Journal of Nematology, 23*, 210–224.

Ambrogioni, L. (1969). Two cases of mixed infections by nematodes of the genera *Heterodera* and *Meloidogyne*. *Redia, 51*, 159–168.

Back, M. A., Haydock, P. P. J., & Jenkinson, P. (2002). Disease complexes involving plant parasitic nematodes and soilborne pathogens. *Plant Pathology, 51*, 683–697.

Bafokuzara, N. D. (1983). Influence of six vegetable cultivars on reproduction of *Meloidogyne javanica*. *Journal of Nematology, 15*, 559–564.

Barceló, P., Sorribas, F. J., Ornat, C., & Verdejo-Lucas, S. (1997). Weed host to *Meloidogyne* spp. associated with vegetable crops in northeast Spain. *IOBC/WPRS Bulletin, 20*, 89–93.

Bello, A. (1998). Biofumigation and integrated crop management. In: Bello, A., González, J. A., Arias, M. &- Rodríguez-Kabana, R. (Eds.), *Alternatives to methyl bromide for the southern european countries* (pp. 99–126). International Workshop, Arona, Tenerife, Spain

Berge, J. B., Dalmasso, A., & Ritter, M. (1972). Studies on *Meloidogyne hapla* found in France. European Society of Nematologists, International Symposium of Nematology (11th), Reading, 2–3.

Bergeson, G. B. (1959). The influence of temperature on the survival of some species of the genus *Meloidogyne*, in the absence of a host. *Nematologica, 4*, 344–354.

Besri, M. (2005). Current situation of Tomato grafting as alternative to methyl bromide for tomato production in the Mediterranean region. Twelfth Annual Conference on Methyl Bromide Alternatives and Emissions Reduction.San Diego California 1–3, Nov-2005.

Boiteux, L. S., & Charchar, J. M. (1996). Genetic resistance to root-knot nematode (*Meloidogyne javanica*) in eggplant (*Solanum melongena*). *Plant Breeding, 115*, 198–200.

Bora, A. (1970). Studies on plant-parasitic nematodes in the Black Sea region and their distribution and possibilities for chemical control. *Bitki Koruma Bulteni, 10*, 53–71.

Bridge, J. (1996). Nematode management in sustainable and subsistance agriculture. *Annual Review Phytopathology, 34*, 201–225,

Bungay, D. P. (1999). Steam sterilisation as alternative to Methyl Bromide. In: Methyl Bromide and Soilborne Diseases. Tenth Annual Interdisciplinary Meeting of the Soil-Borne Plant Diseases interest Group, 8–9 September, Stellenbosch, South Africa.

Busquets, J. O., Sorribas, J., & Verdejo-Lucas, S. (1994). Potencial reproductor del nematodo *Meloidogyne* en cultivos hortícolas. *Investigación Agraria: Producción y Protección Vegetales, 9*, 1–7.

Castagnone-Sereno, P., Bongiovanni, M., & Dalmasso, A. (1993). Stable virulence againts tomato resistance Mi gene in the parthenogenetic root-knot nematode *Meloidogyne incognita*. *Phytopathology, 83*, 803–805.

Castagnone-Sereno, P., Bongiovanni, M., & Djian-Caporalino, C. (2001). New data on the specificity of the root-knot nematode resistance genes Me1 and Me3 in pepper. *Plant Breeding, 120*, 429–433.

Castillo, P., Vovlas , N., Subbotin, S., & Troccoli, A. (2003). A new root-knot nematode, *Meloidogyne baetica* n. sp (Nematoda : Heteroderidae), parasitizing wild olive in Southern Spain. *Phytopathology, 93*, 1093–1102.

Ciancio, A., Leonetti, P., & Alba, G. (2002). Studies on field application of the hyphomycete *Verticillium chlamydosporium* for biological control of root-knot nematodes. *Nematologia Mediterranea, 30*, 78–88.

Colyer, P. D., Kirkpatrick, T. L., Vernon, P. R., Barham, J. D., & Bateman, R. J. (1998). Reducing *Meloidogyne incognita* injury to cucumber in a tomato-cucumber double-cropping system. *Journal of Nematology, 30*, 226–231.

Crump, P. (2001). Steam sterilization in chrysanthemums. In: Vick, K. W. (Ed.), Methyl bromide alternatives (pp. 1). Beltsville: USDA-ARS.

Cuadra, R., Cruz, X., & Fajardo, J. F. (2000). Cultivos de ciclo corto como plantas trampas para el control del nematodo agallador. The use of short cycle crops as trap crops for the control of root-knot nematodes. *Nematropica, 30*, 241–246.

Dabaj, K. H., & Jenser, G. (1987). List of plants infected by root-knot nematodes in Libya. *International Nematology Network Newsletter, 4*, 28–33.

Dalmasso, A. (1980). *Meloidogyne* nematodes and canning tomatoes. Le nematode *Meloidogyne* et la tomate de conserve. *Pepinier-Hortic-Maraich, 205*, 29–32.

Dao, F. D. (1970). Climatic influence on the distribution pattern of plant parasitic and soil inhabiting nematodes. Mededelingen LandbouwHogesnappen. Wageningen, 70–72.

Davis, R. F., Johnson, A. W., & Wauchope, R. D. (1993). Accelerated degradation of fenamiphos and its metabolites in soil previously treated with fenamiphos. *Journal of Nematology, 25*, 679–685.

Di Vito, M., Cianciotta, V., & Zaccheo, G. (1992). Yield of susceptible and resistant pepper in microplots infested with *Meloidogyne incognita*. *Nematropica, 22*, 1–6.

Di Vito, M., Greco, N., & Carella, A. (1986). Effect a *Meloidogyne incognita* and importance of the inoculum on the yield of eggplant. *Journal of Nematology, 18*, 487–490.

Di Vito, M., & Saccardo, F. (1979). Resistance of *Capsicum* species to *Meloidogyne incognita*. In: F. Lamberti & E. E. Taylor (Eds.), *Root-Knot nematodes (*Meloidogyne *sp.): Systematics, Biology and Control* (pp. 455–456). London: Academic Press.

Djian-Caporalino, C., Pijarowski, L., Januel, A., Lefebvre, V., Daubeze, A., Palloix, A., et al. (1999). Spectrum of resistance to root-knot nematodes and inheritance of heat-stable resistance in pepper (*Capsicum annuum* L.). *Theoretical and Applied Genetics, 99*, 496–502.

Dropkin, V. (1969). The necrotic reaction of tomatoes and other hosts resistant to *Meloidogyne*: Reversal by temperature. *Phytopathology, 59*, 1632–1637.

Duncan, L. W. (1991). Current options for nematode management. *Annual Review of Phytopathology, 29*, 469–490.

Eddaoudi, M., Ammati, M., & Rammah, A. (1997). Identification of resistance breaking populations of *Meloidogyne* on tomatoes in Morocco and their effect on new sources of resistance. *Fundamental and Applied Nematology, 20*, 285–289.

EPPO (2006). Data Sheets on Quarantine Pests. *Meloidogyne chitwoodi*. EPPO A2 List of Pests Recommended for Regulationas Quarantine Pests, No. 227, 6.

FAO (2006). FAOSTAT. Citation Database Results.FAOSTAT,

FAO, Grupo de cultivos hortícolas. (2002). (Eds.), *El Cultivo Protegido en Clima Mediterráneo* (pp. 320) Roma: Organización para las Naciones Unidas y para la Agricultura y la Alimentación.

Ferris, H. (1978). Nematode economic thresholds: Derivation, requirements, and theoretical implications. *Journal of Nematology*, 10, 341–350.

Ferris, H., & Noling, J. W. (1987). Analysis and prediction as a basis for management decisions. In: R. H. Brown & B. R. Kerry (Eds.), *Principles and practice of nematode control in crops* (pp. 49–85). Australia: Academic Press.

Ferris, H., Roberts, P. A., & Thomason, I. J. (1985). Nematodes. In: Project, University of California Statewide Integrated Pest Management. (Eds.), *Integrated pest management for tomatoes* (pp. 60–65). Division of Agriculture and Natural Resources. Publication 3274.

Gamliel, A., & Stapleton, J. J. (1993). Characterization of antifungal volatile compounds evolved from solarized soil amended with cabbage residues. *Phytopathology, 83*, 899–905.

Godfrey, G. H., & Oliveira, J. (1932). The development of the root-knot nematode in relation to root tissues of pineapple and cowpea. *Phytopathology, 22*, 325–348.

Goodell, P. B., & Ferris, H. (1989). Influence of environmental factors on the hatch and survival of *Meloidogyne incognita*. *Journal of Nematology, 21*, 328–334.

Greco, N., Brandonisio, A., & Elia, F. (1985). Control of *Ditylenchus dipsaci, Heterodera carotae* and *Meloidogyne javanica* by solarization. *Nematologia Mediterranea, 13*, 191–197.

Greco, N., Brandonisio, A., & Elia, F. (1992). Efficacy of SIP 5561 and soil solarization for management of *Meloidogyne incognita* and *M. javanica* on tomato. *Nematologia Mediterranea, 20*, 13–16.

Greco, N., & Esmenjaud, D. (2004). Management strategies for nematode control in Europe. In: R. Cook & D. J. Hunt (Eds.), *Nematology Monographs & Perpectives. Proceedings of the Fourth International Congress of Nematology* (pp. 33–43). Spain: Tenerife. The Nederlands: Brill, Leiden.

Grujicic, G. (1975). Root knot nematodes (*Meloidogyne* spp.) on kitchen garden vegetables and possibilities of their control by preparations which are not phytotoxic. *Agronomski Glasnik, 37*, 23–24.

Grujicic, G. (1974). Studies on plant parasitic nematodes of maize plants. *Biljna Zastita, 5*, 193.

Grujicic, G. & Paunovic, M. (1971). A contribution to the study of the root-knot nematode (*Meloidogyne hapla* Chitwood). *Zastita Bilja, 22*, 112–113.

Hallmann, J., Hänisch, D., Braunsmann, J., & Klenner, M. (2005). Plant-parasitic nematodes in soil-less culture systems. *Nematology, 7*, 1–4.

Hanna, H. Y., Colyer, P. D., Kirkpatrick, T. L., Romaine, D. J., & Vernon, P. R. (1994). Feasibility of improving cucumber yield without chemical control in soils suceptible to nematode buildup. *Hortscience, 29*, 1136–1138.

Hare, W. W. (1956). Resistance in pepper to *Meloidogyne incognita acrita. Phytopathology, 46*, 98–104.

Hartman, K. M., & Sasser, J. N. (1985). Identification of *Meloidogyne* species on the basis of differential host test and perineal pattern morphology. In: K. R. Barker, C. C. Carter & J. N. Sasser (Eds.), *An advanced treatise on Meloidogyne. Volume II: Methodology* (pp. 69–77). Raleigh, North Carolina, USA: North Carolina State University Graphics.

Hashim, Z. (1979). A preliminary report on the plant-parasitic nematodes in Jordan. *Nematologia Mediterranea, 7*, 177–186.

Hendy, H., Pochard, E., & Dalmasso, A. (1985). Transmission héréditaire de la résistance aux nématodes *Meloidogyne* Chitwood (Tylenchida) portée par deux lignées de *Capsicum annuum* L.: étude de descendances homozygotes issues d'androgenése. *Agronomie, 5*, 93–100.

Hirschmann, H. (1986). *Meloidogyne hispanica* n. sp. (Nematoda: Meloidogynidae), the "Seville root-knot nematode". *Journal of Nematology, 18*, 520–532.

Hooper, D. J. & Evans, K. (1993). Extraction, identification and control of plant parasitic nematodes. In Plant parasitic nematodes in temperate agriculture. CAB International, Wallingford, UK, pp. 1–59.

Hussey, R. S., & McGuire, J. M. (1987). Interactions with other organisms. In: R. H. Brown & B. R. Kerry (Eds.), *Principles and practice of nematode control in crops* (pp. 294–328). Australia: Academic Press.

Ibrahim, I. K. A., Ibrahim, I. A., & Rezk, M. A. (1972). Pathogenicity of certain parasitic nematodes on rice. *Alexandria Journal of Agricultural Research, 20*, 175–181.

Ibrahim, I. K. A., & Rezk, M. A. (1988). The root-knot nematode - a major problem in crop production in Egypt. In: M. A. Maqbool, A. M. Golden, A. Ghaffar & L. R. Krusberg (Eds.), *Advances in Plant Nematology* (pp. 81–98). Karachi, Pakistan: Proceedings of the U.S.-Pakistan International Workshop on Plant Nematology, April 6–8, 1986.

Ibrahim, Y. K. A. (1985). The status of root-knot nematodes in the middle east, region VII of the international *Meloidogyne* project. In: K. R. Barker, C. C. Carter & J. N. Sasser (Eds.), *An advanced treatise on Meloidogyne. Volume II: Methodology* (pp. 373–378). Raleigh, North Carolina, USA: North Carolina State University Graphics.

Jacquet, M., Bongiovanni, M., Martinez, M., Verschave, P., Wajnberg, E., & Castagnone-Sereno, P. (2005). Variation in resistance to the root-knot nematode *Meloidogyne incognita* in tomato genotypes bearing the Mi gene. *Plant Pathology, 54*, 93–99.

Jiménez-Millan, F., Bello, A., Arias, M., & López Pedregal, J. M. (1964). Morfología de las especies del género *Meloidogyne* (Nematoda) de varios focos de infección de cultivos españoles. *Boletin De La Real Sociedad Española De Historia Natural*, 143–153.

Kaplan, D. T., & Keen, N. T. (1980). Mechanisms conferring plant incompatibility to nematode. *Revue de Nématologie, 3*, 123–124.

Karajeh, M., Abu-Gharbieh, W., & Masoud, S. (2005a). Virulence of root-knot nematodes, *Meloidogyne* spp., on tomato bearing the Mi gene for resistance. *Phytopathologia Mediterranea, 44*, 24–28.

Karajeh, M., Abu-Gharbieh, W., & Masoud, S. (2005b). First report of the root-knot nematode *Meloidogyne arenaria* race 2 from several vegetable crops in Jordan. *Plant Disease, 89*, 206.

Karpouzas, D. G., Giannakou, I. O., Walker, A., & Gowen, S. R. (1999). Reduction in biological efficacy of ethoprophos in a soil from Greece due to enhanced biodegradation: comparing bioassay with laboratory incubation data. *Pesticide Science, 55*, 1089–1094.

Karpouzas, D. G., Hatziapostolou, P., Papadopoulou-Mourkidou, E., Giannakou, I. O., & Geogiadou, A. (2004). The enhaced biodegradation of fenamiphos in soils from previosly trated sites and effect of soil fumigants. *Environmental Toxicology and Chemistry, 23*, 2099–2107.

Katan, J., Greenberger, A., Laon, H., & Grinstein, A. (2007). Solar heating by polyethylene mulching for the control of diseases caused by soil-borne pathogens. *Phytopathology, 66*, 683–688.

Khan, M. W. (1982). State of knowledge of root-knot nematodes in Libyan Jamahiriya. Proceedings of the Second Research and Planning Conference on Root-Knot Nematodes *Meloidogyne* spp. Region VII. Raleigh, NC: North Carolina State University Graphics.

Khan, M. W., & Dabaj, K. H. (1980). Some preliminary observations on root-knot nematodes of vegetable crops in Tripoli region of Libyan Jamahiriya. *Libyan Journal of Agriculture, 9*, 127–136.

Kirkegaard, J., & Sarwar, M. (1998). Biofumigation potential of Brassicas - I. Variation in glucosinolate profiles of diverse field-grown Brassicas. *Plant and Soil, 201*, 71–89.

Koliopanos, C. N. (1982). Contribution to the study of the root-knot nematode (*Meloidogyne* spp.) in Greece. Proceedings of the Second Research and Planning Conference on Root-Knot Nematodes *Meloidogyne* spp. Region VII. Raleigh, NC: North Carolina State University Graphics.

Kyrou, N. C. (1976). New records of nematodes in Greece. *Plant Disease Reporter, 60*, 630.

Lagüe, C., Gill, J., & Péloquin, G. (2001). Thermal control in plant protection. In: C. Vincent, B. Panneton & F. Fleurat-Lessard (Eds.), *Physical control methods in plant protection* (pp. 35–46). Berlin, Germany: Springer-Verlag.

Lahtinen, A. E., Trudgill, D. L., & Tiilikkala, K. (1988). Threshold temperature and minimum time requirements for the complete life cycle of *Meloidogyne hapla* from northern Europe. *Nematologica, 34*, 443–451.

Lamberti, F. (1981). Plant nematode problems in the Mediterranean region. *Helminthologial Abstracts, Serie B, Plant Nematology. CAB, 50*, 145–166.

Lopez-Llorca, L. V., Jansson, H. B., Macià, J. G., & Salinas, J. (2006). Nematophagous fungi as root endophytes. In: B. Schulz, C. Boyle & T. Sieber (Eds.), *Microbial root endophytes* (pp. 191–206). Berlin, Germany: Springer.

Macaron, J. H., Laterrot, P., Davet, K., Makkouk, K., & Revise A. (1975). A study of the behaviour in the Lebanon of varieties and hybrids of *Lycopersicon esculentum* Mill. Resistant to nematodes, tobacco mosaic virus and the chief parasitic fungi. *Poljoprivredna Znanstvena Smotra, Agriculrurae Conspectus Scientificus 1976, 39*, 113–119.

Madulu, J. D., & Trudgill, D. L. (1994). Influence of temperature on the development and survival of *Meloidogyne javanica*. *Nematologica, 40*, 230–243.

MBTOC. (2006). Alternatives to methyl bromide for soils uses. Montreal protocol on substances that deplete the ozone layer. United Nations Environment Programme (UNEP). Report of the Methyl Bromide Technical Options Committee.

McSorley, R. & Dickson, D. W. (1991). Determining consistency of spatial dispersion of nematodes in small plots. *Journal of Nematology*, 23, 65–72.

McSorley, R., & Parrado, J. L. (1982). Estimating relative error in nematode numbers from single soil samples composed of multiple cores. *Journal of Nematology*, 14, 522–529.

McSorley, R., & Phillips, M. S. (1993). Modeling population dynamics and yield losses and their use in nematode management. In: K. Evans, D. L. Trudgill & J. M. Webster (Eds.), *Plant parasitic nematodes in temperate agriculture* (pp. 61–85). Wallingford, Oxon, UK: CAB International.

Melakeberhan, H., Xu, A., Kravchenco, A., Mennan, S., & Riga, E. (2006). Potential use of arugula (*Eruca sativa* L.) as a trap crop for *Meloidogyne hapla*. *Nematology, 8*, 793–799.

Minz, G. (1956). How the potato root nematode was discovered in Israel. *Plant Disease Reporter*, 40, 688–699.

Moens, M. G. (1985). Disinfestation of tomato nurseries. *International Nematology Network Newsletter, 2*, 14–15.

Mojtahedi, H., Santo, G. S., & Pinkerton, J. (1991). Efficacy of ethoprop on *Meloidogyne hapla* and *M. chitwoodi* and enhanced biodegradation in soil. *Journal of Nematology, 23*, 372–379.

Netscher, C. (1976). Observations and preliminary studies on the occurrence of resistance breaking biotypes of *Meloidogyne* spp. on tomato. *Cahier Orstom Série Biologiques, 11*, 173–178.

Netscher, C., & Luc, M. (1974). Nématodes associés aux cultures maraïchères en Mauritanie. *Agronomie Tropicale (Nogent-sur-Marne), 29*, 697–701.

Netscher, C., & Mauboussin, J. C. (1973). Resultats d'un essai condernant l'efficacité comparée d'une varieté resistante et de certains nematicides contre *Meloidogyne javanica. Cahier Orstom Série Biologiques, 21*, 97–102.

Netscher, C., & Sikora, R. A. (1990). Nematode parasites of vegetables. In: M. Luc, R. Sikora & J. Bridge (Eds.), *Plant parasitic nematodes in subtropical and tropical agriculture* (pp. 237–283). Wallingford, Oxon, UK; CAB International.

Nico, A. I., Rapoport, H. F., Jiménez-Díaz, R. M., & Castillo, P. (2002). Incidence and population density of plant-parasitic nematodes associated with olive planting stocks at nurseries in southern Spain. *Plant Disease, 86*, 1075–1079.

Noe, J. P. (1985). Analysis and interpretation of data from nematological experiments. In: K. R. Barker, C. C. Carter & J. N. Sasser (Eds.), *An advanced treatise on Meloidogyne. Volume II: Methodology* (pp. 187–196). Raleigh, North Carolina, USA: North Carolina State University Graphics.

Noling, J. W. (1997). Relative lethal dose, a time-temperature model for relating soil solarization efficacy and treatment duration for nematode control. 3-11-1997. Annual International Research Conference on Methyl Bromide Alternatives and Emissions Reductions. San Diego, California, (17) 1–4.

Nusbaum, C. J., & Ferris, H. (1973). The role of cropping systems in nematode population management. *Annual Review of Phytopathology, 11*, 423–440.

Orion, D., Nessim-Bistritsky, B., & Hochberg, R. (1982). Using color infrared aerial photography to study cotton fields infested with *Meloidogyne incognita. Plant Disease, 66*, 105–108.

Ornat, C. (1998). Epidemilogía de *Meloidogyne* en cultivos hortícolas. PhD; Thesis. Facultat De Biologia. Universitat De Barcelona, España.

Ornat, C., Sorribas, F. J., Verdejo-Lucas, S., & Galeano, M. (2001a). Effect of planting date on development of *Meloidogyne javanica* on lettuce in northeastern Spain. *Nematropica, 31*, 148–149.

Ornat, C., & Verdejo-Lucas, S. (1999). Distribución y densidad de población de *Meloidogyne* spp. en cultivos hortícolas de la comarca de El Maresme (Barcelona). Distribution and population density of *Meloidogyne* spp. on vegetable crops in El Maresme county (Barcelona, Spain). *Investigación Agraria: Producción y Protección Vegetales, 14*, 191–201.

Ornat, C., Verdejo-Lucas, S., & Sorribas, F. J. (1997). Effect of the previous crop on population densities of *Meloidogyne javanica* and yield of cucumber. *Nematropica, 27*, 85–90.

Ornat, C., Verdejo-Lucas, S., & Sorribas, F. J. (2001b). A population of *Meloidogyne javanica* in Spain virulent to the Mi resistance gene in tomato. *Plant Disease, 85*, 271–276.

Ornat, C., Verdejo-Lucas, S., Sorribas, F. J., & Tzortzakakis, E. A. (1999). Effect of fallow and root destruction on survival of root-knot and root-lesion nematodes in intensive vegetable cropping systems. *Nematropica, 29*, 5–16.

Philis, J. (1983). Occurrence of *Meloidogyne* spp. and races on island of Ciprus. *Nematologia Mediterranea, 11*, 13–19.

Philis, J., & Vakis, N. (1977). Resistance of tomato varieties to the root-knot nematode *Meloidogyne javanica* in Ciprus. *Nematologia Mediterranea, 5*, 39–44.

Ploeg, A. T., & Maris, P. C. (1999). Effects of temperature on the duration of the life cycle of a *Meloidogyne incognita* population. *Nematology, 1*, 389–393.

Prot, J. C. (1984). A naturally occurring resistance breaking biotype of *Meloidogyne arenaria* on tomato. Reproduction and pathogenicity on tomato cultivars Roma and Rossol. *Revue de Nématologie, 7*, 23–28.

Prot, J. C., & Ferris, H. (1992). Sampling approaches for extensive surveys in nematology. *Journal of Nematology*, 24 (Supplement), 757–764.

Pyrowolakis, E. (1975). Studies on the distribution of the genus *Meloidogyne* on the island of Crete. *Zeitschrift Fur Pflanzenkrankheiten Und Pflanzenschutz, 82*, 750–755.

Rich, J. R., & Olson, S. M. (2004). Influence of Mi-gene resistance and soil fumigant application in first crop tomato on root-galling and yield in a succeeding cantaloupe crop. *Nematropica, 34*, 103–108.

Rich, J. R., & Olson, S. M. (1999). Utility of *Mi* gene resistance in tomato to manage *Meloidogyne javanica* in North Florida. *Journal of Nematology, 31*, 715–718.

Riggs, R. D., & Winstead, N. N. (1959). Studies on resistance in tomato to root-knot nematodes and on the occurrence of pathogenic biotypes. *Phytopathology, 49*, 716–724.

Roberts, P. A. (1995). Conceptual and practical aspects of variability in root-knot nematodes related to host plant resistance. *Annual Review of Phytopathology, 33*, 199–221.

Roberts, P. A. (2002). Concepts and consequences of resistance. In: J. L. Starr, R. Cook & J. Bridge (Eds.), *Plant resistance to parasitic nematodes* (pp. 23–41). Wallingford, Oxon, UK; CABI Publishing.

Roberts, P. A., & Thomason, I. J. (1989). A review of variability in four *Meloidogyne* spp. measured by reproduction on several hosts including *Lycopersicon. Agricultural Zoology Reviews, 3*, 225–252.

Roberts, P. A., Van Gundy, S. D., & McKinney, H. E. (1981). Effects of soil temperature and planting date of wheat on *Meloidogyne incognita* reproduction, soil populations, and grain yield. *Journal of Nematology, 13*, 338–345.

Robertson, L., Lopez-Perez, J. A., Bello, A., Diez-Rojo, M. A., Escuer, M., Piedra-Buena, A., et al. (2006). Characterization of *Meloidogyne incognita, M. arenaria* and *M. hapla* populations from Spain and Uruguay parasitizing pepper (*Capsicum annuum* L.). *Crop Protection, 25*, 440–445.

Ros, C., Guerrero, M. M., Martínez, M. A., Barceló, N., Martínez, M. C., Rodríguez, I., et al. (2005). Resistant sweet pepper rootstocks integrated into the management of soilborne pathogens in greenhouse. *Acta-Horticulturae, 698*, 305–310.

Runia, W. T. (2000). Steaming methods for soils and substrates. *Acta Horticulturae, 532*, 115–123.

Saad, A. T., & Tanveer, M. (1972). *FAO Plant Protection Bulletin, 20*, 31–35.

Sanz, R., Escuer, M., & López-Pérez, J. A. (1998). Alternatives to Methyl Bromide for root-knot nematode control in cucurbits. In: A. Bello, M. González, M. Arias & R. Rodríguez-Kabana (Eds.), *Alternatives to Methyl Bromide for the Southern European Countries* (pp. 73–84). Madrid: DG XI, EU, CSIC.

Sasanelli, N., Di Vito, M., & Zaccheo, G. (1992). Population densities of *Meloidogyne incognita* and growth of cabbage in pots. *Nematologia Mediterranea, 20*, 21–23.

Sasser, J. N. (1954). Identification and Host parasite relationships of certain root-knot nematodes (*Meloidogyne* spp.). Technical Bulletin College Park, Maryland Agricultural Experiment Station, A-77, 21.

Sasser, J. N. (1966). Behavior of *Meloidogyne* spp. from various geographical locations on ten host differentials. *Nematologica, 12*, 97.

Sasser, J. N., & Carter, C. C. (1985). Overview of the international *Meloidogyne* project 1975-1984. In: J. N. Sasser & C. C. Carter (Eds.), *An advanced treatise on Meloidogyne. Volume I: Biology and control* (pp. 19–24). Raleigh, North Carolina, USA: North Carolina State University Graphics.

Scotto La Massese, C. (1961). Overview of problems posed by phytoparasitic nematodes in Algeria. In: *Les nématodes* (pp. 1–27). Paris, France: ACTA.

Sellami, S., Lonici, M., Eddoud, A., & Besenghir, H. (1999). Distribution et plantes hotes associées aux *Meloidogyne* sous abris plastiques en Algerie. Distribution and host plants of *Meloidogyne* in plastic houses in Algeria. *Nematologia Mediterranea, 27*, 295–301.

Shepherd, J. A., & Barker, K. R. (1990). Nematode parasites of Tobacco. In: M. Luc, R. A. Sikora & J. Bridge (Eds.), *Plant parasitic nematodes in subtropical and tropical agriculture* (pp. 493–517). Wallingford, UK: CAB International.

Sikora, R. A. (1992). Management of the antagonistic potential in agricultural ecosystems for the biological control of plant parasitic nematodes. *Annual Review of Phytopathology, 30*, 245–270.

Sikora, R. A., Bridge, J., & Starr, J. L. (2005). Management practices: an overview of integrated nematode management technologies. In: M. Luc, R. A. Sikora & J. Bridge (Eds.), *Plant parasitic nematodes in subtropical an tropical agriculture* (pp. 793–825). Wallingford, UK; CAB Internacional 2005.

Sikora, R. A., & Fernandez, E. (2005). Nematode Parasites of vegetables. In: M. Luc, R. A. Sikora & J. Bridge (Eds.), *Plant parasitic nematodes in subtropical an tropical agriculture* (pp. 319–391). Wallingford, UK; CAB Internacional 2005.

Smith, P. G. (1944). Embryo culture of a tomato species hybrid. *Proceedings of American Society Horticultural Sciences, 44*, 413–416.

Sorribas, F. J. (1996). Incidencia de *Meloidgyne* en el área de producción hortícola del Baix Llobregat. PhD Thesis. Facultat De Biologia. Universitat De Barcelona, España.

Sorribas, F. J., Ornat, C., Galeano, M., & Verdejo-Lucas, S. (2003). Evaluation of a native and introduced isolate of *Pochonia chlamydosporia* against *Meloidogyne javanica. Biocontrol Science and Technology, 13*, 707–714.

Sorribas, F. J., Ornat, C., Verdejo-Lucas, S., Galeano, M., & Valero, J. (2005a). Effectiveness and profitability of the Mi-resistant tomatoes to control root-knot nematodes. *European Journal of Plant Pathology, 111*, 29–38.

Sorribas, F. J., Ornat, C., Verdejo-Lucas, S., Talavera, M., Valero, J., Torres, J., et al. (2005b). Development of predictive models for managment of *Meloidogyne* on tomato crops. *Nematropica, 35*, 99.

Sorribas, F. J., & Verdejo-Lucas, S. (1994). Survey of *Meloidogyne* spp. in tomato production fields of Baix Llobregat county, Spain. *Journal of Nematology, 26*, 731–736.

Sorribas, F. J., & Verdejo-Lucas, S. (1999). Capacidad parasitaria de *Meloidogyne* spp. en cultivares de tomate resistente. *Investigación Agraria: Producción y Protección Vegetales, 14*, 237–247.

Southards, C. J., & Priest, M. F. (1973). Variation in pathogenicity of seventeen isolates of *Meloidogyne incognita*. *Journal of Nematology, 5*, 63–67.

Stapel, L. H. M., & Amsing, J. J. (2004). Populations dynamics and damage potential of the root-knot nematode *Meloidogyne hapla* on roses. Proceeeding of the XXVII ESN International Symposium, Rome, 14–18 June, 86–87.

Stapleton, J. J. (2000). Soil solarization in various agricultural production systems. *Crop Protection, 19*, 837–841.

Starr, J. L., Bridge, J., & Cook, R. (2002). Resistance to plant parasitic nematodes: History, current use and future potential. In: J. L. Starr, R. Cook & J. Bridge. *Plant resistance to parasitic nematodes* (pp. 1–22). Wallinfors, Oxon, UK; CABI Publishing.

Stirling, G. R. (1991). (Eds.), *Biological control of plant parasitic nematodes* (pp.) Wallingford, UK.

Tarjan, A. C. (1953). Geographic distribution of some *Meloidogyne* species in Israel. *Plant Disease Reporter, 37*, 315–316.

Tayar A. (1982). Seed treatment for control of *M. incognita* on cotton. Proceedings of the Second Research and Planning Conference on Root-Knot Nematodes *Meloidogyne* spp. Region VII Raleigh, NC: North Carolina State University Graphics.

Taylor, A. L., & Sasser, J. N. (1978). Biology, identification and control of root-knot nematodes. Coop. Publ. Deps. Plant Pathology, North Carolina State University and U.S. Agency for International Development, Raleigh, North Carolina State University Graphics, 111 pp.

Taylor, A. L., Sasser, J. N., & Nelson, L. A. (1982). (Eds.), *Relationship of climate and soil characteristics to geographical distribution of Meloidogyne species in agricultural soils* (pp. 65). North Carolina USA: Department of Plant Pathology, North Carolina State University & US Agency for International Development Raleigh.

Taylor, C. E. (1979). *Meloidogyne* interrelationships with micro-organisms. In: F. Lamberti & C. E. Taylor (Eds.), *Root-knot nematodes (*Meloidogyne *species): Systematics, biology and control* (pp. 375–398). London. UK: Academic Press.

Thies, J. A., Davis, R. F., Mueller, J. D., Fery, R. L., Langston, D. B., & Miller, G. (2004). Double-cropping cucumbers and squash after resistant bell pepper for root-knot nematode management. *Plant Disease, 88*, 589–593.

Thies, J. A., & Fery, R. L. (2000). Characterization of resistance conferred by the N gene to *Meloidogyne arenaria* Races 1 and 2, *M-hapla*, and *M-javanica* in two sets of isogenic lines of *Capsicum annuum* L. *Journal of the American Society for Horticultural Science, 125*, 71–75.

Thomason, I. J., Van Gundy, S. D., & Kirkpatrick, J. D. (1964). Motility and infectivity of *Meloidogyne javanica* as affected by storage time and temperature in water. *Phytopathology, 54*, 192–195.

Towson, A. J., & Apt, W. J. (1983). Effect of soil water potential on survival of *Meloidogyne javanica* in fallow soil. *Journal of Nematology, 15*, 110–114.

Trivedi, P. C., & Barker, K. R. (1986). Management of nematodes by cultural practices. *Nematropica, 16*, 213–236.

Trudgill, D. L. (1995a). Host and plant temperature effects on nematode development rates and nematode ecology. *Nematologica, 41*, 398–404.

Trudgill, D. L. (1995b). An assessment of the relevance of thermal time relationships to nematology. *Fundamental and Applied Nematology, 18*, 407–417.

Trudgill, D. L., & Perry, J. N. (1994). Thermal time and ecological strategies: a unifying hypothesis. *Annals of Applied Biology, 125*, 521–532.

Tyler, J. (1933). Development of the root-knot nematode as affected by temperature. *Hilgardia, 7*, 391–413.

Tzortzakakis, E. A. (2000). The effect of *Verticillium chlamydosporium* and oxamyl on the control of *Meloidogyne javanica* on tomatoes grown in a plastic house in Crete, Greece. *Nematologia Mediterranea, 28*, 249–254.

Tzortzakakis, E. A., Adam, M. A. M., Blok, V. C., Paraskevopoulos, C., & Bourtzis, K. (2005). Occurrence of resistance-breaking populations of root-knot nematodes on tomato in Greece. *European Journal of Plant Pathology, 113*, 101–105.

Tzortzakakis, E. A., & Gowen, S. R. (1994). The evaluation of *Pasteuria penetrans* alone and in combination with oxamyl, plant resistance and solarization for control of *Meloidogyne* spp. on vegetables grown in greenhouses of Crete. *Crop Protection, 13*, 455–462.

Tzortzakakis, E. A., & Gowen, S. R. (1996). Occurrence of a resistance breaking pathotype of *Meloidogyne javanica* on tomatoes in Crete, Greece. *Fundamental and Applied Nematology, 19*, 283–288.

Tzortzakakis, E. A., & Petsas, S. E. (2003). Investigation of alternatives to methyl bromide for management of *Meloidogyne javanica* on greenhouse grown tomato. *Pest Management Science, 59*, 1311–1320.

Tzortzakakis, E. A., Phillips, M. S., & Trudgill, D. L. (2000). Rotational management of *Meloidogyne javanica* in a small scale greenhouse trial in Crete, Greece. *Nematropica, 30*, 167–175.

Tzortzakakis, E. A., Trudgill, D. L., & Phillips, M. S. (1998). Evidence for a dosage effect of the Mi gene on partially virulent isolates of *Meloidogyne javanica*. *Journal of Nematology, 30*, 76–80.

Tzortzakakis, E. A., Verdejo-Lucas, S., Ornat, C., Sorribas, F. J., & Goumas, D. E. (1999). Effect of a previous resistant cultivar and *Pasteuria penetrans* on population densities of *Meloidogyne javanica* in greenhouse grown tomatoes in Crete, Greece. *Crop Protection, 18*, 159–162.

Van Gundy, S. D. (1985). Ecology of *Meloidogyne* spp. emphasis on environmental factors affecting survival and pathogenicity. In: J. N. Sasser & C. C. Carter (Eds.), *An advanced treatise on Meloidogyne Volume I: Biology and control* (pp. 177–182). Raleigh, North Carolina, USA: North Carolina State University Graphics.

Verdejo-Lucas, S. (1999). Nematodes. In: R. Albajes, M. L. Gullino, J. C. Van Lanteren & Y. Elad (Eds.), *Integrated pest and disease management in greenhouse crops* (pp. 61–68). The Netherlands: Kluwer Academic.

Verdejo-Lucas, S., Buñol, J., Sorribas, F. J., & Ornat, C. (2004). Eficacia del porta-injerto de tomate frente a cultivares portadores del gen Mi para el manejo del nematodo *Meloidogyne*. *Phytoma España, 158*, 13–18.

Verdejo-Lucas, S., Español, M., Ornat, C., & Sorribas, F. J. (1997). Occurrence of *Pasteuria* spp. in northeastern Spain. *Nematologia Mediterranea, 25*, 109–112.

Verdejo-Lucas, S., Ornat, C., Sorribas, F. J., & Stchiegel, A. (2002). Species of root-knot nematodes and fungal egg parasites recovered from vegetables in Almeria and Barcelona, Spain. *Journal of Nematology, 34*, 405–408.

Verdejo-Lucas, S., Sorribas, F. J., Ornat, C., & Galeano, M. (2003). Evaluating *Pochonia chlamydosporia* in a double-cropping system of lettuce and tomato in plastic houses infested with *Meloidogyne javanica*. *Plant Pathology, 52*, 521–528.

Verdejo-Lucas, S., Sorribas, J., & Puigdomènech, P. (1994). Pérdidas de producción en lechuga y tomate causadas por *Meloidogyne javanica* en invernadero. *Investigación Agraria: Producción y Protección Vegetales, 2*, 395–400.

Waldmann, H. (1971). A new method of controlling the root knot nematode? *Gesunde Pflanzen, 23*, 227–232.

Walters, S. A., Wehner, T. C., & Barker, K. R. (1999). Greenhouse and field resistance in cucumber to root-knot nematodes. *Nematology, 1*, 279–284.

Walters, S. A., Wehner, T. C., & Barker, K. R. (1996). NC-42 and NC-43: Root-knot nematode-resistance cucumber germplasm. *Hortscience, 31*, 1246–1247.

Wang, M., & Goldman, I. L. (1996). Resistance to Root Knot Nematode (*Meloidogyne hapla* Chitwood) in Carrot Is Controlled by Two Recessive Genes. *Journal of Heredity, 87*, 119–123.

Webster, J. M. (1985). Interaction of *Meloidogyne* with fungi on crop plants. In: J. N. Sasser & C. C. Carter, (Eds.) *An Advanced Treatise on Meloidogyne Volume I: Biology and Control* (pp. 183–192). Raleigh, North Carolina, USA: North Carolina State University Graphics.

Williamson, V. M. (1998). Root-knot nematode resistance genes in tomato and their potential for future use. *Annual Review of Phytopathology, 36*, 277–293.

Zasada, I. A., & Ferris, H. (2004). Nematode suppression with brassicaceous amendments: application based upon glucosinolate profiles. *Soil Biology & Biochemistry, 36*, 1017–1024.

15

AURELIO CIANCIO

MODELING NEMATODES REGULATION BY BACTERIAL ENDOPARASITES

Istituto per la Protezione delle Piante, CNR, Bari, Italy

Abstract. Some aspects of nematodes regulation by *Pasteuria penetrans* and other endoparasitic Gram-negative bacteria are revised, together with application modeling tools, in reference to their biocontrol potentials. A review is given about general and more detailed epidemiological models and their applications. The models constants accounting for basic biological factors of the parasites and hosts biology and interactions, are also discussed. Some properties of applied models, including the phase plane representation, the identification of equilibrium points and their cyclic relationships are revised, in reference to the study of field and time series data. A modeling scheme for *Pasteuria* and nematode dynamics, accounting for the host life cycle and including its developmental stages, is also proposed. Finally, experimental and practical issues concerning nematodes biological control are also discussed.

1. INTRODUCTION

The attention of producers and consumers for organic productions increased in recent years. Organic productions are characterized by the exclusive use of natural resources or of compounds already present in nature. In Italy, the surfaces cultivated with these technologies progressively increased in the last decade, reaching in the year 2000 almost one million ha, with further increments expected in the subsequent period. Horticultural and industrial crops represent approx. 10% of these surfaces, reflecting a significant component of the market and consumers demand for organic food. The expansion of these productions requires the development of new tools, among which new products and procedures based on biological control agents, as practical alternatives to pesticides.

Plant parasitic nematodes are naturally controlled by several biological antagonists. Among them several fungi are known since the end of the XIX century, thanks to the pioneering observations carried out in agricultural or uncultivated soils (Woronin, 1870; Drechsler, 1934; Duddington, 1957). These studies were subsequently and progressively integrated by observations focusing on the parasitic and predatory activities displayed by nematophagous species commonly isolated from soil (Gray, 1988; Stirling, 1991). Fungi were the first group of antagonists studied, probably because they can easily be cultured *in vitro* and because of the simple microscopy procedures required for recognition of the hyphal structures involved in parasitism or predation (see Chapters 2 and 3 in this volume for revision of nematophagous fungi). A second group of nematode antagonists is represented by soil bacteria, which are the focus of this chapter.

A. Ciancio & K. G. Mukerji (eds.), Integrated Management and Biocontrol of Vegetable and Grain Crops Nematodes, 321–337.

There is today a general, increasing concern about the role, biodiversity and protection of the microbial components of soil. This view arose after the advent of DNA-based technologies. The number of species which can be recovered from soil with traditional methods (i.e. culturing) are known to be several orders of magnitude lower than the real number of species inhabiting soil trophic niches. When using soil DNA extraction, an estimate of $2{\cdot}10^3$ bacterial species, for example, was estimated to be present in each g of soil (Torsvik, Goksøyr, & Daae, 1990). The role and impact of all these organisms on soil functioning and productivity are, indeed, largely underestimated. It is now clear that culturable species represent only a fraction of the soil microbial biodiversity on earth, since several groups, including unculturable symbionts, parasites, endophytes and other decomposers, may remain undetected in a biodiversity census based on traditional identification, due to their trophic biology and obligate behaviour (Amman, Ludwig, & Schleifer, 1995).

Plant parasitic nematodes adapted through a long and selective evolutive process to survive and reproduce in a complex environment such as soil. In this system they are capable of multiplying in spite of a cohort of natural enemies including (apart of bacteria and fungi) aquatic fungi, amoebae or other invertebrate predators like nematodes (see Chapter 1, this volume), tardigrades or mites (Sayre & Starr 1988; Gray, 1988; Stirling, 1991). This complex of species is a fundamental component of the rizhosphere, playing a key role in sustaining fertility through the mobility of mineral elements.

Although the majority of soil microbial species has functions related to the decomposition processes and soil nutrients recycling, it is commonly found that soils with high densities of plant parasitic nematodes show a high diversity of antagonistic microorganisms and invertebrate predators. The indiscriminate use of nematicides and fumigants often induces a significant reduction of these organisms, either as concerns their densities and biodiversity.

In this chapter we will discuss some aspects of the bacterial regulation of nematodes, including some modeling tools. *Pasteuria penetrans* and other bacteria (Fig. 1A, B) are promising biological control agents for management of plant parasitic nematodes. Some aspects of their biology and application are already revised in Chapters 4 and 10 of this volume. Herein we consider some issues related to the activity and ecology of *P. penetrans* and other nematode antagonistic bacteria. Modeling is expected to provide general, theoretical guidelines embracing the study of nematodes regulation in natural conditions. This broad view is needed for the development of biocontrol agents as ordinary products, suitable for the biological or integrated management of most important plant parasitic nematodes. Particular attention is given to models which may describe the role and efficacy of bacteria in natural host regulation, revealing how these species can be exploited, on the basis of parasitism biology and prevalence data.

2. NEMATODE PARASITIC BACTERIA

Among Gram-positive parasites of nematodes, *Pasteuria* spp. (Bacillaceae) are characterized by infective and durable endospores, typically cup shaped (Fig. 1A,

B). They are associated to phytoparasitic or free living nematode species with the only exception of *P. ramosa*, which is found in *Daphnia* spp. (Metchnikoff, 1888; Ebert, Rainey, Embley, & Scholz, 1996). The species mainly studied is *Pasteuria penetrans*, parasitic in root knot nematodes of the genus *Meloidogyne* (Mankau, 1975; Stirling, 1984; Sayre & Starr, 1985, 1988; Anderson et al., 1999).

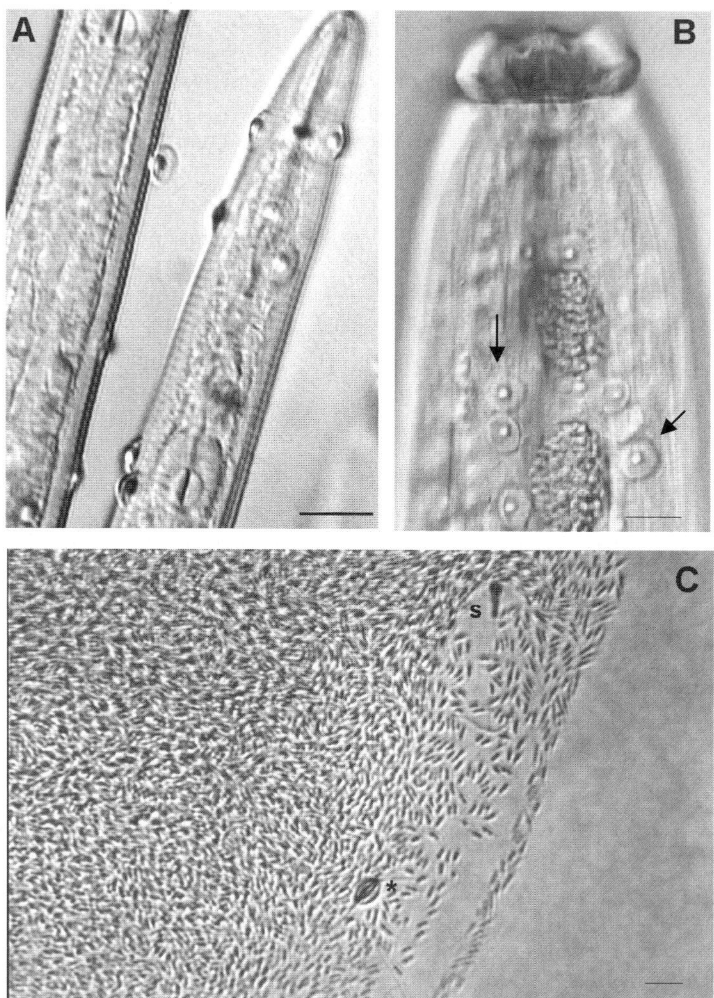

Figure 1. Bacterial parasites of nematodes include Gram + and Gram – species. Pasteuria penetrans *(Bacillaceae), parasitic on* Meloidogyne incognita *(A), is a member of an evolutive radiation of G+ species associated to widely differentiated hosts, including predatory nematodes (B, arrows). Other undescribed G – bacteria also attack* M. incognita *juveniles (C). Also in this case, the bacterial cells are released as the host nematode dies and its body eventually collapses, leaving a few remnants like the stylet (s) and median valve (asterisk). Scale bars: A, B = 10 µm; C = 5 µm.*

The genus *Pasteuria* shows a wide diversity of forms (Starr & Sayre, 1988; Stirling, 1991; Ciancio, Bonsignore, Vovlas, & Lamberti, 1994; Sturhan, Winkelheide, Sayre, & Wergin, 1994) with species sporulating in adult hosts (*P. penetrans*, *P. nishizawae*), as well as species whose endospores were observed in the host juvenile stages only (Giblin-Davis, McDaniel, & Bilz, 1990; Ciancio et al., 1994; Sturhan et al., 1994) or in both host stages (Ciancio, 1995; Galeano, Verdejo-Lucas, & Ciancio, 2003). Actually, the genus is considered to include a high number of species, whose identification is possible thanks to DNA sequencing of the 16S ribosomial gene (Ebert et al., 1996; Anderson et al., 1999; Ciancio, Leonetti, & Finetti Sialer, 2000; Preston et al., 2003; Giblin-Davis et al., 2003; Atibalentja, Noel, & Ciancio, 2004) or of some sporulation genes (Schmidt, Preston, Nong, Dickson, & Aldrich, 2004).

Table 1. Effect of P. penetrans *and efficacy of applied treatments.*

Nematode	*Efficacy**	*Reference*
M. incognita	G>90	Stirling, 1984
M. javanica	E 49 E 40–90[a]	Gowen and Tzortzakakis, 1992
Meloidogyne spp.	G 57–67 G 38–82[a] E 0–49 and 99–43[a]	Tzortzakakis and Gowen, 1994
M. incognita	G 25–31	Jonathan, Barker, Abdel-Alim, Vrain, and Dickson, 2000
M. arenaria *M. javanica*	F 56[b]	Cetintas et al., 2003

*Reduction expressed as % of corresponding controls. Variables and stages: E = eggs · g roots^{-1}; G = root gall index; F = females · g roots^{-1}.
[a]In combination with oxamyl treatments.
[b]Efficacy observed in the field for *M. arenaria* only.

Pasteuria spp. endospores are provided with parasporal fibers, responsible for host adhesion and specificity (Davies et al., 2001; Davies & Williamson, 2006). The endospore has the contemporary function of a durable and infective propagule, which is very resistant to high temperatures and dessiccation and may remain viable for a decade or more (Mankau, 1975; Stirling, 1991). Parasitic specificity is a typical trait of *Pasteuria* spp., due to a preferential adhesion shown towards the nematode species or population which they are associated to in nature (Sayre & Starr, 1985, 1988; Davies et al., 2001; Davies & Williamson, 2006). These properties appear as useful traits for the exploitation of *P. penetrans* as a root-knot nematode biological control agent, provided its mass production is achieved at a low cost. Literature data concerning the application of *P. penetrans* show potentialities for this species, which is actually considered as the most efficient biological control agent of nematodes (Table 1).

Pasteuria spp., however, is not the only bacterial group attacking nematodes. Recently, undescribed Gram-negative bacteria were observed in Southern Italy in several *Meloidogyne* spp. populations. These bacteria are parasitic in juveniles, in which a large number of cells develop after infection, which takes place by germination of adhering bacteria through the nematode cuticle. The developing disease is lethal to nematodes, which release large number of cells at death (Fig. 1C). Showing a similarity with *Pasteuria* spp. biology, also these bacteria appear unculturable. Further investigations, including the DNA sequencing of the 16S ribosomial gene, are required to elucidate their biology and phylogenetic position.

Finally, several other bacterial species, including *Pseudomonas* spp. and other Gram-negative species, were reported to control nematodes in soil, attacking different life stages, including eggs (Esnard, Potter, & Zuckerman, 1995; Siddiqi & Mahmood, 1999; Couillault & Ewbank, 2002; Hamid, Siddiqui, & Shaukat, 2003; Nour et al., 2003; Aksoy & Mennan, 2004). It is hence possible that a deeper insight into the composition and structure of the bacterial soil microflora will reveal further bacterial species or populations, capable of regulating nematodes density or inducing suppressivity.

In the next section we will review some aspects of the ecology of nematodes regulation, with particular attention to models descriptive of the behaviour of *Pasteuria* or other Gram-negative spp., with similar endoparasitic behaviour.

3. MODELING NEMATODES REGULATION

In order to check nematodes regulation in soil by associated bacterial endoparasites, it is useful to rely on a general framework concerning the mechanisms deployed in nature by the antagonists identified. For nematodes, this reference framework is not yet complete, due to the complexity of the edaphic environment and of the relationships therein occuring. A number of experimental data are, however, already available on nematodes regulation by endoparasitic fungi, providing a first insight on some general rules accounting for the basic ecology of microbial regulation (Jaffee, 1992, 2000, 2003; Jaffee, Muldoon, Phillips, & Mangel, 1990; Jaffee, Phillips, Muldoon, & Mangel, 1992; Jaffee & McInnis, 1991; Jaffee & Muldoon, 1994).

The ensemble of nematodes, antagonists and microbial soil components and arthropods (together with the roots as affected by pedologic, climatic or environmental factors), produce what we may consider as a typical complex system. These systems are common in nature, and have chaotic components which make their evolution difficult to predict, in particular for variables like the population density of one or more of their components (Ciancio & Queneherve, 2000). Some interpretative models, however, may reveal key features of the regulation mechanisms occurring in the rhizosphere, and in this view they are herein treated.

Nematodes and parasites modeling received some help from theoretical and applied studies previously carried out for the ecology and control of other pests, in particular insects. Also, the efforts deployed to monitor antagonists or parasitoids, after their introduction in the environment, provided a first basis useful to construct or evaluate already existing models. Some prudence, however, should be taken in applying models constructed for organisms having habits and behaviours different from nematodes, and in particular adapted to describe the dynamics of insects or of their parasitoids (i.e. Nicholson-Bailey's model, not treated herein). These indeed differ from nemtodes for the dimension of the corresponding microcosms, their behaviour and motility, and their environmental spread (Hassel, 1978; Jaffee et al., 1992).

3.1. Lotka–Volterra Model (LV)

A general population regulation model was described last century by Lotka and Volterra, who independently discovered a system of two equations, complex enough to be applied to a wide range of situations and targets, including competition and predation of wild vertebrate species. This model has a broad ecological application range and essentially relies on four constants accounting for some basic relationships.

The LV system represents general antagonistic effects between two species, one of which (X) acts as a prey/host whereas the second (Y) may be a predator or parasite. In general, it may also be applied to describe competition or mutual exclusion between species. In this application, nematode densities are referred to a microcosm volume (i.e. 100 cm^3 or one liter of soil, for nematodes in the plant rhizosphere, depending on the scale used) whereas prevalence (% of true parasitism or % of infected specimens) is used for the antagonist changes in time. In its simplest form at the differences used herein, each value of X and Y may be calculated through sums or differences at time intervals t, which may be days, weeks or months, depending on the time scale used when monitoring both populations. LV model equations (1) and (2) yield values fluctuating with regular cycles in time:

$$X_{t+1} = X_t + a\,X_t - b\,X_tY_t \tag{1}$$

$$Y_{t+1} = Y_t + c\,X_t\,Y_t - d\,Y_t \tag{2}$$

When applying this model to a nematode and bacterial parasite system, the constants are: a = the hosts growth rate; b = the rate of hosts decline due to prevalence; c = the rate of prevalence increase and d = the rate of prevalence decrease due to natural mortality of the parasite.

In this model it is possible to show the relationships linking two species on a single plot, called the phase plane (Fig. 2). In the only case of stable relationships, the calculated points produce a cycle with a "satellite orbit" shape which may be observed when data (real or simulated) are plotted on this plane. The cycle is produced by the observations changes in time, and runs counter clockwise (Christiansen & Fenchel, 1977).

The shape of the cycle varies in function of the initial points used for the simulated dynamics (Figs. 2, 3). Simulated density and prevalence values tend to close the cycle around a single point (called equilibrium point), as much as the initial values of the two variables approach its coordinates. In dynamical terms, at the equilibrium values the prevalence and density changes in time are 0, and their fluctuations are reduced to two continuous straight lines. On the phase plane, the equilibrium value corresponds to a single point, at which no change in the host and parasite densities occur in time ($dx/dt = dy/dt = 0$). The coordinates of the LV equilibrium points (shown by an asterisk) are given by the ratios of the constants used in the model: $X^* = d/c$ and $Y^* = a/b$.

This model was applied to the study of a population of *Xiphinema diversicaudatum* in rhizosphere of peach and of a population of the citrus nematode *Tylenchulus semipenetrans*, each associated to a specific *Pasteuria* form (Ciancio, 1995, 1996; Ciancio & Rocuzzo, 1992). A difference from the general model is that prevalence was considered instead of the parasite true density.

Although the phase plane representation of the population dynamics fits some LV cycles calculated for field populations, the model does not provide too many informations about the inner mechanisms of regulation, due to its lack of analytical details. Because of its regularity, it cannot explain too the effect of the several variables involved in nematode parasitism in nature and other stochastic effects due to external factors. Equations (1) and (2), however, improved the interpretation of data providing a better fit than other models applied to insects (i.e. Nicholson-Bailey's model), because of the instabilities produced by the latter system (Atibalentja, Noel, Liao, & Gertner, 1998).

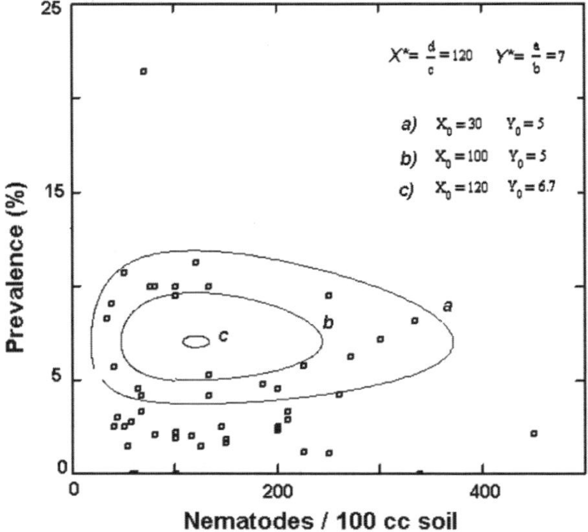

Figure 2. Density/prevalence phase plane showing the effect of different starting points (a, b, c) applied to fit a LV model to spatial sampling data (squares) from a Xiphinema diversicaudatum *field population and an associated* Pasteuria *sp. The effect of the different inital values used in the model is shown by the closure of the corresponding cycles around the equilibrium point identified by coordinates X*, Y* (from Ciancio, 1995).*

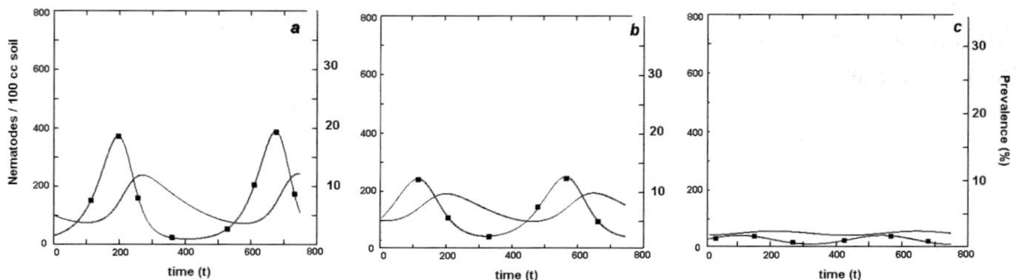

Figure 3. Time series plots (in arbitrary time units) of density (squares) and prevalence, obtained using the same inital values (a, b, c) of the LV model shown in figure 2. The starting points have an effect on the cycles, which approaches a uniform line as initial values approach the equilibrium point.

3.2. Anderson and May Model G

A series of detailed epidemiological models was developed by Anderson and May (1981), which are more complex than the LV or Nicholson-Bailey systems. These models provide also a basis useful to construct models *ad hoc* for nematodes and their associated antagonists. They allow a more detailed description of the relationships between host and parasite, and are based on parasitism transmission and densities of healthy and infected hosts. Some applications already provided a good description of nematodes regulation by the endoparasitic fungus, *Hirsutella rhossiliensis* (Jaffee, 1992, 2000, 2003; Jaffee et al., 1990).

Among others, Anderson and May Model G (AM-G) may yield a deeper insight on the nematodes dynamics, relying on propagules transmission and on the presence of two components of the infected host population, including the infected, but not yet transmissive, hosts. One feature of this model is that it may provide/forecast the densities of the bacterial propagules free in soil, since thay may be treated as distinct units (cells). In fact, and differing from fungi (due to their mycelial nature), bacterial cells are more suitable to represent the real parasite units used for the quantitative density simulations of these models.

AM-G results by three equations accounting for densities, in a microcosm volume, of healthy (X) and infected hosts (Y), and on the numbers of the parasite free propagules (W, i.e. endospores or cells free in soil). A fourth equation accounts for the total host population numbers (H = X+Y). In its simpler, non derivative form, the system is as follows (with t = time):

$$H_{t+1} = H_t + r\,H_t - \alpha Y_t \tag{3}$$

$$X_{t+1} = X_t + a\,(X_t + Y_t) - bX_t - v\,W_t X_t + \gamma Y_t \tag{4}$$

$$Y_{t+1} = Y_t + v\,W_t\,X_t - (\alpha + b + \gamma)\,Y_t \tag{5}$$

$$W_{t+1} = W_t + \lambda\,Y_t - (\mu + vH_t)\,W_t \tag{6}$$

AM-G provides, for any time step t, the variations of the cited populations through eight constants, which account for some basic biological factors governing the parasite and host biology and interactions. They are:

a = host multiplication rate
b = host mortality rate
r = growth rate of the host population ($a–b$)
α = parasitism induced host mortality
v = rate of host variation (from infected to infective)
γ = host recovery rate
λ = number of parasite's propagules produced per host
μ = mortality rate of the parasite

AM-G may be applied to the study of time series of nematodes and parasites densities, obtained through the study of their population dynamics in field or controlled conditions. It may also yield prevalence data (prevalence = Y/H). This system, however, requires the direct determination of the bacterial propagules densities in soil, a task that is not yet fully accomplished for i. e. *Pasteuria* spp., although antibody-based techniques provided the first estimation of the bacterium density in soil (Fould, Dieng, Davies, Normand, & Mateille, 2001).

AM-G represents a reliable quantitative basis needed for the analysis of the density-parasitism relationships in time and/or in space (Jaffee & McInnis, 1991) or for the identificatin of density dependent factors linking two or more organisms (Jaffee et al., 1990; Kasumimoto, Ikeda, & Kawada, 1993). Also in this model the relationships among variables may be represented using phase planes, in which equilibrium points may be calculated. As stated by Anderson and May (1981), the equilibrium point (as usual, shown by an asterisk), of the total host population H is

$$H^* = \frac{\Gamma}{\beta\,[1-(r/\alpha)-(1/\Lambda)]} \tag{7}$$

H* depends on the rate of hosts loss from the infected class $\Gamma = \alpha + b + \gamma$, on the coefficient of propagules transmission $\beta = v\lambda/\mu$ and on the total number of infective stages produced per host $\Lambda = (\alpha + b + \gamma) / \lambda$.

For the density of the antagonist propagules, the equilibrium value is

$$W^* = \frac{r\,\Gamma}{v\,(\alpha-r)} \tag{8}$$

W* depends on the growth rate of the host population $r = a–b$, on the rate of hosts loss from the infected class Γ previously described, on the mortality induced by parasitism (α), and on the rate of the host variation, from infected to infective (v).

One of the advantages offered by modeling concerns the possibility of evaluating medium and long-term effects of inundative treatments or of simple inocula. This evaluation may be simulated by increasing the densities of one or both organisms during the model runs and/or changing the levels of parasite transmission within the host population. Also, the application of AM-G and more complex models offers the possibility to study the effect and role of the biological parameters describing the host-parasite interaction or their basic biology, thus providing a first analytical tool for the investigation of the biological requirements or suitability of one or more biological control agents.

AM-G may offer a reliable interpretation of the effects of regulation between parasites like *Pasteuria* spp. and nematodes. It is, however, limited as concerns the capability to describe different life stages of the host population, which for sedentary nematodes include eggs, juveniles, pre-adult stages and females. The developement of a further model class, closer to the nematode host biology, and suitable for *Pasteuria* and other bacterial species, is described in the next section.

3.3. Modeling Pasteuria

Nematodes are characterized by different stages in their life-cycles. Modeling their density changes in time requires the inclusion of the delays related to stages development and the description of the behaviour of the specific fraction targeted by the bacterial parasite. It is known, for example, that some *Pasteuria* spp. adhere and parasitize host's juvenile stages, which do not penetrate roots and remain in soil where they die (Davies, Flynn, Laird, & Kerry, 1990). Other species, i.e. *P. penetrans* or *P. nishizawae*, allow the development of the sedentary female nematode, developing colonies inside their host body during the moulting and maturity phases, even allowing a small number of eggs to be produced (Sayre & Starr, 1988; Starr & Sayre, 1988; Noel, Atibalentja, & Domier, 2005).

For any detailed application of modeling to the *Pasteuria*-nematodes dynamics, the life cycle of the host must be accounted for and described in detail by the model, including the developmental stages within roots and the eggs densities, as in the case of sedentary species.

On the other side, although the specificity of the *Pasteuria*-nematode interaction simplifies modeling because of the bacterium obligate parasitism, the reproductive rate of the parasite must be also carefully evaluated. A number of *Pasteuria* cells is "lost" during the sporulation phase, since not all the bacterial vegetative stages within infected hosts reach maturity, in order to yield durable endospores. Other factors should also be accounted for, i.e. the time spent in soil by the endospore and required for parasporal fibers exposure; the probabilistic nature of transmission; the time period required for endospores activation and germination; the parasite specificity levels and the genetics underlying the parasporal fibers and cuticle interactions; the removal of propagules by wind or soil water; the loss of propagules due to adhesion in large numbers to J2, reducing their motility and root penetration capacity (Davies, Laird, & Kerry, 1991); the possible feeding of other soil organisms on resting endospores. Finally, also some external factors governing the energy flow proceeding from the plant through the nematode and up to the bacterium (i.e. climates, temperature, plant development and nutrition, other erbivorous and plant pathogens effects, etc.) should be included in a descriptive model.

A possible model describing the relationship of a bacterial nematode parasite like *P. penetrans* and a root knot nematode is shown in Fig. 4. At this regard, to construct the model we can start from the basic nematode life-cycle, based on eggs hatching (at rate h), followed by J2 moulting (at rate m), which eventually yield females, producing eggs (at rate α). All these stages should be introduced into the model with their corresponding mortality rates (d or df for adult females). Due to the matching (at transmission rate β) of the J2 with the *P. penetrans* endospores released in soil (at rate σ) by infected females, some nematodes are unable to enter the root system due to endospore encumberence (uncapable to move and lost from the microcosm, at rate φ) whereas a larger fraction of J2 reaches the roots (at rate m), in which they will complete the parasite cycle, yielding infected females producing new endospores. A small fraction of infected females (with mortality rate dfp) may finally be allowed to produce a few eggs, at rate γ. Also the propagules introduced into the model should display a corresponding natural mortality (μ, endospores mortality rate). In synthesis, although the nematode-bacterium relationship may appear simple, the description of the life cycles of both organisms, (without inclusion of other external ecological factors, i.e. roots development, temperature, effects due to other parasites and predators) requires the quantification of a wide array of constants. At this regard it is worth to note that constants always represent a "simplification" of a natural system, in which real functions take place.

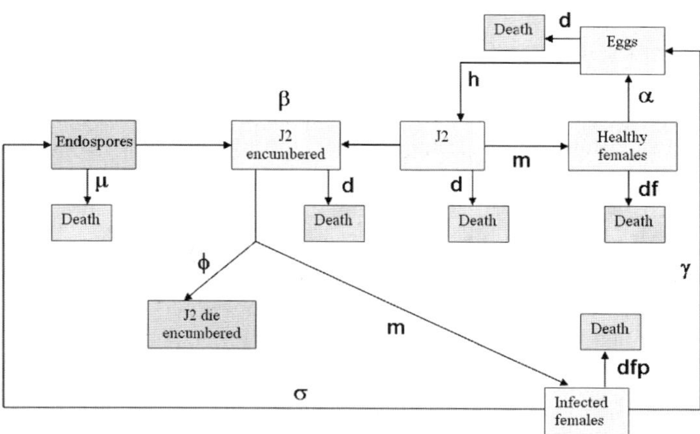

Figure 4. A model for a Pasteuria penetrans *or similar antagonists and a root-knot nematode population. Letters show constant rates accounting for changes of the model components.*

4. EXPERIMENTAL AND PRACTICAL ISSUES

To determine the extent and potentials of a biological antagonist, an appropriate series of time samplings and replications must be planned (Jaffee & McInnis, 1991; Jaffee, 1992; Verdejo-Lucas, 1992; Ciancio, 1995). Sedentary nematodes are confined within a "microcosm" corresponding, in the majority of cases, to the volume of soil explored by the plant roots. In this space, J2 mobility is functional to the search of a root penetration site, often the apex, whereas movements on longer distances and dispersion in other parcels or on a wider surface are mainly due to the action of man or to passive trasport (machinery and soil movements, irrigation, wind etc.). The likelihood of a local extintion as estimated by modeling must be referred, hence, to this rhizosphere microcosm.

It is also worth to note that samplings describing all microcosm changes in time, should remain as close as possible to the volume initially identified (i.e. a plant root system). This microcosm should be considered as a single observation, replicated in other parcels or field areas, depending on the nematode spatial dispersion and distribution. However, the same sampling action introduces a source of variation in the study, since a fraction of the population is removed, together with soil, from the microcosm. In this way, time sampling alters the population dynamics, which should follow a different path, in absence of any experimental assay. This factor must be taken into account, also considering the effects of density and prevalence values on the subsequent dynamics, as evidenced by the cycles variations experienced when changing the starting simulation points.

Since sampling is of a destructive kind, in order to analyse density dependent relationships it is useful to measure the densities of both organisms in their phase space, possibily through a single sampling plan or scheme, carried out with several replicated samples. These may then be collected to obtain a clear quantification of a density-parasitism relationship, without affording a long term temporal study (Jaffee & McInnis, 1991; Ciancio, 1995). The rationale behind this action is that by this way we can eliminate the variable "time", through the analysis of samples on a plane formed by two variables (i.e. host density and antagonist prevalence or propagules density), measured at a single moment. This procedure is based on the assumption that data from i.e. 40–50 sampling sites will show asynchronous observations (microcosms) representative of different moments of their cycle. Their contemporary projection on a single phase plane may thus allow the reconstruction, by inference, of the original cycle's path.

To determine the number of samples (N) to collect in a spatial sampling, with a given standard error to mean ratio E (i.e. 0.05 or 5%), Taylor's power law (Taylor, 1961) relating mean and variance, ($s^2 = a \cdot \overline{x}^b$) may be used, based on different combinations of observations from previous explorative samplings or time series. Parameter b is an index of aggregation, whereas parameter a is related to the sample size (McSorley, Dankers, Parrado, & Reynolds, 1985). McSorley et al. (1985) provided the relationship to determine N, once the parameters of Taylor's power law are known

$$N = (1/E)^2 \cdot a \cdot \overline{x}^{(b-2)}$$

As stated previously, a property of the simulated orbits including the majority of the observations, is that they may produce several possibile "cycles", varying in function of their starting points, which are the first initial values used for computations. This property is worth further investigations and experimental testings, since it suggests that the population dynamics observed in the field may be affected not only by the basic biological parameters of the organisms involved, but also by their mutual quantitative relationships. As shown, minimal changes of prevalence and densities values in time are found when observations approach a particular region of their phase space. The conditions leading towards the equilibrium points and/or the region of their contour, require special attention in field applied studies, since they may possibly represent field effects, concerning soil suppressivity or natural nematodes regulation.

5. CONCLUSIONS

Simulations, even if cannot "forecast" the evolution in time of a natural system due to its external perturbations and to the presence of its own chaotic components, may allow the understanding of some details of the mechanisms of nematodes natural regulation or suppressivity. Simulations show that the behaviour and dynamics of a simple system including a host and i.e. a bacterial parasite population is not only affected by the biological constants characterizing the two organisms, but also by the densities at which they occur. In some cases, changes in one or more constant/components of a model during a simulation (including the initial points used to start the model) may yield a cycle path leading to the extinction of one or both components (i.e. a local extinction may be considered when the cycle orbit becomes wide enough to reach one of the axes). Furthermore, by this way it is possible to estimate the doses and the time required to reach an equilibrium between host and parasite, or to induce a local extinction, when routine treatments with biocontrol agents are possible. This may be achieved, in the real system, through the introduction of a biological antagonist or by increasing its density, if it is already present in the microcosm (Jaffee, 1992; Ciancio, 1995; Atibalentja et al., 1998; Ciancio & Quenehervé, 2000).

A second factor to consider in the modeling approach concerns the detailed knowledge required about the antagonists biology and specificity. For the practical purpose previously cited, monitoring an isolate after its introduction in the environment represents a key issue: technologies based on PCR amplification are today available, exploiting specific genes and/or allowing the detection of particular regions of DNA. Through these techniques it is already possible to identify a microbial species or even a single isolate after its release (Hirsh, Mauchline, Mendum, & Kerry, 2000; Hirsh et al., 2001; Mauchline, Kerry, & Hirsch, 2002; Ciancio et al., 2000; Ciancio, Loffredo, Paradies, Turturo, & Finetti-Sialer, 2005) and it is expected that their application will become routine in field populations ecology studies.

Expanding this view, these technologies are expected to produce further developments when they will be integrated with methods of "precision farming" in crops biological protection. In consideration of the progress of electronic devices and of the integration of systems based on information technologies, it is possible to conceive future scenarios based on real time monitoring with data transmission and analysis. These systems, connecting the field soil microcosm to producers and consumers, will result informative about the status of a biological crop protection procedure and the related added value.

However, the benefits expected by the exploitation of soil microbial organisms for the economy, the environment and the society as well as the implications linked to the development of industrial products and processes finalized to biological control, are several. Thanks to these microorganisms and the knowledge acquired through their environmental study and modeling, new production sectors may arise. If the market expectations and the demand of bionematicides will grow, thanks to *Pasteuria* spp. and other similar bacteria, it is possible to expect a reduction of the environmental impact of nematicides with higher safety levels for farmers and consumers and a parallel reduction in the greenhouse gases release.

REFERENCES

Aksoy, H. M., & Mennan, S. (2004). Biological control of *Heterodera cruciferae* (Tylenchida: Heteroderinae) Franklin 1945 with fluorescent *Pseudomonas* spp. *Journal of Phytopathology, 152*, 514–518.

Amman, R. I., Ludwig, W., & Schleifer, K. H. (1995). Phylogenetic identification and in situ detection of individual microbial cells without cultivation. *Microbiological Reviews, 59*, 143–169.

Anderson, R. M., & May, R. M. (1981). The population dynamics of microparasites and their invertebrate hosts. *Philosophical Transactions of the Royal Society of London, 291*, 451–524.

Anderson, J. M., Preston, J. F., Dickson, D. W., Hewlett, T. E., Williams, N. H. & Maruniak, J. E. (1999). Phylogenetic analysis of *Pasteuria penetrans* by 16S rRNA gene cloning and sequencing. *Journal of Nematology, 31*, 319–325.

Atibalentja, N., Noel, G. R., Liao, T. F., & Gertner, G. Z. (1998). Population changes in *Heterodera glycines* and its bacterial parasite *Pasteuria* sp. in naturally infested soil. *Journal of Nematology, 30*, 81–92.

Atibalentja, N., Noel, G., & Ciancio, A. (2004). A Simple method for the extraction, PCR-amplification, cloning, and sequencing of *Pasteuria* 16S rDNA from small numbers of endospores. *Journal of Nematology, 36*, 100–105.

Cetintas, R., Lima, R. D., Mendes, M. L., Brito, J. A., & Dickson, D. W. (2003). *Meloidogyne javanica* on peanut in Florida. *Journal of Nematology, 35*, 433–436.

Christiansen, F. B., & Fenchel, T. M. (1977). Theories of populations in biological communities. In *Ecological studies series* (Vol. 20, 144 pp.). Springler-Verlag, Berlin, Heidelberg, New York.

Ciancio, A. (1995). Density dependent parasitism of *Xiphinema diversicaudatum* by *Pasteuria penetrans* in naturally infested soil. *Phytopathology, 85,* 144–149.

Ciancio, A., & Roccuzzo, G. (1992). Observations on a *Pasteuria* sp. parasitic in *Tylenchulus semipenetrans*. *Nematologica, 38*, 403–403.

Ciancio, A. (1996). Time delayed parasitism and density-dependence in *Pasteuria* spp. and host nematode dynamics. *Nematropica, 26*, 251 [Abstract].

Ciancio, A. & Quenehervé, P. (2000). Population dynamics of M*eloidogyne incognita* and infestation levels by *Pasteuria penetrans* in a naturally infested field in Martinique. *Nematropica, 30*, 77–86.

Ciancio, A., Leonetti, P. & Finetti Sialer, M. M. (2000). Detection of nematode antagonistic bacteria by fluorogenic molecular probes. *EPPO/OEPP Bulletin, 30*, 563–569.

Ciancio, A., Bonsignore, R., Vovlas, N., & Lamberti, F. (1994). Host records and spore morphometrics of *Pasteuria penetrans* group parasites of nematodes. *Journal of Invertebrate Pathology, 63*, 260–267.

Ciancio, A., Loffredo, A., Paradies, F., Turturo, C., & Finetti-Sialer, M. (2005). Detection of *Meloidogyne incognita* and *Pochonia chlamydosporia* by fluorogenic molecular probes. *OEPP/EPPO Bulletin, 35*, 157–164.

Couillault, C., & Ewbank, J. J. (2002). Diverse bacteria are pathogens of *Caenorhabditis elegans*. *Infection and Immunity, 70*, 4705–4707.

Davies, K. G., Flynn, C. A., Laird, V., & Kerry, B. R. (1990). The life-cycle, population dynamics and host specificity of a parasite of *Heterodera avenae* similar to *Pasteuria penetrans*. *Revue de Nématologie, 13*, 303–309.

Davies, K. G., Laird, V., & Kerry, B. R. (1991). The motility, development and infection of *Meloidogyne incognita* encumbered with spores of the obligate hyperparasite *Pasteuria penetrans*. *Révue de Nematologie, 14*, 611–618.

Davies, K. G., Fargette, M., Balla, G., Daudi, A. Duponnois, R., Gowen, S. R., et al. (2001). Cuticle heterogeneity as exhibited by *Pasteuria* spore attachment is not linked to the phylogeny of parthenogenetic root-knot nematodes (*Meloidogyne* spp.). *Parasitology, 122*, 11–120.

Davies, K. G., & Williamson, V. M. (2006). Host specificity exhibited by populations of endospores of *Pasteuria penetrans* to the juvenile and male cuticles of *Meloidogyne hapla*. *Nematology, 8*, 475–476.

Drechsler, C. (1934). Organs of capture in some fungi preying on nematodes. *Mycologia, 26*, 135–144.

Duddington, C.L. (1957). *The friendly fungi* (p. 188). London: Faber & Faber.

Ebert, D., Rainey, P., Embley T. M., & Scholz, D. (1996). Development, life cycle, ultrastructure and phylogenetic position of *Pasteuria ramosa* Metchnikoff 1888: rediscovery of an obligate endoparasite of *Daphnia magna* Straus. *Philosophical Transactions of the Royal Society of London B, 351*, 1689–1701.

Esnard, J., Potter, T. L., & Zuckerman, B. M. (1995). *Streptomyces costaricanus* sp. nov., isolated from nematode-suppressive soil. *International Journal of Systematic Bacteriology, 45*, 775–779.

Fould, S., Dieng, A. L., Davies, K. G., Normand, P., & Mateille, T. (2001). Immunological quantification of the nematode parasitic bacterium *Pasteuria penetrans* in soil. *FEMS Microbiology Ecology, 37*, 187–195.

Galeano, M., Verdejo-Lucas, S., & Ciancio, A. (2003). Morphology and ultrastructure of a *Pasteuria* form parasitic in *Tylenchorhynchus cylindricus* (Nematoda). *Journal of Invertebrate Pathology, 83*, 83–85.

Giblin-Davis, R. M., McDaniel, L. L., & Bilz, F. G. (1990). Isolates of the *Pasteuria penetrans* group from phytoparasitic nematodes in bermudagrass turf. *Journal of Nematology*, (Suppl.), *22*, 750–762.

Giblin-Davis, R. M., Williams, D. S., Bekal, S., Dickson, D. W., Brito, J. A., Becker, J. O., et al. (2003). Candidatus '*Pasteuria usgae*' sp. nov., an obligate endoparasite of the phytoparasitic nematode *Belonolaimus longicaudatus*. *International Journal of Systematic Evolutive Microbiology*, *53*, 197–200.

Gowen, S. R., & Tzortzakakis, E. (1992) Biological control of *Meloidogyne* spp. with *Pasteuria penetrans*. *EPPO/OEPP Bulletin, 24*, 495–500.

Gray, N. F. (1988). Fungi attacking vermiform nematodes. In G. O. Poinar & H. B. Jansson (Eds.), *Diseases of nematodes* (Vol. 2, pp. 3–8). CRC Press, Boca Raton, FL.

Hamid, M., Siddiqui, I. A., & Shaukat, S. S. (2003). Improvement of *Pseudomonas fluorescens* CHA0 biocontrol activity against root-knot nematode by the addition of ammonium molibdate. *Letters in Applied Microbiology, 36*, 239–244.

Hassel, M. P. (1978). *The dynamics of arthropod predator-prey systems* (p. 237). USA: Princeton University Press.

Hirsh, P. R., Mauchline, T. H. Mendum, T. H., & Kerry, B. R. (2000). Detection of the nematophagous fungus *Verticillium chlamydosporium* in nematode-infested plant roots using PCR. *Mycological Research, 104*, 435–439.

Hirsh, P. R., Atkins, S. D., Mauchline, T. H., Morton, O. C., Davies, K. G., & Kerry, B. R. (2001). Methods for studying the nematophagous fungus *Verticillium chlamydosporium* in the root environment. *Plant and Soil, 232*, 21–30.

Jaffee, B. A. (1992). Population biology and biological control of nematodes. *Canadian Journal of Microbiology, 38*, 359–364.

Jaffee, B. A. (2000). Augmentation of soil with the nematophagous fungi *Hirsutella rhossiliensis* and *Arthrobotrys haptotyla*. *Phytopathology*, *90*, 498–504.

Jaffee B. A. (2003). Correlations between most probable number and activity of nematode-trapping fungi. *Phytopathology, 93*, 1599–1605.

Jaffee, B. A., & McInnis, T. M. (1991). Sampling strategies for detection of density-dependent parasitism of soil-borne nematodes by nematophagous fungi. *Revue de Nematologie, 14*, 147–150.

Jaffee, B. A. & Muldoon, A. E. (1994). Susceptibility of root-knot and cyst nematodes to the nematode-trapping fungi *Monacrosporium ellipsosporum* and *M. cionopagum*. *Soil Biology and Biochemistry, 27*, 1083–1090.

Jaffee, B.A., Muldoon, A. E., Phillips, R., & Mangel, M. (1990). Rates of spore transmission, mortality, and production for the nematophagous fungus *Hirsutella rhossiliensis*. *Phytopathology, 80*, 1083–1088.

Jaffee, B., Phillips, R., Muldoon, A., & Mangel, M. (1992). Density-dependence host-pathogen dynamics in soil microcosms. *Ecology, 73*, 495–506.

Jonathan, E. I., Barker, K. R., Abdel-Alim, F. F., Vrain, T. C. & Dickson, D. W. (2000). Biological control of *Meloidogyne incognita* on tomato and banana with rhizobacteria, actinomycetes, and *Pasteuria penetrans*. *Nematropica, 20*, 231–240.

Kasumimoto, T., Ikeda, R., & Kawada, H. (1993). Dose response of *Meloidogyne incognita* infecting cherry tomatoes to application of *Pasteuria penetrans*. *Japanese Journal of Nematology, 23*, 10–18.

Mankau, R. (1975). *Bacillus penetrans* n. comb. causing a virulent disease of plant parasitic nematodes. *Journal of Invertebrate Pathology, 26*, 333–339.

Mauchline, T.H., Kerry, B. R., & Hirsch, P. R. (2002). Quantification in soil and the rhizosphere of the nematophagous fungus *Verticillium chlamydosporium* by competitive PCR and comparison with selective plating. *Applied and Environmental Microbiology, 68*, 1846–1853.

McSorley, R., Dankers, W. H., Parrado, J. L. & Reynolds, J. S. (1985). Spatial distribution of the nematode community on Perrine Marl soil. *Nematropica, 15*, 77–92

Metchnikoff, E. (1888). *Pasteuria ramosa* a répresentant des bacteries a division longitudinale. *Annales de l' Institut Pasteur, 2*, 165–170.

Noel, G. R., Atibalentja, N. & Domier, L. L. (2005). Emended description of *Pasteuria nishizawae*. *International Journal of Systematic Evolutive Microbiology, 55*, 1681–1685.

Nour, S. M., Lawrence, J. R., Zhu, H., Swerhone, G. D. W., Welsh, M., Welacky, T. W., et al. (2003). Bacteria associated with cysts of the soybean cyst nematode (*Heterodera glycines*). *Applied and Environmental Microbiology, 69*, 607–615.

Preston, J. F., Dickson, D. W., Maruniak, J. E., Nong, G., Brito, J. A., Schmidt, L. M., et al. (2003). *Pasteuria* spp.: systematics and phylogeny of these bacterial parasites of phytopathogenic nematodes. *Journal of Nematology, 35*, 198–207.

Sayre, R. M., & Starr, M. P. (1985) *Pasteuria penetrans* (ex Thorne, 1940) nom. rev., comb. n., sp. n., a mycelial and endospore-forming bacterium parasitic in plant-parasitic nematodes. *Proceedings of the Helminthological Society of Washington, 52*, 149–165.

Sayre, R. M., & Starr, M. P. (1988). Bacterial diseases and antagonism of nematodes. In G. O. Poinar & H. B. Jansson (Eds.), *Diseases of nematodes* (Vol. 1, pp. 69–101). CRC Press, Boca Raton, FL.

Schmidt, L. M., Preston, J. F., Nong, G., Dickson, D. W., & Aldrich, H. C. (2004). Detection of *Pasteuria penetrans* infection in *Meloidogyne arenaria* race 1 in planta by polymerase chain reaction. *FEMS Microbiology Ecology, 48*, 457–464.

Siddiqi, Z. A., & Mahmood, I. (1999). Role of bacteria in the management of plant parasitic nematodes: a review. *Bioresource Technology, 69*, 167–179.

Starr, M. P., & Sayre, R. M. (1988). *Pasteuria thornei* sp. nov. and *Pasteuria penetrans* sensu stricto emend., mycelial and endospore-forming bacteria parasitic, respectively, on plant parasitic nematodes of the genera *Pratylenchus* and *Meloidogyne*. *Annales de l' Institut Pasteur/ Microbiologie, 139*, 11–31.

Stirling, G. R. (1984). Biological control of *Meloidogyne javanica* with *Bacillus penetrans*. *Phytopathology, 74*, 55–60.

Stirling, G. R. (1991). *Biological control of plant parasitic nematodes: progress, problems and prospects* (p. 282). Oxon, UK: CAB International.

Sturhan, D., Winkelheide, R., Sayre, R. M., & Wergin, W. P. (1994). Light and electron microscopical studies of the life-cycle and developmental stages of a *Pasteuria* isolate parasitizing the pea cyst nematode, *Heterodera goettingiana*. *Fundamental and Applied Nematology, 17*, 29–42.

Taylor, L. R. (1961). Aggregation, variance and the mean. *Nature, 189*, 732–735.

Torsvik, V., Goksøyr, J., & Daae, F. L. (1990). High diversity in DNA of soil bacteria. *Applied and Environmental Microbiology, 56*, 782–787.

Tzortzakakis, E. A., & Gowen, S. R. (1994). Evaluation of *Pasteuria penetrans* alone and in combination with oxamyl, plant resistance and solarization for control of *Meloidogyne* spp. on vegetables grown in greenhouses in Crete. *Crop Protection, 13*, 455–462.

Verdejo-Lucas, S. (1992). Seasonal population fluctuations of *Meloidogyne* spp. and the *Pasteuria penetrans* group in kiwi orchards. *Plant Disease*, *76*, 1275–1279.

Woronin, M. (1870). *Sphaeria limaneae, Sordaria coprophila, S. firmiseda, Arthrobotris oligospora; Eurotium, Erysiphe, Cicinnobolus; nebst Bemerkungen über die Geschlechtsorgane des Ascomyceten.* In A. de Bary & M. Woronin (Eds.), *Beiträge zur Morphologie und Physiologie der Pilze.* Verlag von C. Winter Frankfurt a. M., Germany. (Vol. 7, pp. 325–360).

INDEX